UNDERSTANDING TRACK ENGINEERING

Edited and published by

The Permanent Way Institution

Printed in Great Britain by The Lavenham Press

Permanent Way Institution

The Institution for Rail Infrastructure Engineers

ISBN 978-0-903489-05-8

PWI 20150414

INTRODUCTION

The Permanent Way Institution was founded in 1884 and since that date it has strived to provide training, development and networking opportunities for its membership. The Institution has a proud history of publishing textbooks which help the industry to share knowledge and experience. This new book has its roots deep in a previous publication - the 6th Edition of British Railways Track (the authors of which are acknowledged in that publication but I also thank them again for the information carried forward to this new book).

There has been significant changes in the rail industry since those authors produced the chapters for the 6th Edition back in 1994 with an industry on the cusp of privatisation. Twenty years on we have a very different landscape and a need to present a book to the industry. Firstly, to aid new entrants understand the basic principles, and secondly, to provide a definitive source of reference for more experienced engineers.

Understanding Track Engineering draws together the key principles of the track engineering discipline, set in the context of today's railway. I commend it to the industry.

Steve Whitmore
President, Permanent Way Institution

FOREWORD

The role track engineering plays in ensuring safe, efficient rail transport is invisible to rail users – and that's how it should be. Track Engineers need a wide variety of skills and comprehensive knowledge of this complex subject to perform at their best, delivering safe and efficient track infrastructure. This book is essential reading for all track engineers covering, as it does, the whole spectrum of track engineering practice.

Understanding Track Engineering starts with the background to track engineering and its place in the railway engineering system. There is then comprehensive coverage of track components; rails, sleepers and fastenings. The book moves on to the design of railway layouts in plain line and switches & crossings, dealing with geometric principles and their practical application.

The support provided by the ballast and earthworks is covered next, followed by off-track management of the lineside. These elements of track engineering are vital to safe and efficient operation but can be overlooked by inexperienced engineers, hence their inclusion here.

Maintenance of the track and its geometry is covered in broad terms as an introduction to the subject. This area is covered comprehensively in the PWI's detailed books 'Plain Line Maintenance' and 'S&C Maintenance'.

The final elements are track renewal and new railway construction, again summarised as an introduction to these two areas.

I am confident that this book will be found widely on the desk of track engineers. It is an excellent introduction for those starting out in the rail industry and an invaluable reference work for engineers who have chosen rail as an exciting and involving career.

Brian Counter
Technical Director, Permanent Way Institution

ACKNOWLEDGEMENTS

The PWI gratefully acknowledges the contribution made by Dr Brian Counter, Technical Director PWI, in the creation of this book.

Starting with 'British Railway Track 6th Edition' as the basis, he has restructured and updated the content to ensure that it is relevant to current railway practice. In this endeavour he has been ably assisted by PWI Fellows Malcolm Pearce (Track Renewals), David Johnson (Gauging), Colin Wheeler (Law and Safety) Member, Vice President Paul Bull (Rails) and Richard Johnson (Welding).

The PWI would also like to acknowledge the contributions of the following to British Railway Track 6th edition on which this book has been based.

A. Blower, P. O. Jennings, T. A. Jipson, D. Hill-Smith, C. F. Bonnett, R. M. Heath, J. McCahill, J. K. Wright, F. J. Roberts, A. J. McCluskey, P. Grisdale, E. A. Labrum, D. R. Gillan, N. J. Duffy, J. S. Bell, A. Cooksey, J. B. Ellis, D. Ingram, L B Foster, M. B. P. Allery, D. Cook, D. Lindsay, J. C. Morgan, D. Stuckey, J. D. Strange, P. Hunt, R. Bance & Co Ltd, Pandrol International Ltd, J. McMorrow, D. J. Ayres, R. W. Collingwood, J. H. B. Cook, M. A. Winn, D. L.Cope, J. C. Morgan, R. B. Lewis, R. Harvey and R. J. Gostling,

Production and printing of the volume has been led by Alison Stansfield, Communications Director PWI with layout and formatting by Holly Hayley and Kerrie Pearson.

Photographs have been kindly provided by :

Cross Country Trains, London Underground, Jay Jaiswal, AREMA, Nexus, Network Rail, Thermit Welding (GB) Ltd, Railtech UK Ltd, Rail Technology.com, Cemex Rail Solutions, Rail One, LCR Ltd, RailPictures.net, Steven Pearson, Balfour Beatty Rail Technologies, D Ratledge, Rail Infrastructure, Jörg, Schnabel, Carillion, Colas Rail, Docbrown.info, Scalefour Society,

CONTENTS

8

10

CHAPTER 1

INTRODUCTION

1.1 THE ORIGINS OF TRACK (THE PERMANENT WAY)

The term "track" is commonly used worldwide to represent guided forms of way as far as transport infrastructure is concerned, however the expression "permanent way" is deep rooted in the development of this specialist engineering subject. The term "permanent way" was created in the early 19th century to distinguish between the final constructed railway and any "temporary ways" or tracks that were used on a temporary basis, usually for hauling of materials or excavated earth and rock.

Wheeled transport before the middle of the 19th Century implied exclusively wheels with rigid tyres. It was found at a very early stage of the development of land transport that most road surfaces and foundations were very quickly damaged by heavy wagons on rigid tyres, and man's inventiveness almost as quickly found that by providing a surface of stone slabs or wooden baulks a much more nearly permanent hauling way could be achieved. Examples of such hauling ways can be found in connection with colliery operations as early as the 16th Century, and were common by the early 18th Century.

The effects of the Industrial Revolution were initially to develop this idea by adding wrought iron plates to reduce wear on the wooden baulks. As shown in Figure 1.1, cast iron plates and later on, edge rails represented an improvement on these ideas. Edge rails (1789) by raising the wheels above ground, enabled the use of flanged iron wheels. Between the late 1770's and around 1825, wrought iron rails were developed further and became strong enough to carry the weight of the vehicles without the support of a longitudinal wooden baulk. A selection of these later developments are shown in Figure1.2

The combination of improved track and steam locomotion was so effective that railways effectively monopolised medium to long distance land transport for the remainder of the 19th Century, spreading from their centre of invention in North East England across most of the world. Only the inventions of the pneumatic tyre, the internal combustion engine, and smooth hard roadway pavements restored the competitive position of roads in the early 20th Century, making possible the much more diverse inland transport scene which is so familiar today, in which "railways" are seen as having the uniquely identifying feature of steel wheels running on steel rails.

1.2   THE STOCKTON AND DARLINGTON RAILWAY AND OTHER EARLY DEVELOPMENTS

The story of railway track in Britain really starts about 1825 with the construction of the Stockton and Darlington Railway. At the opening of this railway as a public, passenger carrying, steam locomotive hauled railway, in 1825, Stephenson used T-shaped rails of a type developed earlier by John Birkinshaw of the Bedlington Ironworks, Northumberland. These rails were 15 feet long, carried in chairs on stone sleeper blocks set at 3 foot spacing, and weighed 28lbs per yard. They were 2¼ ins broad across the head, which was ¾ ins deep. The rails were 2 ins deep at the chairs and bellied to a depth of 3¼ ins in the middle of the space between them. The general features of this track are shown in Figure 1.2.

For the Liverpool and Manchester Railway, Stephenson adopted a somewhat similar form of track construction to that used for the Stockton and Darlington. The need to provide a stronger rail was, however, recognised and the rails first used weighed 17.36 kg per metre (35 lbs per yard). Later a 24.80 kg per metre (50 lbs per yard) rail was introduced and eventually the standard rail weight on the Liverpool and Manchester became 37.20 kg per metre (75 lbs per yard).

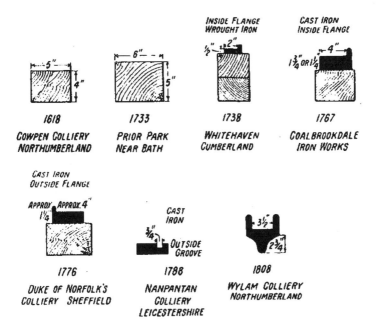

Figure 1.1 Development of early track from the log to the tram plate

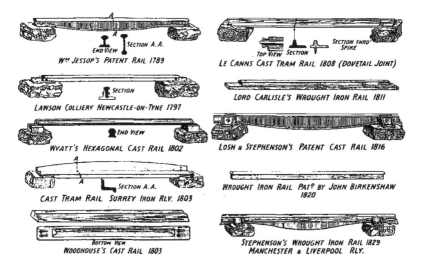

Figure 1.2 The early evolution of the iron rail

The Leeds and Selby Railway used 3962 mm (15 ft) lengths of 20.45 kg per m. (42 lbs per yard) T section rail fixed in small chairs at intervals of 915 mm (6 ft) mounted on stone blocks, or alternatively larch sleepers. On some sections of the line the stone blocks were continuous under each rail, connected by iron ties-to keep the rails to correct gauge.

Figure 1.3 Early Stone blocks forming support to iron rails on the original Pickering – Grosmont railway in the early 19th century (courtesy docbrown.info/docspics.)

As the mileage of track increased it was soon found both impractical and uneconomic to continue the use of individual stone blocks, and these gave way to transverse timbers (sleepers) which not only supported the rails but ensured proper gauge retention. The term "sleepers" had been in general use since at latest the early 17th Century to denote any kind of beam or baulk laid directly on the ground to support other beams, although the term cross tie, generally regarded in UK as an Americanism, is probably more apt in the railway context.

For the Great Western Railway (GWR) Brunel returned to the concept of the combination of 30 foot long, longitudinal timbers of 14 inch by 7 inch cross section supporting light section rails (in fact bridge rails in this case). Gauge retention and correct track spacing was achieved in this type of track by means of transoms extending right across the two tracks at 15 foot intervals, and as shown in Figure 1.3, the whole was held down in the original concept by beech piles. These latter were soon abandoned. This type of track, first introduced in 1838, and subsequently much improved, e.g. by the use of a form of baseplate under the bridge rail (Macdonnell plates), endured until beyond the final change of gauge on GWR in 1892.

1.3   TRACK GAUGE

Many new students of railway track engineering ask the question as to why most countries in the world have adopted and continue to adopt on their high speed lines a standard track gauge of an unusually specific figure. We can trace the historical path that led to the distance between the running edges of the rails being determined as 1435 mm (4 ft 8½ inches).

The gauge of the potential guided "wagonways" was virtually decided this as they had been laid since Roman times to accommodate the wheels of carts and wagons which used the common roads of the country. The original wagons were usually a similar width to a roman chariot which was designed to accommodation the width of two horses side by side.

A dimension of 1524 mm (5 ft) between the centres of wheels was the generally accepted early norm. The Wylam wagonway, afterwards the Wylam plateway, the Killingworth railroad and the Hetton railroad were already practically of the same gauge and since some of the spoil wagons used in the construction of the Stockton and Darlington were brought from Hetton, and others being intended for future use on the completed railway were built to the same dimensions, it was not unnatural that the same gauge was adopted. Most railways in UK adopted a track gauge more or less identical with that chosen by Stephenson for the Stockton and Darlington Railway. However, this gauge was not universal, and gauges both broader and narrower than this were adopted. The most celebrated UK example of broader gauges was the GWR, for which Brunel persuaded his Directors to agree to the adoption of a gauge of 2140 mm (7 ft ¼ inches).

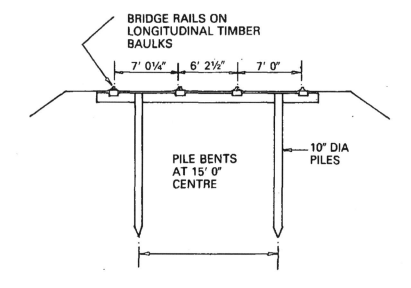

Figure 1.4 Great Western Railway Broad Gauge Track

The technological and economic implications of Brunel's visionary innovation in the face of an already substantial mileage of Standard Gauge track were such that a violent controversy developed which was finally resolved by a UK Royal Commission in 1845 which decided in favour of Stephenson's gauge, which to this day has been universally known as STANDARD GAUGE.

Nevertheless, the GWR did not finally abandon its broad gauge until 1892, and the Royal Commission's ruling did not prevent other countries from adopting gauges either broader or narrower than standard. Notably in the context of the PWI the main line railways of Ireland are to this day to the significantly broader gauge of 5 ft 3 ins, whilst the world's most extensive railway system (that of Russia) has over 87000 route miles of track of 5 ft gauge. A small number of railway concerns in this country (mainly associated with mineral workings) were developed with gauges down to around 2 ft, and very extensive systems were constructed in developing countries with gauges of 3 ft to 3 ft 6ins. As a result taken worldwide, "Standard Gauge" actually accounts for about 60% of all track mileage.

1.4   THE DEVELOPMENT OF BULL-HEAD TRACK

In the early years of the railways three types of rails emerged as possibilities. These were:

a) Tee-section rails where the wheels were carried on the upper bar of the tee, whilst the vertical leg or web of the tee was supported at discrete intervals by chairs carried on transverse sleepers.

b) Bridge rails or similar, carried on longitudinal timbers.

c) Flat bottom rails (or "Vignoles" rails, after the engineer who patented the design) which like Tee-section rails behaved like girders in supporting the wheels between sleepers. However, the flat foot of the rail enabled the rail to be carried directly on a sleeper without the use of a chair.

1.4.1   The Invention of Bull-Head Rail

The development which had by far the greatest influence upon the subsequent history of track design in UK was due to Joseph Locke, a former pupil of Stephenson, who had been involved in the construction of the Liverpool and Manchester. In 1835 he designed and put into service on the Grand Junction Railway some double headed rails with both heads of the same dimension. The intention was that these rails would serve a double turn of duty. When the uppermost head was worn to its limit the rail was to be turned upside down and the part which had been seated in the chair would become the rail head. However, under the constant hammering of heavy dynamic loads, which caused movement between the lower head and the base of the chair, the underside of the rail became so indented, or galled, that the rail was unfit for further service after it had been turned.

Since the rail could not be inverted it was natural to extend the idea by enlarging the head, which suffered the worst wear, and thus the concept of the bull-head (BH) rail was born.

Within a few years wrought iron rails of true bull-head section were being laid. The Shropshire Union Railway, opened in 1848, was probably the first line to use bull-headed track. Although the double headed rail did not fulfil the function for which it was originally designed, the chaired road with its rails securely keyed in position proved very satisfactory. Indeed, it was reckoned to be much more satisfactory than the type of tracks being laid at that time in Europe and North America where the practice was to use flat bottomed rails dog spiked directly onto rough timber cross ties.

By the middle of the 19th Century, standard gauge railways in UK were laying their new tracks almost exclusively with BH rail.

1.4.2   Development of Bull Head Rail

Rail weights which varied widely in the early years increased appreciably during that decade and by 1849 it is recorded that of the 705 km (438 miles) managed by the London and North Western Railway (L.N.W.R) about 241 km (150 miles) were laid with 37.20 kg per metre (75 lbs per yard) rails, about 161 km (100 miles) with 32.24 kg per metre (65 lbs per yard) rails, and the remainder with rails varying from 29.76 kg per metre (60 lbs per yard) to 42.16 kg per metre (85 lbs per yard). It is noteworthy that before 1850 a 42.16 kg per metre (85 lbs per yard) rail had been in use for some time.

As the weight and speed of trains increased it was found that rails of wrought iron could not stand up to the demands placed upon them and in 1857 the first steel rails were laid on the Midland Railway at Derby. They are recorded as having been in use for 16 years at a location where previously it had been necessary to relay every few months. The first steel rails on the L.N.W.R. were laid at Rugby, Stafford and Crewe in 1862 and in comparative tests at Camden they were found to have a life of up to thirty times that of the best quality iron rails.

As heavier rail sections were finding favour so too were longer rails, primarily so as to reduce the number of rail joints. By the turn of the century the bulk of the mileage was laid with 45ft rails.

**Figure 1.5   Jointed Bull Head Track on LUL Central Line
(courtesy Scalefour Society)**

The introduction by the L.N.W.R. of the 18.29m (60ft) rail in 1910 created the standard rail length which still exists worldwide today on jointed track. By the early 1900s rail weights had also increased to the 47.13 to 49.61 kg per metre (95 to 100 lbs per yard) range. In 1914, for new rails the Midland was using 49.61 kg (100 lb) rails, and the L.N.W.R. was using 51.1 kg (103 lb) rails in certain locations.

As evidence of the increased demands upon the track, in 1939 many expresses were loading up to 700 tonnes and attaining speeds up to 145 km (90 miles) per hour. Steam traction still predominated, and with it was its marked detrimental effect upon the track due to the extreme difficulty, or even impossibility, of obtaining a mechanical balance at varying speeds. At the turn of the century a heavy express passenger train weighed only about 350 to 400 tonnes and its maximum speed was around 113 km (70 miles) per hour.

In their final development, two BH rail sections became standard. These British Standard (BS) rails weighed 95 lb/yd (BS95R, illustrated in Figure 1.6), for all main line renewals, and 85 lb/yd (BS85), for branch lines and other tracks carrying light traffic.

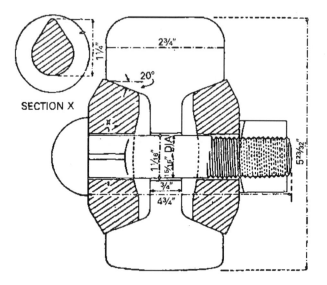

Figure 1.5 95 lb Bull Head Rail, Fishplates and Fishbolt

Other rails whose use persisted until the conversion to flat bottom (FB) rail included the "00" section of the G.W.R. with a weight of 97.5 lbs per yard, which that railway standardised as early as 1894, and some rail weighing as much as 100 lbs per yard was rolled, mainly for use in the Severn Tunnel.

The length of rail supplied from the rolling mills stabilised at 60ft, corresponding to the deck length of the available rail carrying wagons, and they were normally rolled from medium manganese steel. The design of rail joint was a continual source of debate but ultimately the "suspended" joint in which the rail ends cantilever out from the adjacent chairs and are joined by a fishplate having four holes became the standard.

BH rails were supported in cast iron chairs which normally had three tapering holes in the base to accommodate the coach screws used for fixing them to the sleepers. The chair had two lugs or jaws and the web of the rail was held tight up against the inner jaw by a key, usually of oak or teak, and, later, of steel which was driven in between the outer jaw and the rail.

1.4.3   Sleepers

It has already been mentioned that stone blocks gave way to cross sleepers very early in the history of railways. However, the original concept, that sleepers supported the ends of very short rails, implied sleeper spacings of around 5 ft. With experience, it became necessary to close up the sleepers to 4ft, and later 3ft spacing. By the time the UK went over to flat bottom rail, it had become normal to provide 24 sleepers per 60 ft rail length, with a graduated spacing so that the sleepers nearest to the joints were less than 2ft apart, whilst those in the centre of the rail length were at 3ft spacing. Needless to say, the coming of mechanised relaying, long welded rail, and automatic tamping, has brought an end to that practice.

Since by the early 19th Century, Britain's deciduous forests had largely been felled to produce timber for battleships and charcoal for iron smelting, softwood (i.e. timber from coniferous trees), either locally grown or imported, figured largely as the raw material of sleepers from early days. This is in sharp contrast to France, where even today, the use of locally grown oak and beech for sleeper wood is not unknown. If softwood is used, it is essential to use some form of timber preservative, if rapid decay of sleepers is to be avoided.

Treatment with water soluble processes such as "Kyanizing" was attempted but by 1840 the use of creosote or tar was fairly universal. As time went by the importance of penetration of the timber by preservative was realised, and by the 1880's several of the railways had established creosoting plants where sleepers could be impregnated with creosote under pressure before being distributed for relaying.

1.4.4   Ballast

Bearing in mind that railways came into existence as a means of providing a hard bearing surface on an existing roadway which consisted of compacted earth and stones, it is not perhaps surprising that it was not immediately realised that it was important to provide a thick layer of hard material under the sleepers to distribute wheel loads. Thus the early railways were ballasted with all manner of material which today would be regarded as wholly unsuitable, e.g. ashes, chalk, burnt clay etc., but which was cheaply and easily available locally. It is easy to be critical with hindsight, but it has to be borne in mind that much of South East England has no deposits of hard stone, and until the railway actually existed, transport of thousands of tons of crushed stone from say Dartmoor to Surrey, would have been virtually impossible.

However, experience proved the necessity of a layer of good quality ballast under and around the sleepers, and harder materials, such as gravel, blast furnace slag, crushed gravel, and ultimately crushed limestone or igneous rock came into use. The transformation only came slowly, however, and it may be noted that even as late as the formation of the London and North Eastern Railway in 1922, 90% of the mileage of the former North Eastern Railway was ballasted with ash. On the Southern Railway, similarly, large scale distribution of crushed metamorphic rock from Meldon Quarry did not start until the early 1930's.

1.4.5 Drainage

There is no reason to believe that the engineers who designed and built the first railways did not understand the importance of good drainage. On the other hand there is plenty of evidence that it was not always achieved, or at least was not good enough, and it is clear from the records that many serious accidents resulted from the drainage not being good enough in the first place, or else of its being neglected. As the weight and speed of traffic increased, so did the importance of drainage and formation work.

1.5 RECENT DEVELOPMENTS

It seems a reasonable generalisation to state that improvements in track design are responses either to an economic stimulus or to changes in axle weights, speeds and overall tonnages of traffic resulting in a perceived inability of an existing trackform to support the new loadings. No account, however brief, of recent developments in track engineering can be understood without reference to this latter feature, which has exerted a profound influence over all the developments which have taken place.

1.5.1 Developments in Motive Power

The UK has moved over the last 100 years from an exclusive reliance on steam to its complete abandonment in favour of diesel and electric traction.

The first electric trains appeared in UK in the 1890's, and after some experimentation with overhead current collection, the high capital costs involved in either raising overbridges or lowering the track appeared unattractive, and most electrification up until 1939 used the third rail principle, with the current being collected by a shoe sliding along the top surface of a conductor rail fixed to the sleeper ends. Current was supplied at 600 volts, later raised to 750. The Medway towns, Sevenoaks, Brighton and Portsmouth had electric services to London by 1939, and most track south of the Thames is now electrified on this system. Some features of this system will be described in Chapter 3 (Rails). As a result of the obvious hazards associated with the exposed conductor rail, its tendency to ice up in cold weather, and some anxiety about its performance at very high speeds, this system has come to be regarded as somewhat archaic, and some safer and more technically advanced examples of third rail traction are described in Chapter 3.

Where it is physically possible to do so, and/or where the rate of return on investment is expected to be adequate, there are considerable advantages in electrical engineering terms to be gained from using much higher voltages for the traction current than the 750 or so which is practical with third rail supply, and consequently the world wide tendency has been towards current collection from overhead wires supported by catenaries strung between gantries or cantilevered masts. Some early examples were installed in the UK using 1500 to 3000 volts DC, but with advances in electrical engineering technology the use of alternating current (AC) at up to 25000 volts has become practical. The major electrification programmes North of London during the 1960's and since have used this technique.

From the track engineer's point of view, the significant features of electrification are:

1) Electrification changes the track environment and imposes new disciplines on track maintenance and renewal staff.
2) Due to the absence of the reciprocating parts the hammer blow associated with steam traction is absent, and hence electrically driven axles are potentially kinder to the track than those driven by steam. Unfortunately this effect is partially countered in

practice by the heavy unsprung mass produced by some forms of suspension, and by the reduction in diameter of the driving wheels.

3) The increased availability of tractive effort at high speed, coupled with the increase in the number of driven axles, particularly when multiple unit stock is employed, produces greater longitudinal traction forces in the rails, which must be resisted by the track fastenings.

4) Increased tractive effort at high speed has increased both maximum and average speeds out of all recognition, and to take advantage of this enhanced capability, every curve on the routes affected has had to be assessed for its maximum potentiality. Frequently this has meant redesigning the curve, and even, in recent years, revising the basic curve design rules

The powers and axle weights of electric and of diesel locomotives are rather similar, but of course the impact of that form of traction on the track man is nowhere near so dramatic as it is with electric traction. Indeed it might almost be said that the main effect of the change from steam to diesel traction was that instead of the ballast being fouled with ashes it became fouled with spilt fuel oil.

1.5.2 Developments in Rolling Stock

Passenger coaches have become heavier and longer during the period under review. However, passenger rolling stock axle weights are not generally seen as a limiting factor in track design. Changes in length, and also changes in the kinematic behaviour of coaching stock call for continual review of the route availability situation.

Changes in freight vehicle design on the other hand have had a dramatic effect on track design. For the reason for this, it is only necessary to compare typical train weights at the beginning and end of the review period:

A typical train of 1900 might have comprised a locomotive with 3, 17 tonne driving axles. The remainder of the train would consist of 40 or so wagons totalling 80, 7 tonne axles. Only the loco and the guard's van would have been braked, and the train speed would not exceed about 60 km/h.

A typical train in the early 1980's will have comprised a locomotive with 6, 21 tonne driving axles. The remainder of the train would consist of 10 wagons totalling 40, 25 tonne axles, all of which will be braked, and the train speed may reach 125 km/h.

Thus the train will have increased in all up weight by 84%, the number of heavy axles will have increased fifteen fold from 3.6% of the total to 100%, and the speed has doubled.

The modern train imposes very heavy longitudinal forces on the track fastenings when stopping, and experience has demonstrated the inadequacy of earlier standard track forms, ballast depths, and formations to support the greatly increased static and dynamic loadings.

1.5.3   Rail developments in the 20th Century

A particular problem relating to the track in the middle of the 20th century was the absence of suitable indigenous supplies of sleeper timber and difficulties in obtaining timber from the Baltic countries. This led to considerations of alternative sources of supply and alternative materials, namely, steel and concrete, neither of which had previously found favour with track engineers. Engineers also resumed the investigation of US and European practice as regards alternative rail sections, resilient track fastenings, and long welded rail.

1.5.4   The introduction of Flat Bottom Rail

Recognition of the desirability of providing a form of track construction less prone to lateral distortion and capable of carrying heavy trains at high speed at reduced maintenance cost was evidenced by the trials of flat bottomed rails. The main advantage of the FB rail over the BH rail of equivalent weight is its greater lateral stiffness. It was also recognised that the fastenings available for FB track were more elaborate and effective than the crude dog spikes of earlier years.

In 1948 the decision was taken in the UK to change from BH to FB rails. It was accepted that traffic demands on main lines, where high speed trains ran in the years just prior to the war, were more than BH rails would stand without excessive maintenance costs, and since the extensive trials with FB track had clearly proved that a saving in maintenance was achievable by the adoption of the stiffer rails, the change in policy was agreed. It was decided not to use any of the FB rail sections which had been on trial, but to introduce two entirely new sections as the new standards. A heavy rail with a weight of 54 kg per m (109 lbs per yard) to replace the BS95R BH section and a lighter rail of 48.6 kg per m (98 lbs per yard) in place of the BS85 BH section (cf section 1.4.2 above). The features of the new rails are described in Chapter 3.

In 1959 it was decided to discontinue the use of the 54 kg (109 lb) rail, and instead to use the B.S. 110A rail. This is practically identical with the U.I.C. 54 kg per m rail. The decision to use it was influenced by export considerations and the economy to be gained by aligning UK requirements with the potentially much bigger export market. In 1968 the 110A rail was modified by thickening the web to form the present standard 56 kg 113A rail section.

1.5.5 Continuously Welded Track

The earliest trials with Continuously Welded Rail Track (CWR) date back to when the then London Passenger Transport Board made its first tentative steps toward putting CWR into its tunnels, following in the steps of continental and the United States railways. UK trials commenced in 1955 and installation of CWR became the standard in all suitable main line locations in the mid-1960's.

The advantages of CWR compared with jointed track are:

(i)   up to one third longer rail life
(ii)   cost of maintenance is about half that of jointed track
(iii)   reduces the number of rail breakages by eliminating the potential
        for star cracks emanating from fish bolt holes
(iv)   saves around 5 per cent of traction energy costs
(v)   increases sleeper life
(vi)   gives a better quality ride
(vii)   permits higher speeds
(viii) reduces wear and tear on vehicles
(ix)   eliminates the 'clickety-clack' of jointed track thereby reducing
        noise to lineside inhabitants.

The change to CWR was accompanied by the effective abandonment of softwood sleepers for new work in locations of fast or heavy traffic, in favour of either prestressed monobloc concrete sleepers, or hardwood (usually Jarrah) sleepers. UK standard concrete sleepers are pretensioned on the "long line" system. At the same time the fastening spikes came to be regarded as obsolete, and after many trials, spring fastenings became standard.

1.5.6   The Challenge of Higher Speeds

Long distance passenger rail traffic has to compete on journey time with the best that can be done, either by motorists over medium distances, or by air transport over the longer runs such as London to Glasgow. This challenge led to the concept of a tilting train in the 1980's, which although it was claimed to be able to achieve shorter journey times over existing track, by virtue of its ability to tilt, still imposed increased track forces when traversing curves at what were, from the track point of view, cant deficiencies considerably exceeding anything then contemplated. The research work arising from this challenge led to an improved appreciation by track engineers of the detailed interactions between vehicle and track. Tilting trains are now common throughout the world.

To conclude, developments in the field of track engineering in the last 20 years have taken place on a scale not attained in any similar period in UK railway history. The ultimate aim of successive generations of track engineers has been a "fit and forget" track with a life span of several times that of the conventional tracks of a mere fifty years ago. With traffic speeds of 200 kph (120 mph) now commonplace in the UK and high speed development over 320 kph (200 mph) already in use between London and Europe and more lines in prospect, the railway as a means of land transport is still unsurpassed and that superiority is in no small measure due to the excellence of modem track.

1.6 RAILWAY SYSTEMS IN THE UK

There are a varied number of guided systems of way in the world, the major common features being steel wheels moving over steel rails. This section describes the systems in the UK.

1.6.1 Main Line Railways

In the UK, there are around 37,800km of standard gauge track (1432 or 1435mm) which comprises of main strategic routes, to and from London, cross-country routes, secondary lines and branch lines. This forms a comprehensive inter-urban network for high speed passenger and freight traffic with slower speed connecting services into urban and rural locations. Axle loads vary across the network but the main routes are cleared for 25 tonne axle loads and in some locations 30 tonne axle loads are permitted.

Figure 1.6 Photo courtesy of Cross Country Trains

In Ireland, there are 1940km in the South and 330km in the North with a gauge of 1600mm. There are similar structural clearance restrictions to mainland Britain but axle loads have been limited to 17 tonnes on many routes, although some developments towards higher axle loads are in progress.

In Europe and most of the world, the standard gauge of 1435mm is adopted. There are notable exceptions such as Spain who use 1600mm and there are many 1000mm gauge lines in existence.

1.6.2   Urban Railway Systems

There are a number of light railway systems in the UK, which have differing characteristics including gauge. The largest urban system in the UK is London Underground (LUL). The total LUL network, including all surface, sub-surface and tube routes adds up to 394km, or 826 track km of running line, to which must be added 253km of depots and sidings.

**Figure 1.7 London Underground Track Open Junction tube in background
(Courtesy LUL)**

The split between tube, sub-surface and surface lines is:

In the open:	725km
Sub-surface:	73km
Tube:	281km

Maximum speeds (110km/h in the open and from 56 to 80km/h in tunnel) and axle loads (16.74 tonnes) are lower than on mainlines but services, particularly in the central areas, are very frequent so that there are high maximum annual gross tonnages.

There are other light railway systems in the UK which are similar to LUL and use dedicated tracks with open and underground sections. The oldest of these is the Glasgow subway built in 1896 and has a 4ft gauge. Docklands Light Railway is an automatic rapid transit system with a standard gauge but has many tram features. Other independent metro systems exist such as the Liverpool Loop and Link and the Tyne and Wear Metro (Nexus). These use a main line route enhanced with tunnels and electrical power systems.

1.6.3   Tram Systems

There are a number of tram systems in the UK which utilise a traditional street running system typical in Europe. However, many systems in the UK adopt open running utilising ex-standard gauge railways in the suburbs. The type of trams used does vary, depending upon the need to use normal height platforms. The following list is not exhaustive but reflects systems operating in the UK:

    a.   Manchester Metrolink
    b.   Sheffield Supertram
    c.   Midland Metro (Birmingham)
    d.   Croydon Tramlink
    e.   Edinburgh Tram System
    f.   Nottingham Express Transit
    g.   Blackpool Tram System

**Figure 1.8 Tram traversing a junction
(courtesy Croydon Tramlink Ltd.)**

1.6.4   Heritage Railways

Standard gauge private railways may or may not be connected to main lines and include around forty stretches of line totally some 400 or so route miles rescued by Preservation Societies after abandonment. These latter are usually passenger carrying and are subject to the same type of regulation, being classified as Light Railways. Typically the preserved railway came into existence with the dual object of rescuing a stretch of railway line seen by its supporters as having some special scenic or historic features worthy of retention, and of restoring steam locomotives to working, after (in many cases) they had been sold for scrap. Wherever possible, such railways took over the track which happened to be in situ at the time of abandonment which was mostly obsolete combinations of BH and FB rail and sleepers of all types. Maintenance of these railways tends to be kept to a minimum consistent with safe operation at an acceptable standard of comfort, and is carried out to a considerable extent by volunteer labour using manual techniques. Renewables and extensions are carried out often using reconditioned materials.

Figure 1.9 Heritage Railway showing the use of Bull Head Track and semaphore signals (Courtesy Bluebell Railway)

1.6.5   Freight Railways

There are also a number of railways which are purely freight orientated, and these are the descendants of the industrial railways which existed in various forms in mines and quarries long before the arrival of steam locomotion. Where these privately owned railways have a connection to mainline tracks, their load and structure gauges, and construction and operating practices must be acceptable to the railway authorities. Such railways were common and very extensive during the 19th Century, when heavy haulage using road vehicles was totally underdeveloped, and undertakings such as steelworks, gasworks, power stations, and docks used rail almost exclusively for internal transport.

1.6.6   Narrow Gauge Railway

There were, during the heyday of railway building, many railway undertakings where there was no possible link with any main line railway, and in these cases both the track gauge, and the loading and structure gauges, became parameters which could at the inception of the undertaking be determined to suit the particular conditions anticipated for it.

If tunnelling is a feature of the undertaking, as for instance in coal mines, then a narrow gauge railway has distinct economic advantages in that it allows a larger vehicle body to pass through an access tunnel of very small diameter. Such considerations will still be very important even today where the tunnel has to be driven through overburdened or some other unproductive strata, and must have been even more so where the engineering techniques available for tunnelling and roof support were primitive. The strongest shape for a tunnel lining is circular, and clearly, in such a section, the narrower the trackform can be made then, other things being equal, the lower down it can be placed in the tunnel, and the more room is left above it for payload. For example, if the tunnel diameter is limited to 3 metres for constructional reasons, then the usable height (and therefore the usable cross section of the tunnel) is increased by at least 16% by using a 762 mm track gauge instead of standard gauge. The precise increase depends upon the overall width chosen for the track form, and the shape of the completed tunnel structure. Similar considerations apply to bridge construction.

Considerations such as these led to hauling ways in mines generally being laid to gauges down to 600 mm or so, and in some cases, particularly in difficult and lightly populated terrain, where construction of a standard gauge railway would be both expensive and unlikely to pay its way, it became convenient to extend the hauling way into a surface railway. Very few of these railways have survived. The best known survivors in North Wales are the 597 mm gauge Blaenau Ffestiniog Railway, which extends 21.7 km from Blaenau Ffestiniog to Porthmadog and the 686 mm gauge Talyllyn Railway with a route length of 11.6 km.

There are a number of narrow gauge railways in Wales with gauges of around 600mm, together with a further twenty or so railways built mainly as tourist attractions in various parts of the country, to gauges varying from as small as 260 mm to 1067 mm. Extensive narrow gauge branch line systems are found in France, Switzerland and Germany. The track of these railways varies from light section FB rails welded to steel cross-ties, to smaller editions of main line track, with quite heavy section FB rail with PANDROL brand rail clips laid on hardwood sleepers. All such railways are again subject to regulation if they carry passengers, and it is worth observing that the different relationship existing between the track gauge and the rolling stock parameters imposes a discipline of line and level which may in its own way be no less stringent than that required on a standard gauge line.

1.6.7   Specialist Railway Systems

There are a number of specialist systems which often utilise additional technology and are mainly for tourist attractions. Examples of these include mountain railways and extended rollercoaster rides in theme parks.

The UK boasts one mountain railway requiring rack and pinion traction, in the shape of the Snowdon Mountain Railway. This 800 mm gauge, steam and diesel operated railway extends 7.4 km from Llanberis, on the edge of Llyn Padarn to the summit of the mountain. The line rises 957 m with an average gradient of 1 in 7.7 and has a maximum gradient of 1 in 5.5. The track comprises 40 lb/yd running rails with PANDROL brand rail fastenings on steel sleepers. The traction is applied solely to the central rack which has a double row of teeth with angled side "girders" to provide contact with the stabilising grippers on the vehicles.

The Isle of Man has a number of specialist railways which include a mountain railway and an electric railway.

1.7   LEGAL RESPONSIBILITIES OF TRACK ENGINEERING

The railway engineer's responsibility for safety can be divided into two main aspects:

i.   the construction and maintenance of a railway system that is safe for the passage of trains

ii.   undertaking work in a way that is safe for those doing the work

The engineer's responsibility in the UK derives from both Common Law and Statute Law. In Common Law, if an accident occurs resulting in a loss or injury to anyone, that person may have a Common Law claim for compensation. If, for example, it can be shown that the accident was due to negligence on the part of the Railway or its employees, then the injured person would have a claim against the Railway Authority.

However, it is up to the injured party to bring a legal action to support his claim for compensation. If the railway denies the claim, the subsequent court hearings and possibly appeals, mean that the claimant may have a long wait and no certainty of redress, to say nothing of the cost involved. Since the Common Law operates on the principle of "precedent" (which means that a Judge in determining a case, will so far as possible, follow the principles established in previous cases), a measure of "accident prevention" is derived from the knowledge that failure to take a particular precaution will be regarded by the Court as neglect to act with reasonable care in the particular circumstances. This is especially recognised in the UK legislation known as the Health and Safety at Work Act, 1974. This specified measures aimed at reducing the risks of accidents. The engineer is required to abide by the Law and failure to do so is punishable.

In the early days of UK railway regulation, three very important safety standards had to be met, namely:

- The block system of signalling to be used
- Points and signals to be interlocked
- All passenger trains to be equipped with continuous brakes which would be self-applying and therefore 'fail-safe' in the event of the train brake pipe becoming severed and resulting in loss of air pressure or vacuum

These apparently simple requirements were far-reaching at the time, and are still valid in the context of today's far more sophisticated signalling and train control systems.

It is usual for the government to lay down the requirements for the basic essentials of the safe running of any railway, including:

- Loading and structure gauges
- Clearances
- Electrification
- Design of structures and stations
- Arrangements at level crossings

1.8   TRACKSIDE SAFETY

The railway presents numerous dangers as a working environment. Some of these, such as uneven surfaces, the lifting of heavy and perhaps badly balanced equipment, and contamination from oil or other substances which can be hazardous to health, are common to civil engineering sites generally and have been for decades. More recently the rapid increase in the use of machinery including rail mounted vehicles and road/railers on sites which are by their very nature long and narrow, together with night working has increased some risks. Persons who work in this environment must understand the risks involved to preserve health and safety. The most serious categories of risk which are particular to the railway industry are:

    i.   The risk of being struck by moving trains
    ii.   The risk of electrocution by the traction current or associated equipment

It is from these hazards that the majority of fatal accidents to employees of the railway engineering departments arise. Railways have developed safety systems and instructions, the objective of which, is to help staff identify when a danger is likely to arise, and to define suitable protective procedures. These are legally required for compliance with current safety legislation.

Since a train moving fast can neither change direction nor stop to avoid people, it behoves the latter to get out of the way of the train in time. In order to do so they must first become aware that the train is approaching. This necessitates the planning of work to ensure their protection ideally by working when trains are not running (a usual requirement at speeds of more than 100mph) or the planning and use of a protection method which either separates workers from moving trains or if this cannot be achieved the provision of a warning system giving the people adequate time to move into a place of safety before a train passes.

A suitable system of safe working must be set up to protect track engineering staff which usually involves an assessment of the time needed to clear the track. An example of a sighting distance chart is shown below.

Permissible speed (mph)	Sighting distances in METERS to provide minimum warning time						
	15 sec	20 sec	25 sec	30 sec	35 sec	40 sec	45 sec
140	1000	1300	1600	1900	2200	2600	2900
125	900	1200	1400	1700	2000	2300	2600
100	700	900	1200	1400	1600	1800	2100
90	700	900	1100	1300	1500	1700	1900
75	600	700	900	1100	1200	1400	1600
60	500	600	700	900	1000	1100	1300
40	300	400	500	600	700	800	900
20	200	200	300	300	400	400	400

It is normally an essential legal requirement to carry out a risk assessment to determine whether any track engineering work can be carried out whilst trains are running.

CHAPTER 2

TRACK ENGINEERING CONCEPTS AND DEFINITIONS

2.1   INTRODUCTION

In Chapter 1 the development of railways was outlined in an historical and generalised perspective and the basic concepts of track engineering such as rails, sleepers, ballast, formation, and junctions were introduced. The processes of design, manufacture, and construction of railway infrastructure cannot however, be effectively described in detail without using the language of engineering. This chapter provides the basic concepts related to track engineering.

2.2   UNITS OF MEASUREMENT

In order to describe any object unambiguously, it is essential to have universally agreed standards by which its properties can be measured. We need to be able to describe size and weight in such a way that the person with whom we are communicating, can visualise and if necessary, reproduce the object under discussion, in identical form to the original. To do this, we need a system of weights and measures, and the bases of any such system are called "units". In general terms, a unit is an agreed standard which is capable of accurate reproduction by everyone.

Track (or permanent way as it was known originally) was developed using imperial units in the UK such as inches, feet, yards and miles, however due to the sophisticated inter-relationships between different units and their non-universality, metric units are now used. This is known as the SI system.

One metre is equal to 3.2808 feet (39.37 inches). It is usually represented by the lower case letter "m". For convenience, a number of multiples and subdivisions of the metre are defined. The commonest of these are:

- The kilometre which is 1000 metres. It is usually represented by "km".
- The millimetre which is 0.001 metre. It is usually represented by "mm" (ie, 1000 mm = 1 m).
- The centimetre, which is 10mm, and is usually referred to as "cm".

Although the centimetre is not a recommended term for use in engineering, it is used occasionally in this book.

2.2.1 Area and Volume

The base units for areas and volumes are square metres (m^2) and cubic metres (m^3) respectively. A square metre is the area of a square whose sides are one metre long. Similarly a cubic metre is the volume of a cube whose sides are one metre long. The commonest subdivisions of a square metre (m^2) are square millimetres (mm^2) and square centimetres (cm^2); $1m^2 = 10,000cm^2 = 1,000,000\ mm^2$

The usual multiple of a square metre is the HECTARE, which is the area of a square whose sides are 100 m long. 1 hectare (ha) = 10,000 m^2. 1 hectare is equivalent to approximately 2.47 acres.

The commonest subdivision of the cubic metre (m^3) is the cubic centimetre (cm^3); $1m^3 = 1,000,000\ cm^3$. Of importance also, the LITRE (l) is the volume contained in a cube whose sides measure one tenth of a metre (10cm or 100mm). One litre contains 1000 cm^3. This relationship gives rise to the other and nowadays more commonly encountered name for a cubic centimetre, the MILLILITRE or *MIL* (ml). $1\ m^3 = 1,000,000\ cm^3 = 1000l$ and also $1l = 1000\ cm^3 = 1000$ ml.

2.2.2   The Unit of Mass

The term MASS is used in science and engineering as distinct from
the term "weight", to denote the quantity of matter which an object
contains. The distinction between "weight" and "mass" in scientific
terms will be further discussed below. The unit of mass in the Metric
system is the GRAM. The gram is defined as "the mass of a cubic
centimetre of water at the temperature of melting ice".
In engineering it is not often necessary to refer to subdivisions of
grams, but since the gram is quite small, it is common to use
multiples, the most common of which are the KILOGRAM (kg) and
TONNE (t). The relations between these quantities are:

1 kg = 1,000 g and 1t = 1,000 kg.

The following conversions from "Imperial" measures may be useful:
1lb = 453.6 g, 1kg = 2.205 lb, 1tonne = 2,205 lb = 0.9842 ton. (lb is a
pound weight, ton is an imperial ton which is 2240 lb.)

2.2.3   Time

In order to describe concepts such as speed it is necessary to refer to
a third basic measure, that of TIME. "Time" has the standard
dictionary meaning of the continuum within which a series of events
take place, and is measured in seconds (s/sec), minutes (min), hours
(h/hr), and years (yr).

2.2.4   The Measurement of Angles

Most readers will be familiar with the measurement of angles in
degrees, using a protractor. A degree is an arbitrary measure of angle,
and may be readily comprehended as being one ninetieth of a right
angle. Measuring an angle with a protractor is however not much
practical use for work on site, and some readers will be familiar with
the practice of referring to angles by a number (eg a "one in ten"

crossing). This is often called the UNIT METHOD of angular measurement.

Since it is necessary to be able to relate this method of measuring angles to the measurement of angles by degrees, it is necessary to be precise in defining exactly how the distances involved are measured. The method is called the CENTRE LINE METHOD (CLM), and this will now be described.

Refer now to Figure 2.1(a). In this figure, the angle we wish to measure is that between BA and BD, which we shall call angle ABD. To make the measurement, we set out two equal distances "c" along the two lines BA and BD, and measure the distance AD, which we will call "b". Next, we find point C, midway between A and D, and measure the length BC, which we denote by "a". Then in CLM, the angle ABD is one in N, where:

$$N = \frac{a}{b} \dots\dots\dots\dots 2 - (1)$$

Clearly, if we arrange the measurements so that the length "b" of line BD is equal to one unit of length, then N is numerically equal to the length "a" of line BC.

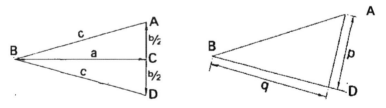

Figure 2.1 (a) Centre Line Measure   (b) Right Angle Measure

(c) Isosceles Measure

There is a practical difficulty in measuring the length of line BC in a real life layout, due to the fact that the rails do not end in fine points. To overcome this, it is usual to measure the spread across from rail to rail at two places on either side of the crossing nose, where the spread from running edge to running edge is, say 100mm, and to measure the distance between these two places along the central axis of the vee. The N value will then be this distance in mm divided by 200.

Example: We find that the distance between the two places where the spread is 100mm is 2150mm. Calculate N.

$$N = \frac{2150}{200} = 10.75$$

2.2.4   Dimensional Analysis

The measures used in engineering are built up in a systematic fashion from the abstract concepts of length, mass, and time described above. In abstract, as distinct from literal terms, these concepts are known as "DIMENSIONS". When writing about "dimensions" in the abstract, it is customary to write the concept under consideration in square brackets.

The dimension "length" is written as LENGTH, abbreviated as L
The dimension "mass" is written as MASS, abbreviated as M
The dimension "time" is written as TIME, abbreviated as T

It is important to distinguish between the abstract dimensional concept, in which the UNITS are not required to be specified, and the real life concept which only exists alongside statements of both magnitude and also the units in which it is measured. Thus for example LENGTH stands by itself but the length of a rail must be stated as eg, 18.3 metres.

These concepts have already been discussed in terms of the units used to measure them. However, each can be subjected to dimensional analysis, and this elementary example should serve to show the principles of the method:

$$AREA = L \times L = L^2 \quad and \quad VOLUME = L \times L \times L = L^3$$

As explained in Section 2.5 above, the measurement of an angle in the CLM system is obtained by dividing one length by another. By following the principles described above, the dimensions of the quantity "N" are:

$$[N] = \frac{[L]}{[L]} = 1$$

Therefore "N" is a dimensionless quantity because any unit divided by itself is equal to unity. This kind of relationship is called a "RATIO".

2.2.5 Relationships Between Different Methods Of Measuring Angles

The correct way of measuring an angle involves determining a ratio between two lengths. The most familiar method is to measure an angle in degrees using a protractor. What is actually done here is to measure the length of the portion of the circumference of a circle described from the intersection of two lines as centre and which is intercepted between the two lines. A "degree" is the angle subtended by one-360th part of the circumference of a circle. However, the actual length of circumference corresponding to a degree, varies according to the radius of the circle. Hence the magnitude of the angle can only be described by relating the length of the portion of the circumference of the circle embraced by the two arms forming the angle, to its radius. The selection of the factor 360 for a degree, is related to historical methods of measuring time and position in navigation. It is in a sense arbitrary, and does not fit very well with decimal systems of calculation. For this reason, another arbitrary measure, the "grad", is sometimes used. A grad is the angle subtended at the centre of a circle by an arc of whose length is one-400th part of its circumference.

If the angle is measured by the direct ratio between arc and radius, the name given to the unit is the RADIAN (rad). In this system, if the length of arc is equal to the radius, the angle subtended is one radian.

When working in radians therefore, an angle is defined by the equation:

$$\theta = \frac{L}{R} \ \text{- - - - - - - - - - - - - - - - - -} \ 2 - (2)$$

where θ is the angle subtended in radians (NB; $360° = 2\pi$ rad).

L is the length of arc
R is the Radius.

Angles can also be defined by the ratios between the sides of a right angled triangle. Consider the right angled triangle ABC, shown in Figure 2.1(a), in which the angle to be measured is angle ABC and angle BCA is a right angle. No matter how large or small the triangle is made, the ratios between the lengths of the three lines AC, AB and BC will always be the same.

The three lines can be paired in three ways, as follows:
AB with AC
AB with BC
BC with AC

The ratios derived from these pairings form a series of functions which are of very wide application throughout surveying and engineering design generally.

The ratio AC/AB is called the SINE of angle ABC.
The ratio BC/AB is called the COSINE of angle ABC.
The ratio AC/BC is called the TANGENT of angle ABC.

These words are usually abbreviated to SIN, COS, and TAN.

If we now apply the TANGENT ratio to Figure 2.1(a), we have if the angle ABD is θ :

$$tan \frac{\theta}{2} = \frac{AC}{BC} = \frac{b/2}{a} = \frac{b}{2a}$$

If as before we make b = 1, then a = N so we can write:

$$tan \frac{\theta}{2} = \frac{b}{2a} = \frac{1}{2N} \text{-------------} 2-3$$

Transposing

$$N = \frac{1}{2 \, tan \, \theta/2} \text{---------------} 2-4$$

It should also be noted that there are two other ways of defining "N" besides CLM. One of these should be obvious from the foregoing. It is known as "Right Angle Measure" (RAM) as shown in Figure 2.1(b). If "N" in RAM is called Nr, then:

$$Nr = \frac{1}{tan \, \theta} \text{---------------} 2-5$$

The third method is called the "Isosceles Measure" (IM), since as shown in Figure 2.1(c), it involves measuring the length of one of the two equal sides of an isosceles triangle. If "N" obtained by this method is called $N1$ then:

$$N1 = \frac{c}{b} = \frac{1}{2 \, sin \, \theta/2} \text{-----------} 2-6$$

The small differences between the values of N for the same value of θ, can be determined by using equations 2-(3), 2-(4), and 2-(6).

For example, if N in CLM is 8, then:

θ = 7.15266875 degrees

Nr = 7.96875

$N1$ = 8.015610

Whilst these differences are apparently small (incidentally they become more significant as θ increases), and it may appear easier to apply the Isosceles Method, particularly on the ground, the advantage of the CLM method is that it lends itself very easily to calculation without resort to trigonometrical ratios.

2.3 FORCES, MOTION AND STRESS

2.3.1   Velocity and Speed

If an object is in motion, it possesses the quality known as velocity. Velocity has both magnitude and direction. If the velocity does not change from moment to moment, then the object is described as having a constant velocity. The magnitude element of velocity is called speed. In general, a quality having magnitude and not direction is called SCALAR, whilst one having both magnitude and direction is called a VECTOR, and the latter is not fully described unless both elements are quoted. Thus one cannot fully describe the movement of a train merely by giving its speed. However, the speed at which an object is travelling is expressed in terms of the distance travelled in a unit of time.

Its Dimensional Analysis therefore is:

$$[SPEED] = [L] \times [T]^{-1} \ or \ \frac{[L]}{[T]}$$

Typical units are metres per second or kilometres per hour (m/s, km/h).

2.3.2   Acceleration

If the velocity of a particle changes either in magnitude or direction, then that particle is said to undergo acceleration. Thus acceleration is also a vector, like velocity, and to define it fully, requires a statement of both magnitude and direction. The magnitude of acceleration is described as the change of speed in a unit of time, and has the dimensions:

$$[ACCELERATION] = [SPEED] \times [T]^{-1} = [L] \times [T]^{-2} \text{ or } \frac{[L]}{[T]^2}$$

Typical units are metres per second per second (m/s^2) and millimetres per second per second (mm/s^2).

An example of the concept of acceleration as a vector which is of particular interest in the context of this book, is to consider what happens when an object such as a train is moving around the circumference of a circle such as that shown in Figure 2.2. Its speed is constant, in the sense that its rate of forward motion along the rails does not change. However, if the motion of the train as it approaches and passes point X, is viewed from a point on the tangent, say B, it is seen that as the object approaches X, it will appear to have motion in two directions at the same time. When the object is at Position 1 in Figure 2.2, it will be moving towards B (if it is moving clockwise), but it will also appear to be moving away from the centre of the circle. However, this movement is slowing down and as it passes Position 2, it appears to have no outwards motion at all, as it is moving along in a direction which is tangential to the curve. By the time the object has reached Position 3, it will be moving towards the centre of the circle at ever increasing speed. Thus the object can be seen to be undergoing an acceleration at right angles to its constant forward motion. This is referred to as *radial acceleration.* Its magnitude can be calculated if the forward speed and the radius of the circle are known, from the following formula:

Let the radial acceleration be   a
Let the radius be   R
Let the speed be   V

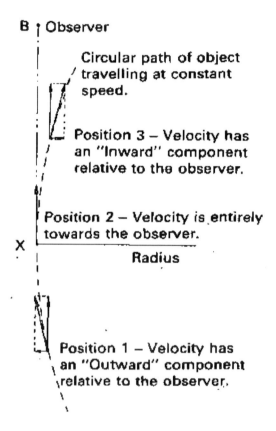

B ↑ Observer

Circular path of object travelling at constant speed.

Position 3 – Velocity has an "Inward" component relative to the observer.

Position 2 – Velocity is entirely towards the observer.

X

Radius

Position 1 – Velocity has an "Outward" component relative to the observer.

Fig. 2.2 Acceleration

Then

$$a = \frac{V^2}{R} \quad \dots\dots\dots\dots\dots\dots\dots\dots\dots\dots 2-7$$

For the proof of this formula, the reader should consult any textbook on applied mathematics.

2.3.3 Newton's Laws of Motion and The Dimension of Force

Isaac Newton, is generally accredited as being the first person to realise the relationships between force and motion. He expressed these relationships in three statements which are generally referred to as Newton's Laws of Motion, upon which effectively the whole of engineering science is based. The first of these laws states that unless acted upon by an external force of some kind, an object will continue in a state of rest or uniform motion in a straight line. Thus the word "Force" in engineering terms is formally defined as that which tends to disturb the state of rest or uniform motion of an object. The second law states that when a force P is applied to an object of mass M, it undergoes an acceleration a, and P, M, and a are related by the equation:

$$P = M \times a \quad\text{................................} 2-8$$

This applies either if the object is at rest and free to move and it is desired to set it in motion, or if the object is moving at constant speed in a straight line, and it is desired to change its velocity in either magnitude or direction, or both. We apply the principles of Dimensional Analysis to the parameters M and "a" to determine the dimension of [FORCE] thus:

$$[FORCE] = [MASS] \times [ACCELERATION]$$
$$= [M] \times [L] \times [T]^{-2}$$
$$= [P]$$

The unit of Force used in the SI system is the *newton,* after the aforesaid mathematician. It is defined as that force which will cause a mass of 1kg to accelerate by $1m/s^2$, and it is given the symbol N. It should be obvious from its engineering definition, that a force is a vector just like velocity and acceleration. Newton's Third Law states that every force is accompanied by an equal and opposite reaction.

2.3.4 Centrifugal force

Since as explained above, an object travelling in a circle undergoes a continuous acceleration as defined in equation 2-7, we can substitute this value for "a" in equation 2-8, and we then see that the object is subject to an inwards acting force whose magnitude is given by:

$$Centripetal force = M \times \frac{V^2}{R} \quad - - - - - - - - \quad 2-9$$

When a wheelset is going round a curve, this centripetal force must be applied by the track.

Hence, by Newton's Third Law, the track experiences a reaction which is equal and opposite to the centripetal force, that is to say that the track feels as if it were being pushed outwards by the wheels. This is the so-called centrifugal force. Its magnitude is the same as the centripetal force.

The centrifugal force acts horizontally at the centre of gravity of the vehicle.

2.3.5 Mass and weight

The formal definition of the word "force" uses the words "tend to" because as already explained, force may be applied in circumstances where the body does not actually move or it may be applied to prevent motion. The most obvious example of the use of force to prevent motion is in the action of holding an object in the hand against the force of gravity.

Imagine that you are holding in your hand, an object which in common parlance is said to weigh one kilogram. The correct engineering statement about the quantity of material contained within that object is that its mass is 1kg, and that the force you are applying to support it is equal and opposite to the effect of gravitational attraction on that mass. If you ceased to apply a force to the object (eg, you drop it), then the force of gravity would cause it to move downwards with an acceleration of 9.81m/s², and it follows that the force you are applying, which you perceive as the "Weight" of the object, is:

$$\{1kg\} \quad @ \{9.81m/s\} \quad = 9.81 \ newtons$$

If you were out in space, free of all gravitational pull, then the object would have no weight at all, that is to say that you could hold it in your hand without applying any force at all to hold it in place, but it would still have a mass of 1kg. The difference between "Mass" and "Weight" is that whereas "Mass" refers to the quantity of matter contained within an object, "Weight" refers to the effect of gravitational attraction on that object. Because of the all-pervading nature of the Earth's gravitational attraction, it is common to use another unit for force, the "Kilogram Force" (kgf), which is the force due to gravity on a mass of lkg.

$$1 \ kgf = 9.81 \ newtons$$

This explains why the vertical force due to the vehicle mass, dealing with the relationships between centrifugal force, cant, and wheel weight transfer, includes the term "g" for the acceleration due to gravity.

2.3.6   Equilibrium of Forces and Vectors

Newton's Third Law introduces the concept of equilibrium of a force system. For example, a rail stretched by a tensor is in equilibrium, because the force applied by the tensor is balanced by the tension in the rail, which is in turn balanced by the restraint applied by the rail fastenings in the anchor length. In the above example, we considered only two forces, and it is fairly obvious that to be in equilibrium, a pair of forces must be equal and opposite, and they must line up with one another. If a set of three or more forces act upon a body, the conditions for equilibrium are somewhat more complex, and depend firstly upon whether the lines of action of all the forces meet at a single point. If the lines of action of the forces do indeed meet at a point, and a state of equilibrium exists, then the relations between the forces can be determined by the rule known as the Triangle of Forces and Equilibrium of Forces. (See Figures 2.5 and 2.6)

a) Representation of Vectors

A vector can be represented on a drawing by a line of a specified length and direction. Thus in Figure 2.3(a), lines AB, AC, and AD are all vectors. Their magnitude is indicated by the length of each line, and the direction by both the orientation of the lines and by the arrow. Thus vector AD in Figure 2.4(a) is equal and opposite to vector EF in Figure 2.3(b). Note that the position of the line on the drawing does not necessarily indicate the position of the vector in space.

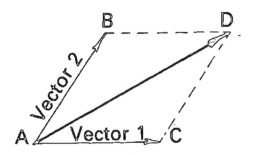

Figure 2.3 Parallelogram of Vectors

b) The Parallelogram of Vectors

The rule for combining two vectors is called the PARALLELOGRAM OF VECTORS. This is illustrated in Figure 2.3. In this figure, AB represents vector (1) in both magnitude and direction, and AC represents vector (2) similarly. The parallelogram is completed by drawing CD equal in length to AB and parallel to it, and BD equal and parallel to AC. Then the single vector which is the aggregate of vectors (1) and (2), is represented in both magnitude and direction by the diagonal AD. This vector is called the RESULTANT of vectors (1) and (2).

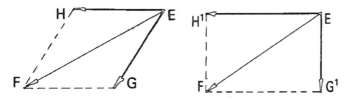

Figure 2.4 (a) Resolution of a Vector   **(b) Resolution of a Vector into components at Right Angles**

c) Resolution of Vectors

Conversely, it is possible to represent one vector by any two others, provided that the latter form two sides of a parallelogram with the original force as diagonal. This is shown in Figure 2.4. In Figure 2.4(a), the shape of the parallelogram is the same as in Figure 2.3. Vector EF which is equal and opposite to vector AD, is divided into two forces EG and EH, which are equal and opposite to vectors AB and AC. However, any convenient parallelogram will serve, and the commonest use of the device is to replace the original vector by two vectors which are mutually at right angles. This is shown in Figure 2.4(b). By means of this device, known as RESOLUTION of forces, a single force can be replaced by two other forces acting at right angles to one another. When a force is replaced in this way, it is said to be resolved into components. This concept is used in Figure 2.2 to show the change of direction of an object travelling around a circular curve at uniform speed.

d) Equilibrium of forces

If the vectors of the previous paragraphs are identified as forces, we can state that the resultant of two forces AB and AC is one force AD. We can further see that the force required to balance forces AB and AC, is a force of the same magnitude as the resultant AD, and acting in the same direction as the resultant, but in the opposite sense. The lines of action of the three forces must also meet at a single point. This balancing force is shown as force No 3 (which is line AD[1]) in Figure 2.5. Generalising this statement: If three forces acting upon the same point are in equilibrium, then if we take any two of these forces and combine them, the resultant will be equal in magnitude and opposite in direction to the third force.

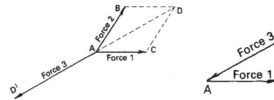

Figure 2.5 Equilibrium of Forces   **Figure 2.6 Triangle of Forces**

e)   Triangle of Forces

Another way of expressing the situation described in above, is shown in Figure 2.6. Here in Triangle ACD, vector AC represents force No 1 as before, but force No 2 is now represented by vector CD, which is identical in magnitude and direction to line AB in Figure 2.5. The third side of the triangle is vector DA, which is clearly identical with vector AD[1] and is hence a complete representation of force No 3. Hence triangle ACD represents the three forces AC, AB, and AD[1] which are in equilibrium. Thus we have another rule, which states that if three forces acting upon the same point can be represented in magnitude and direction by the three sides of a triangle taken in order, then those three forces are in equilibrium. This is called the TRIANGLE OF FORCES.

2.3.7 Stress of forces

If two equal and opposite forces are applied to stretch a metal bar (eg, a rail), the bar is said to be in a state of direct tensile stress. The degree of stress in the bar is obtained by dividing the force applied, by the cross-sectional area of the bar. If this example is generalised to include compressing as well as stretching, we can say that direct stress is defined as "The force acting per unit area affected". In dimensional terms:

$$[STRESS] = [FORCE] + [AREA]$$
$$= [M] \times [L] \times [T]^{-2} + [L]^2$$
$$= [M] \times [L]^{-1} \times [T]^{-2}$$

Stress is usually given the symbol "f" and its basic units are newtons per square metre. The unit "newton per square metre" is termed the *pascal* and is given the symbol Pa. This unit is very small, and the unit more commonly used in the applications with which we are concerned, is the Megapascal, or MPa, which is equal to one million Pa. 1 MPa is the same as 1 N/mm^2.

To connect this measure with more familiar notions, atmospheric pressure in Imperial units is 14.7 lbf/in^2. In SI units it is 0.1013 MPa. The breaking strength of Grade 'A' rail steel is 900 MPa. The standard conversion factor between Imperial and SI units is 1 tonf/in^2 = 15.444 N/mm^2 (MPa).

When a metal bar is stretched, it extends, and when compressed, it shortens. These statements are obvious, but certain important relationships have to be noted in connection with these changes of length.

If the bar is marked out in equal divisions of unit length before the force is applied, it will be found (assuming that the bar is of constant cross-sectional area) that the length of each and every segment of the bar changes by the same amount under the action of the forces. The amount of this change in length per unit length of the bar is called STRAIN.

This can be expressed as a formula:

$$STRAIN = \frac{Change\ in\ Length}{Original\ Length}$$

Since the "Dimension" of both "Change in Length" and "Original Length" is [L], STRAIN is a dimensionless RATIO. If the force applied to the bar is small compared with the force required to break it, it will be found that the strain caused by the force is proportional to the stress. This is expressed as:

$$\frac{STRESS}{STRAIN} = E$$

Where E remains constant whatever the stress. This relationship is often referred to as HOOKE'S LAW. The constant of proportionality, E, is termed the MODULUS OF ELASTICITY, or YOUNG'S MODULUS. Since the term "strain" is a non-dimensional ratio, it follows that the units in which E is expressed are the same as those for "stress", e.g., N/mm^2. The value of E for rail steel is 2.069×10^5 N/mm^2. For concrete, the value of E is about 7% of that for steel.

Provided again, that the force applied to the bar is not excessive, the bar will return to its original shape precisely, when the load is released. Such behaviour is described as ELASTIC, and classic methods of analysis of engineering structures assume ELASTIC BEHAVIOUR throughout.

As the stress in the bar is increased, there comes a point when any further increase causes a change in the behaviour of the metal. It starts to change shape at a greater rate as stress increases, and when the load is released, the bar no longer returns to its original shape. The stress at which this starts to happen is called the YIELD STRESS. It occurs at the YIELD POINT, or LIMIT OF PROPORTIONALITY. When structures are designed assuming elastic behaviour, it is necessary to ensure that at no point in the structure does the stress in a member exceed its yield point. In order to give security against accidental overload, effects of corrosion etc, it is usual to limit the stress to a value significantly below the yield stress. This limiting stress is called the ALLOWABLE STRESS or SAFE WORKING STRESS.

If the bar is stretched beyond its yield point, it starts to become thinner, and eventually, as stress increases still further, a neck starts to form at some point in the middle of the bar, and immediately after this happens, the bar breaks. The stress at which this occurs is called the ULTIMATE TENSILE STRESS (UTS). If the bar is in compression, failure occurs either by buckling, if the bar is slender, or it becomes steadily more barrel shaped, if it starts off fairly squat in shape.

2.3.8 Torsion - couples, and moments

We saw above that three forces whose lines of action meet at a point, can be in equilibrium under certain conditions. We now consider what happens if the forces do not all meet at the same point.

A person driving home a chairscrew with a Tee-headed box spanner, grasps the ends of the cross-bar and turns it by simultaneously pulling with one hand and pushing with the other. The forces applied by his hands are equal and opposite and act parallel to one another. Such a combination of forces is extremely common and is called by engineers, a COUPLE. The magnitude of a couple is described as its MOMENT and is defined by multiplying together the applied force and the distance between the two forces, as shown in Figure 2.7. The distance between the forces is called the MOMENT ARM. Applying the principles of dimensional analysis again, we have:

$$[MOMENT] = [FORCE] \times [MOMENT\ ARM]$$
$$= [MLT^{-2}] \times [L]$$
$$= [ML^2T^{-2}]$$

The units of MOMENT are "newton-metres" or some multiple thereof. The symbol for newton-metres is Nm. It is important to note that the moment produced by a couple is independent of the point at which it is measured. However, if the point at which the moment is calculated does not lie on the line of action of one of the forces making up the couple, then both forces must be counted. Thus in Figure 2.7, the magnitude of the couple can be obtained by multiplying either force "P" by the moment arm "A", or by adding together the product of both the forces "P" and the distance between the point of application of the force and the axis of the spanner shaft, which would be "A/2". The result is in each case the same, ie, "PA". The twisting effect of the couple upon the vertical rod connecting the crossbar of the spanner and the bolt box is referred to as torsion.

2.3.9 Shear stress and shear strain

The effect which torsion has on the metal of the spanner shaft is to strain it in a way which is different from that described above. As shown in Figure 2.7, the effect is that a small square drawn on the surface of the rod becomes diamond shaped. This type of strain is called SHEAR STRAIN and the stress associated with it is called SHEAR STRESS.

Shear stress can be caused in other ways. For example, when a train is braking or accelerating, it tends to pull the rail through the fastenings, and in so doing the upper surfaces of the rail pads are subject to a force in one direction, and the lower surfaces to an equal and opposite force, thus applying a couple to the pad, and causing the pad to become parallelogram-shaped. The shear stress in this case can be easily calculated, being simply the force applied to one face of the pad, divided by the area of the pad, and its units are the same as those quoted above.

There are only two types of stress to which a material can be subjected, direct and shear, and these have now been described. We now show that there is a connection between them.

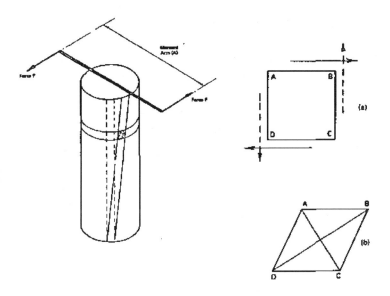

Figure 2.7 A couple acting to produce Torsion in a bar.   **Figure 2.8 Shear Strain and Principal Stresses.**

When an element (ie, a very small square such as that shown shaded in Figure 2.7 of material is placed in shear, equal and opposite forces produced by the externally applied couple, operate along two opposite faces (eg, faces AB and DC in Figure 2.8(a)). By Newton's Third Law, these forces are opposed by two other forces along the complementary faces (also as shown in Figure 2.8(a)).

When this happens, the square element is distorted and becomes a parallelogram with sides of equal length (the geometrical term for which is a RHOMBUS) (as shown in Figure 2.8(b)).

In this diagram, the diagonals of the rhombus have been drawn. Obviously they are of unequal length. If these diagonals are compared with the diagonals of the original square, it is seen that one has become longer and the other shorter. Hence the effect of shearing the element has been to extend the material of which it is made, in the direction BC (ie, it is put into tension) and at the same time the material has been compressed in direction AD. Furthermore, the two diagonals BC and AD are at right angles, just as the diagonals of the original square were. The tensile and compressive stresses in directions BC and AD are called PRINCIPAL STRESSES, and the planes represented by BC and AD are the PRINCIPAL PLANES, since the stresses are at their maximum across those planes.

(NOTE: A simple demonstration of this behaviour can be obtained by drawing a small square on the surface of an india-rubber eraser and distorting the latter by hand, when the square will change shape as described).

If the direction of shear reverses, as happens in a rail when a wheel moves from one rail to the other over a fishplated joint, then clearly the stress across each of the principal planes will also reverse, and it can be seen how the potential for fatigue arises from fluctuating shear loads. Furthermore, since the principal planes are at 45 degrees to the vertical, it is easy to see why it is that star cracks around fishbolt holes in rails are always at 45 degrees to the vertical.

The example of the Tee-headed spanner was selected because the action imposed by the torque applied to the handle of the spanner on the shank, is pure shear. Very often other forces besides pure shear come into play. For instance if a single-ended spanner is used to turn a bolt, a torque is still applied to the bolt, but the operator only supplies one of the two equal and opposite forces which constitute the twisting couple. The other is supplied by a force applied at the jaws of the spanner, and this force acts transversely across the bolt, as illustrated in Figure 2.9. If the bolt has a long unsupported shank, this force will tend to bend the bolt.

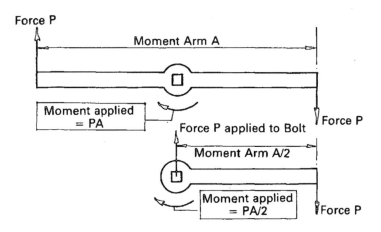

Figure 2.9 Difference between Double ended and Single ended spanners

When as already described, a rail pad is subject to shear due to traction or braking forces, the situation is complicated because the wheel loads also apply pressure. In this case, there will, in Figure 2.8, be a pressure applied to the faces AB and CD of the element, at the same time as the shear along the same surfaces. The principal stresses discussed previously will still be there, but their value will change, and the orientation of the principal planes will change. If in fact the compressive stress is large enough, both principal stresses can become compressive.

These effects are important in understanding the behaviour of elastic rail pads, and their theory will be further developed in Chapter 6.

2.4 BEAMS UNDER A COMBINATION OF EXTERNAL LOADS

Suppose that a beam AB (see Figure 2.10(a)) which is weightless and pinned at A to a rigid support and mounted on rollers at B, is subject to an inclined force P. Clearly there will be forces operating at A and B which taken together with P, are in equilibrium. To determine what those forces are, we first resolve the inclined force P, into horizontal and vertical components H and W respectively, as shown in Figures 2.10(b) and 2.10(c). Consider first the force system shown in Figure 2.10(b). It is immediately obvious that since the beam is not able to move away from A, the horizontal reaction R_H must be equal and opposite to H, and that there will be a tensile force of this magnitude in the beam

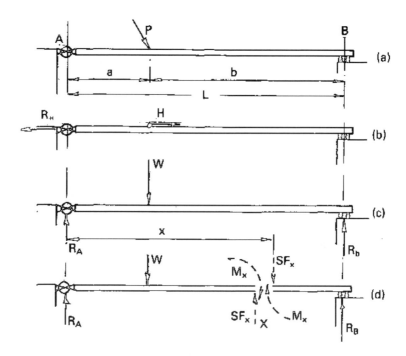

Figure 2.10 Beam under External Loads

The vertical force W, will be resisted by vertical reactions at A and B, as shown in Figure 2.10(c), but it is not possible by inspection, to say how big, forces R_A and R_B must be, in order that the bar shall not move, since we have two "unknowns". The problem is solved as follows:

Since the beam does not rotate, the moment of R_B taken about A must be equal to the moment of W taken about A. We can write:

$$R_B \; x \, L = W \, x \, a$$

Or:

$$R_B = \frac{Wa}{L} \quad \dots\dots\dots\dots\dots\dots\dots\dots\dots\dots\dots\dots\dots\dots.2-10$$

Since by definition, the bar remains stationary, Newton's Laws tell us that reactions R_A and R_B taken together must be just large enough to balance W. Hence we can write:

$$R_A = W - R_B = \frac{(L\text{-}a)W}{L} = \frac{bW}{L} \quad \dots\dots\dots\dots\dots\dots.2-11$$

The force systems represented by Figure 2.10, although much simplified, are in principle, those which act on a simply-supported bridge girder carrying a train. Force P represents a wheel load combined with traction/braking load.

The example is a particular case of the general principle that a body is in equilibrium under a set of external forces if the following conditions are met:

i. The forces on the body are resolved in two directions at right angles and the sums of the resolved forces in each direction must be zero.

ii. The sum of the moments of all the forces acting on the body, taken about any point, must be zero.

2.4.1   Internal Actions in a Beam - Bending Moments and Shearing Forces

We have seen that bars subject to direct and shearing loads, develop internal stresses. The question arises, what kind of stresses develop in a beam such as that shown in Figure 2.10.

It is intuitively evident that the beam will tend to jack-knife downwards under load W, and that if the beam is not strong enough, that is just what will happen. It is perhaps less evident that the same tendency exists, but to a lesser extent, at every point in the span. To see this, imagine what would happen if the beam were cut, say at X, and that all the external forces continue to act. Consider the equilibrium of the right-hand end of the bar as shown in Figure 2.10(d). A downwards vertical force of magnitude R_B must act at the cut end to balance the reaction R_B. Similarly for the portion of the beam to the left of X, the external forces R_A and W must be balanced by a vertical force, whose magnitude can be obtained from equations 2-10 and 2-11 and is clearly equal to R_B. This vertical internal action in the beam is called the SHEARING FORCE.

Applying condition (ii) above to the right-hand portion of the cut bar, and taking moments about X, we find that the moment of R_B about X is:

$$R_B \ x(L\text{-}x)$$

For equilibrium, there must be a moment equal and opposite to this moment in the beam at X. If we refer to this moment as M_x, then by applying equation 2-10:

$$M_x = \frac{W(L-X)a}{L} \quad\ldots\ldots\ldots\ldots\ldots\ldots\ldots\ldots\ldots 2-12$$

whilst for the portion to the left of X, the moments of R_A and W about X are:

$$R_A \ x - W(x-a)$$

Similarly this net moment must be balanced by an equal and opposite moment in the beam at X, and if the value of R_A is substituted from equation 2-11, it will be found that the value of M* so obtained, is equal and opposite to the value obtained from equation 2-12. This internal moment is called the BENDING MOMENT.

This very simple example demonstrates that generally speaking, when a beam supports external loads, internal force systems called BENDING MOMENT (BM) and SHEAR FORCE (SF) are produced, and that these vary along the length of the beam.

To sum up, the Bending Moment is equal to the algebraic sum of the moments of the external forces taken about one side of the section under consideration, whilst the Shear Force is equal to the algebraic sum of the external forces acting on one side of the section. The units of BM are those of MOMENT (newton-metres) whilst those of SF are those of FORCE (newtons).

2.4.2   Shapes for beams or girders

It is a matter of common observation that beams sag under load. This observation provides the key to understanding how the beam resists the bending moments described in the preceding section. As the beam sags, it becomes curved, so the length of the top surface of the beam shortens, whilst the length of the bottom surface becomes slightly longer. The beam can be visualised as a bundle of fibres clamped together, with the top fibres compressed and the bottom ones stretched. Towards the middle of the beam the compression and stretching is less, so that somewhere in the middle of the beam there is a fibre which has no axial stress. This is called the *neutral axis* (NA). The stress "f" (compressive or tensile) in any particular fibre at a distance *"y"* from the NA is a function of the bending moment "M", and of a particular geometrical property of the beam's cross section, known as its "Moment of Inertia" (I), values of which are usually obtained from steel tables. It is given by the expression:

$$\frac{M}{I} = \frac{f}{y} \quad \dots\dots\dots\dots\dots\dots\dots\dots\dots\dots\dots\dots\dots\dots\dots\dots\dots 2-13$$

The aggregate of all the tensile and compressive forces in the fibres forms a couple and it is this which holds the beam up against the BM. This couple is called the Resistance Moment or Moment of Resistance (MOR). From what has already been written, it will be clear that the fibres farthest from the NA will be the most efficient from the point of view of providing this MOR, and therefore a cross section such as that of a railway rail, with substantial top and bottom flanges and a fairly slender web joining them is most efficient. There are other reasons for the particular forms taken by the head, foot, and web of a rail, and these will be discussed in Chapter 3.

The MOR of any beam is the value of "M" obtained from equation 2-13 by substituting for "f" and for "y", the maximum allowable stress in the material from which the beam is made, and the distance of the farthest fibre from the NA, respectively.

The sleepers which carry railway rails are also beams, and although they do not span openings in the sense described in this chapter, the interactions between the rail seat reactions and the supporting, but at the same time, yielding, ground, produce bending and shear forces which the sleeper must be strong enough to resist.

It is emphasised that whatever the shape of the beam, the supporting of loads across an opening or on a yielding foundation is always accompanied by the development of bending and shear as outlined in principle above. Aspects of the subject relevant to rails and sleepers are dealt with in detail in. Anyone wishing to learn more about the subject in general terms is recommended to study a textbook dealing specifically with structural analysis.

2.5 CENTRIFUGAL FORCE, EQUILIBRIUM CANT, CANT DEFICIENCY AND WHEEL WEIGHT TRANSFER

This section deals with the concepts of resolution of forces, equilibrium of forces and moments, to derive the standard equations used to determine the speed potential of curves, and to derive relationships between speed, radius and weight transfer.

2.5.1 Equilibrium Cant

We start by considering the case of an evenly loaded vehicle travelling at constant speed along straight track with no cant. It is immediately evident that since the weight of the vehicle is acting vertically downwards through the longitudinal axis of the track, the weight is evenly divided between the wheels, so that if:

M = Vehicle Mass [kg]
g = acceleration due to gravity [m/s²]
Q = Wheel weight [N]
e_a = applied cant [m]
s = distance between centres of wheel/rail contact patches [m]

Then:

$$Q = \frac{Mg}{2} \quad \text{..} 2 - 14$$

Similarly, the weight force is acting perpendicular to the floor of the vehicle, so that if an object is placed on the vehicle floor there can be no transverse component acting on it. If the track is still straight, but canted over to one side, then the vehicle floor is no longer perpendicular to the line of action of the weight force, and that force will resolve into two components, one in the plane of the floor and one at right angles to it.

The relationship between the weight force and its components, is expressed by the triangle ABC in Figure 2.11 (AC represents M$_g$, AB represents T_m, BC represents P_m). This triangle is geometrically similar to triangle IOF in which:

$$IO \, represents \; s; \quad OF \, represents \; e_a; \quad IF \, represents \; \sqrt{s^2 + e^{2a}}$$

Because the triangles are similar;

$$\frac{IO}{AC} = \frac{OF}{AB} = \frac{IF}{BC}$$

hence:

$$\frac{s}{Mg} = \frac{e^a}{T_m} = \frac{\sqrt{s^2 + e^{2a}}}{P_m}$$

Thus the force in the plane of the rails is T_m where:

$$T_m = Mg.\overset{e_a}{\overline{s}}$$

and the force perpendicular to the plane of the rails is P_m where:

$$P_m = Mg. = \frac{\sqrt{s^2 + e^{2a}}}{s}$$

We now consider the conditions when the track is curved. We know that a centrifugal force will operate. If

$$v = Vehicle \, Speed \, [m/s]$$
$$R = Radius \, [m]$$

then in this terminology:

$$Centrifugal \, Force = \frac{Mv2}{R} \, [N]$$

Its line of action will be horizontally outwards. If the track is uncanted, this force will act on any object placed on the vehicle floor, so that it will tend to slide outwards (it will in fact move if the centrifugal force is large enough to overcome the friction between the object and the floor). If the "Object" is a passenger, he will experience a sensation of discomfort, as if he were being pushed over. If now the track be canted inwards, this centrifugal force will again be resolved into two components, one parallel to and the other perpendicular to, the floor.

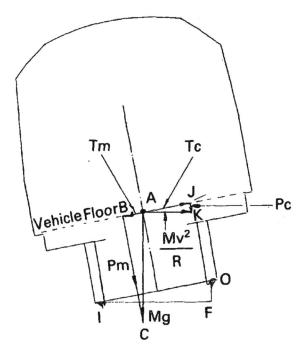

Figure 2.11 Resolution of Weight Force and Centrifugal Force on Canted Track

As before, the relations between these forces are represented by triangle AJK in Figure 2.11, and this triangle, like triangle ABC, is similar to the track triangle IOF, so that the component in the plane of the floor, T will be:

$$T_c = \frac{Mv^2}{R} - \frac{\sqrt{s^2 + e^{2a}}}{s} - \frac{Mge_a}{s}$$

and the component perpendicular to the floor will be:

$$P_c = \frac{Mv^2}{R} - \frac{e_a}{s}$$

All four forces, P_m, P_c, T_m and T_c, act through the object at point A. The forces P_m and P_c are in the same direction and combine to form the "weight" of the object as seen by the vehicle floor, whilst T_m and T_c act in opposite directions and tend to cancel one another out. The aggregate force perpendicular to the floor is P.

The net force parallel to the floor is T, where:

$$T_c = \frac{Mv^2}{R} - \frac{\sqrt{s^2 + e^{2a}}}{s} - \frac{Mge_a}{s}$$

Since "e_a" is small compared with "s", the term $\frac{\sqrt{s^2 + e^{2a}}}{s}$ is very nearly unity (its value departs from unity by less than 0.5% for the largest normal value of e_a). Hence to a close approximation:

$$T = M \left[\frac{v^2}{R} - \frac{ge^a}{s} \right] \dots\dots\dots\dots\dots\dots\dots 2-15$$

Clearly, if v^2/R is equal to ge_a/s, T will become zero, and the object on the vehicle floor will experience no sideways force due to either centrifugal force or to the weight force. By transposition, the value of e_a which gives T equal to zero for any given combination of v, R, and s is given by:

$$[[e^a]] \ _{T=0} \quad = \quad \frac{v^2 s}{R g}$$

This unique value of cant is termed the EQUILIBRIUM CANT and it is denoted by e_q. Thus:

$$e_q = \frac{v^2 s}{R g} \dots \dots \dots \dots \dots \dots \dots \dots \dots \dots \dots \dots \dots \dots 2-16$$

It must be emphasised that even though at equilibrium cant the passenger experiences no sensation of lateral force, the force is still there, and the ballast or other trackbase has to be capable of resisting it, otherwise the rails and sleepers will move bodily outwards.

2.5.2   Cant deficiency

Equilibrium can exist at only one speed for any combination of cant and radius. If as will most often be the case, the speed is higher or lower than this, a deficiency or excess of cant is said to exist. The significance of this statement may be seen from what follows.

By rearranging equation 2-15 we have:

$$T = \frac{Mg}{s} \cdot \left[\frac{v^2}{R g} - e_a \right]$$
$$= \frac{Mg}{s} \cdot \left[e_q - e_a \right]$$

Thus the net transverse force acting on the contents of a vehicle going around a curve is proportional to the difference between the equilibrium cant and the actual cant. This difference is called the CANT DEFICIENCY if the actual cant is less than the equilibrium cant and CANT EXCESS if the opposite is true. If the difference between e_a and e_q be represented by d, we can write:

$$T = \frac{Mgd}{s} \quad .... 2-17$$

Since for comfort and safety, it is necessary to impose limits on the allowable net transverse force, it is convenient to express this limit as a limit on Cant Deficiency or Excess. These limits are defined and discussed in Chapter 8.

2.5.3 A Practical Equilibrium Cant Equation

Equation 2-16 expresses cant in the somewhat inconvenient units of metres, and also involves two terms which for all practical purposes are constant. It is therefore simplified as follows:

The term "s" is for standard gauge approximately 1.502m, whilst "g" is 9.807m/s². Substituting these values, equation 2-16 becomes:

$$e_q = 0.1532 \frac{v^2}{R} \quad ....2-16A$$

This is still not very convenient since e_q is in metres and v in m/s. If we wish to express the speed in km/h, we can use the symbol V and then:

$$v = V/3.6$$

If we wish to express the equilibrium cant in mm, we can use the symbol E_q and then:

$$e_q = E_q/1000$$

Substituting these values in equation 2-16A we have:

$$E_q = 11.82 \ \frac{V^2}{R} \dots\dots\dots\dots\dots\dots\dots\dots\dots \ 2 - 18$$

This result, is the foundation of all the curve design calculations to be found in Chapters 8 and 9.

2.6 THE RAILWAY TRACK AS A STRUCTURE

2.6.1 Structural Analysis of the rail under Vertical Loading

The structural analysis of the rail has to take account of the fact that the sleepers deflect elastically into the ballast as each wheel passes, and (ideally) return to the unloaded position afterwards. The problem of determining exactly how the rails behave under these conditions is quite a difficult one and before the advent of computers, the solution usually accepted was based on the concept that the sleepers are so close together, that they form an effectively continuous resilient support to the rail, i.e., that the rail is a beam of infinite length on a continuous elastic foundation. This problem was solved by engineers in the 19th Century, and the equations involved are sometimes called the ZIMMERMANN equations after one of the workers in this field. Since comprehensive treatments of the derivation of the formulae are available in textbooks on structural engineering only the basic formulae are quoted here.

The terminology used is as follows:

M   =   Bending Moment at a distance x from the point of application of the load [N.mm]

F   =   Shearing Force at point x [N]

Y   =   deflection of the rail at point x

M_0 =   BM under the load

F_0 =   SF under the load

y_0 =   deflection under the load

Q   =   wheel weight [N]

E   =   Young's Modulus for rail steel [N/mm^2] (typically E = 2.069 x 10^5 N/mm^2)

I   =   Moment of Inertia of rail section [mm^4]

K_T =   Track Modulus [N/mm^2]

1c   = $\sqrt[4]{\dfrac{EI}{K_T}}$ $[mm]$ $[The$ Characteristic Length is $l_c\sqrt{2}]$

Using this terminology, the bending moment, shearing force and deflection at a distance x from the point of application of the load are:

$$M = \frac{Ql_c}{2} \cdot e^{\frac{x}{l_c\sqrt{2}}} \cdot \cos\left[\frac{x}{l_c\sqrt{2}} + \frac{\pi}{4}\right] \ldots \ldots \ldots \ldots \ldots 2 - (19)$$

$$M = \frac{Q}{2} \cdot e^{\frac{x}{l_c\sqrt{2}}} \cdot \cos\left[\frac{x}{l_c\sqrt{2}}\right] \ldots \ldots \ldots \ldots \ldots \ldots 2 - (20)$$

$$M = \frac{Ql_c^3}{2\,EI} \cdot e^{\frac{x}{l_c\sqrt{2}}} \cdot \cos\left[\frac{x}{l_c\sqrt{2}} + \frac{\pi}{4}\right] \ldots \ldots \ldots \ldots \ldots 2 - (21)$$

At the point of application of the load, (i.e. when $x = 0$), the values of M, F, and y become:

$$M_0 = \frac{Ql_c}{2\sqrt{2}} \ldots \ldots \ldots \ldots \ldots \ldots \ldots \ldots \ldots \ldots \ldots \ldots \ldots \ldots .2 - (22)$$

$$F_0 = -\frac{Q}{2} \ldots \ldots \ldots \ldots \ldots \ldots \ldots \ldots \ldots \ldots \ldots \ldots \ldots .2 - (23)$$

$$y = \frac{Ql_c^3}{2\sqrt{2EI}} \ldots \ldots \ldots \ldots \ldots \ldots \ldots \ldots \ldots \ldots \ldots \ldots .2 - (24)$$

Equations 2-23, 2-24, and 2-25 represent the maximum values of M, F and y for any given combination of Q, E, I, and K_T, subject to the reservations discussed below. It will be noted that all the terms involving x, in these equations are dimensionless, since both x and l_c

have the dimension of LENGTH. Hence, considering first equation 2- 23, if the equation is transposed to give:

$$\frac{M}{Ql_c} = \frac{1}{2} \cdot e^{\frac{x}{l_c\sqrt{2}}} \cdot cos\left[\frac{x}{l_c\sqrt{2}} + \frac{\pi}{4}\right] \dots\dots\dots\dots\dots 2-(25)$$

then the expression M/Ql_c is also dimensionless and a graph of M/Ql_c against x/l_c will be of universal application and can be used to predict the shape of the bending moment diagram for any set of track constants. Similar dimensionless functions for shearing force and deflection are:

$$\frac{F}{Q} = \frac{1}{2} \cdot e^{\frac{x}{l_c\sqrt{2}}} \cdot cos\left[\frac{x}{l_c\sqrt{2}}\right] \dots\dots\dots\dots\dots\dots\dots 2-(26)$$

$$\frac{yEI}{Q\,l_c^3} = \frac{1}{2} \cdot e^{\frac{x}{l_c\sqrt{2}}} \cdot cos\left[\frac{x}{l_c\sqrt{2}} + \frac{\pi}{4}\right] \dots\dots\dots\dots 2-(26)$$

The curves of M/Q1_c, F/Q, and yEI/ Ql_c^3 against x/l_c are drawn to scale in Figure 2.12. They show how the bending moment, shearing force and deflection peak under the wheel, and show also how all these parameters reverse at a distance from the wheel, thus illustrating the "precession wave" effect which tends to lift the sleepers out of the ballast ahead of and behind the wheel. It is noted that Equations 2-20, 2-21,2-22 are all damped cosine curves with a wavelength of $2\sqrt{2}\,\prod l_c$.

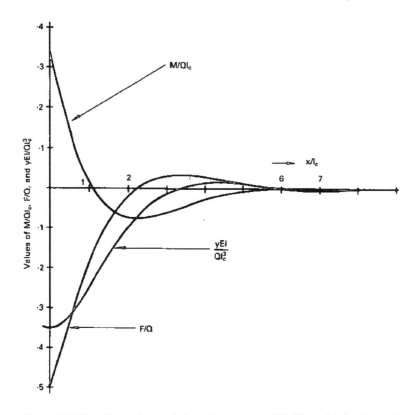

Figure 2.12 Non-dimensional relationships between BM, SF, and deflection, and distance from point of application of load, for a point load supported by a beam on an elastic foundation

When two or more wheels follow one another at a relatively close spacing, it may happen that the spacing corresponds to the distance of the above mentioned uplift from the wheel load. In such cases, the maximum BM produced by the combination will actually be less than the maximum BM produced by either wheel acting alone.

2.6.2   Measures of the stiffness off the track structure
As an alternative to the track modulus, a measure termed the LUMPED TRACK STIFFNESS (K_L) may be used. The lumped track

stiffness is the wheel load in newtons required to cause the track immediately under the wheel to deflect by 1mm. Using the terminology already defined:

$$K_L = Q/y_0$$

Applying equation 2-25, K_T and K_L can be related as follows:

$$K_L = \frac{2\sqrt{2}\,E1}{l_c^{\frac{3}{2}}} = 2\sqrt{2} \cdot K_T^{3/4}\,(EI)^{1/4} \dots\dots\dots\dots\dots 2\text{-}29$$

It should be noted that whereas K_T is a measure which is intrinsic to the trackbed, K_L is partly dependent on the flexural rigidity of the rail, as is shown by equation 2-29.

A third measure of the stiffness of the track is the SLEEPER MODULUS (K_s), which is defined as the rail seat reaction required to produce unit deflection of the sleeper.

If the sleeper spacing is smm then

$$K_s = s.\ K_T \dots\dots\dots\dots\dots\dots\dots 2\text{-}30$$

This last measure is probably the most truly intrinsic measure of track stiffness, since it is clearly related to the support immediately given to the sleeper by the ballast.

On the reasonable assumption that the support given to an individual sleeper by the ballast does not vary with the sleeper spacing, then K_T will be inversely proportional to the sleeper spacing, with consequential effects on l_c, M_0, F_0, and y_0

2.6.3   Typical Range of Trackbed Stiffness

For track (say using new BS113A rail, with sleepers at 654mm spacing) on good firm ballast, elastic deflections of the order of 1mm under a 10-tonne wheel load may be expected. If the trackbed is in very poor condition, deflections of the order of 5mm are conceivable. Values of K_L, K_T and K_s for two typical conditions are given in 2.1.

TABLE 2.1
Track Parameters for varying stiffnesses

Parameter		Firm	Soft
Lumped Track Stiffness	K_L kN/mm	90.4	19.6
Track Modulus	K_T N/mm²	60	7.83
Sleeper Modulus	K_s kN/mm	39.2	5.1
Deflection under 10 tonne wheel mm		1.1	5

2.6.4   Rail Seat Reactions

The rail seat reaction (RSR) can be estimated from the deflection curve, but the more accurate method of estimating it is by computing the change in shear force F from the mid point of the sleeper bay on one side of the sleeper of interest, to the mid point of the sleeper bay on the other side.

The largest rail seat reaction occurs when a wheel is directly over a sleeper, and in this case the RSR is twice the change in shear stress between the value for x equal to zero and the value for x equal to half a sleeper space.

For the firm track bed shown in Table 2.1 about 42% of the wheel load is taken by this sleeper, and 26% by the adjacent sleepers on either side. The third sleepers see rather more than the balancing 3% each owing to the uplift on the 4th and 5th sleepers away from the wheel.

For soft track, the RSR under the wheel is reduced to 25% of the wheel load, the adjacent sleepers seeing 21%, 12% and 6% respectively.

With the wheel midway between the sleepers, the RSR's on the sleepers next to the wheel are, for the stiff trackbed, 37.5% of Qo, and for the soft trackbed, 24.5% - only slightly less, that is, than if the wheel is directly over the sleeper.

By modifying the sleeper spacing the effect on the RSR, (assuming that the sleeper modulus remains unchanged), is to reduce the peak RSR as sleeper spacing is reduced, and vice versa.

For example, if the Sleeper Modulus is 39.2 (see Table 2.1), then reducing the sleeper spacing from 654mm to 610mm (i.e.from 28 to 30 sleepers per rail length) reduces the peak RSR from 42% to 39% of the wheel load. Increasing the spacing to 762mm (24 per length) increases the peak RSR to almost 46% of the wheel load.

2.6.5 Peak Bending Moment in the Rail

Graph (1) in Figure 2.12 shows that for uniform elastic support the BM under the load is:

$$M_{max} = 0.3536 \, Ql_c \quad\text{...} 2\text{-}31$$

and when the wheel is over a sleeper, this value holds good. When the wheel is midway between the sleepers however, there is an additional component due to the need for the rail to act as a bridge between the sleepers.

An estimate of the increased bending moment may be obtained in the following way:

> *Let the wheel load Q be applied at x = 0*
> *Sleeper (1) is at x = s/2 Sleeper (2)*
> *is at x = 3s/2 etc.*
> *M1, M2 etc are the BMs at sleepers (1), (2) etc.*
> $F_0 = SF \text{ between } x = 0 \text{ and } x = s/2 = -\dfrac{Q_0}{2}$
> *F1 = SF between x = s/2 and x = 3s/2*

Then in accordance with the elastic theory of bending:

$$M_0 = M_1 = F_0 \cdot \frac{s}{2}$$
$$= M_1 + \frac{Q \cdot s}{4} \dots \dots \dots \dots \dots \dots \dots \dots \dots \dots \dots \dots .2 - 32$$

But $M_1 = M_2 - F_1 . s$

hence $M_0 = M_2 - \left(F_1 - \dfrac{Q}{4}\right) s \dots \dots \dots \dots \dots \dots \dots \dots 2 - 33$

This process can be extended to any number of bays, so generalising to the nth sleeper:

$$M_0 = M_n - \left(\sum_{j=1}^{j-n-1} F_j - \frac{Q}{4} \right) s \dots \dots \dots \dots \dots \dots .2 - 34$$

M_n and F_j are obtained either from the graphs of Figure 2.1 or by the use of equations 2-20 or 2-21. Since M_x and M_2 (at least) are significantly changed in moving from continuous to discrete supports, it is found in practice that it is necessary to take equation 2-32 to about 6 or 7 terms in order to obtain an estimate of M_0 which does not vary significantly with n. Taking as an example the stiff track of Table 2.1, the BM under a wheel positioned midway between two sleepers is estimated as follows:

Using equation 2-32 ....Mo = 222.8Q
Using equation 2-33 ....Mo = 216.4Q
Using equation 2-34

        with n equal to 3...............................Mo   = 212.8Q
        with n equal to 4...............................Mo   = 212.0Q
        with n equal to 5...............................Mo   = 212.2Q
        with n equal to 6...............................Mo   = 212.4Q
        with n equal to 7...............................Mo   = 212.5Q

These values compare with a peak value of 188.3Q if the wheel is over a sleeper. These coefficients have the dimension of [mm] so that if Q is, say l00kN, the BMs would be 21.25kNm and 18.83kNm respectively.

This calculation has been repeated for a matrix of eight cases, representing the upper and lower sleeper modulus values of Table 2.1 with the four "standard" spacings of 24, 26, 28 and 30 sleepers per length, and the result is plotted in Figure 2.14.

These graphs show how the difference between the BM with the wheel over the sleeper, and the BM with the wheel midway between sleepers, increases (as would be expected) with increasing sleeper spacing, as well as showing the substantial increase in BM associated with a very soft trackbed.

2.6.6   Lateral loads on the rail

Lateral loads arise from:

- centrifugal forces
- curving (frictional) forces
- wind loading
- lurching and nosing associated with track irregularities
- wheelset instability

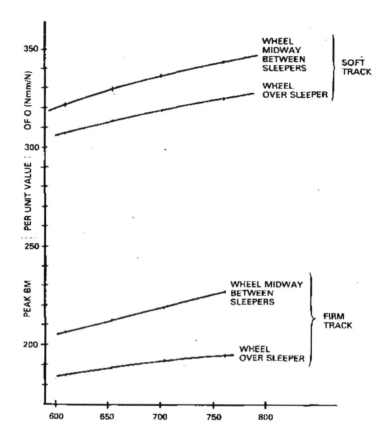

Figure 2.13 Effect of Sleeper Spacing, Wheel Position, and Track Modulus on Peak Bending Moment

The BM due to lateral forces is, in theory, calculable from the "beam on elastic foundation" theory described above, using Iyy for the rail instead of Ixx, and assigning a value to the notional "lateral track modulus". There are however, many uncertainties involved, and it is usually taken that the BM due to a lateral load Y on the rail for a sleeper spacing of s is:

$$M_y = \frac{Ys}{4} \dots\dots\dots\dots\dots\dots\dots\dots\dots\dots\dots\dots .2-35$$

The extreme fibre stresses occur at the side of the rail head and at the outer edge of the rail foot and are obtained in the usual way, using I_{yy} for the rail.

2.6.7   Torsional effects

Neither the vertical nor the lateral loads are applied on an axis of symmetry of the rail, and consequently the rail is subject to torsional forces. The fulcrum about which the moment arms of the vertical and lateral forces are measured is usually taken to be a point on the vertical centre-line of the rail section, level with the contact point of the rail fastenings on the upper surface of the rail foot.

The torsional moment M_T is then

$$M_T = Y.E_Y - Q.E_Q \dots\dots\dots\dots\dots\dots\dots 2\text{-}36$$

where:

        E_Y is the vertical offset of the lateral load Y from the fulcrum
        E_Q is the horizontal offset of the vertical load Q from the vertical axis of the rail

Because of the complex shape of the rail head, the stresses resulting from this torsion are difficult to evaluate and any discussion of the method by which they are derived, is beyond the scope of this book. However, from the work of Timoshenko the stress in the extreme fibre of the rail foot of the BS113A rail is approximately:

$$f_T = 12.4 M_T [N/mm^2]$$

M_T being expressed in kNm.

2.6.8   Selection of a Rail Section, or Determination of Maximum Allowable Axle Load

Rail stresses are determined for the purpose of arriving at the rail section and grade of steel to be specified on a new project, or on an existing railway, to provide a rationale for accepting or rejecting a proposal to increase speeds or axle loads. In line with Eurocodes in structural design, three criteria can be identified. These may be designated respectively "ultimate loading", "serviceability" and "fatigue".

The use of these criteria considerably post-dates the decision to use BS113A as standard, but it is instructive to see how that historic decision stands in the light of these more recent criteria, and in the discussion which follows the use of 25 tonne axle load wagons will be referred to as an example. It is emphasised that this example is not intended to represent the whole decision making process, but purely to illustrate the methods involved.

Ultimate Loading Criterion

A notional ultimate loading criterion would be that the rail must not fail, during derailment due to overturning. Under these conditions, the rail carries:

Vertical load equal to the full 25-tonne axle load so that:

$$Q = Qo = 122.6kN.$$

Lateral load as calculated from equation 2-17 is 96.6kN. Thermal tensile stress as calculated from:

$$f_{Th} = 0.0000115 \times \Delta t \times E$$

For Δt equal to 42°C (i.e. for a neutral temperature of 27°C, and an assumed rail temperature of -15°C), this gives:

$$f_{Th} = 100 \, N/mm^2.$$

Sleeper spacing is assumed to be 654mm, and the trackbed firm, as per Table 2.1. Then the BM due to vertical load M_0 is:

$M_0 = 0.222\,Q = 54.45\,kNm$

$M_L = 15.79\,kNm$

The section moduli for new BS113A rail are taken as:

$Z_{xx} = 3.11 \times 10^5\,mm^3;$

$Z_{yy} = 5.97 \times 10^4\,mm^3$

The offsets E_Y and E_Q respectively, are taken to be 126mm and 10mm giving from Equation 2-36

$M_T = 9.72\,kNm.$

From Equation 5-(15A) the extreme fibre stresses are:

due to bending under vertical load	(f_Q)	175.1 N/mm²
due to bending under lateral load	(f_Y)	264.5 N/mm²
due to torsion	(f_T)	120.5 N/mm²
due to temperature	(f_{Th})	100.0 N/mm²
Total Stress		660.1 N/mm²

These values may be compared with UTS of 700 N/mm² for normal grade rail steel, or 900 N/mm² for Grade A steel.

Serviceability loading criterion

The notional maximum service loading under steady state conditions is assumed to be associated with a cant deficiency of 110mm.

Considering the leading axle of a bogie, a reasonable estimate of the lateral force is a combination of:

- The unbalanced centrifugal force associated with a cant deficiency of 110mm, plus

- The gauge spreading force due to a creepage factor of 0.1 on the wheel weight of the inner wheel.

The thermal stress f_{Th} is as per the previous section.

In this case the wheel weight $Q = Q_0 + \Delta Q$

$Q_0 = 122.6\ kN$

$\Delta Q = 23.92\ kN$

$M_0 = 0.222\ (Q_0 + \Delta Q) = 32.53\ kNm$

whence $f_Q = 104.6\ N/mm^2$.

The centrifugal component of Y is given by equation 2 - (17) and is:

$Y_c = 17.96\ kN.$

The gauge spreading force is:

$Y_G = 0.1\ (Q_0 - \Delta Q) = 9.87\ kN.$

Hence:

$Y = Y_C + Y_G = 27.83\ kN$

$M_L = 4.55\ kNm$

$f_y = 76.2\ N/mm^2$

$M_T = 2.04\ kNm$

$f_T = 25.7\ N/mm^2$

Thus the total stress is 306.1 N/mm², which is well below the yield point of 467 N/mm² for Normal grade rail steel.

Allowance for impact can be made by applying an appropriate factor to one or other of Q or Y.

A factor of 1.85 applied to Q would increase the direct bending stress but would reduce the torsional bending stress leading to an overall maximum extreme fibre stress of 379.5 N/mm².

A factor of 1.5 applied to Y would increase the lateral and torsional bending stresses to give an extreme fibre stress of 366.0 N/mm².

Fatigue strength criterion

This criterion considers the range of bending stress on the longitudinal centre line of the rail (ie, from maximum sagging BM under the wheel, to maximum hogging BM in the uplift area), and compares this range to the allowable range computed from the fatigue strength of the rail, taking account of thermal and residual stress.

The fatigue strength of the rail is approximately 300 N/mm². The residual tensile stress in the surface layers of the rail foot is taken as 80 N/mm², and as before, the maximum thermal tension is taken as 100 N/mm², giving a background stress of 180 N/mm².

Account is taken of this stress by means of what is known as the Smith diagram (see Figure 2.14). In this diagram, for Normal grade rail, the line OU passes through the origin and the point U (700, 700) which represents the ultimate tensile strength of the steel. Lines AU and BU are drawn such that at the point where line BU cuts the x-axis, the intercept in the y direction between the two lines is equal (to scale), to the fatigue strength. For a fatigue strength of 300 N/mm², this implies that point A is at coordinates (150, 0) and B is at (150, 300). From this information, the equations of lines AU and BU can be determined.

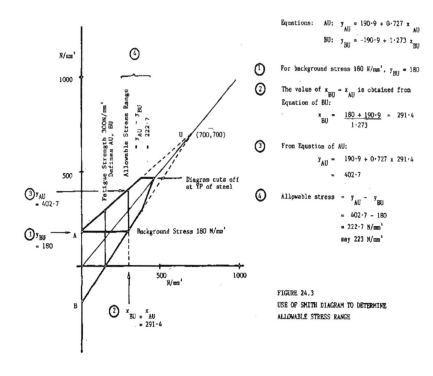

Figure 2.14 Use of Smith diagram to determine allowable stress range

The lines AU and BU cut off as shown at the yield point of the steel. From this diagram, the allowable bending stress range for a given background stress is obtained by determining the x-coordinate of the point on line BU, having a y-coordinate equal to the given background stress. The allowable stress range is the y-intercept between lines AU and BU for that x-coordinate.

For Normal grade rail steel, for a fatigue strength of 300 N/mm², and a background stress of 180 N/mm², the allowable stress range turns out to be 223 N/mm². The rail BM's are computed for a cant deficiency of 110mm. The maximum sagging BM is 0.222Q, and the maximum hogging BM is 0.043Q, giving a stress range of 123.9 N/mm².

2.6.9   Commentary on the Results of Stength Calculations

The figures suggest that whilst well within its fatigue criterion, a
BS113A rail will be working quite hard when carrying maximum axle
loads at maximum permitted speeds on sharp curves with high cant
deficiencies, and that under these conditions, good track geometry
and a firm trackbed would be of great significance in keeping down
maximum stresses in the high rail.

It has to be pointed out that in UK conditions, the full thermal tension
would only be developed for a small percentage of the total service life
of the track. At the same time, the fact that at low temperatures the
thermal stress contributes almost as much to the total stress in the
rail, as the bending under primary vertical loads, under normal service
conditions, is interesting, seen against the observed increased
tendency towards broken rails in cold weather.

It is worth noting also that in sharp curves with high cant deficiencies
the life of the high rail may well be determined by sidewear rather than
by fatigue. As already pointed out, the values obtained above are
illustrations rather than definitive, and a complete appraisal of a rail
section would require the consideration of a number of loading
combinations representing the possible extreme circumstances of the
given project.

2.6.10   Analysis of distribution of longitudinal forces along track

The analysis of the forces associated with traction and braking,
assumes that the fastener system allows the rail to move relative to
the sleeper, in the direction of action of the force. It is a precondition of
ability to resist creep that the rail must return to its original position
when the force is removed. Assuming that the movement of the rail
relative to the sleeper is proportional to the shear force acting across
the pad (ie, the pad behaves "elastically"), it can be shown (using the
usual rules of elasticity which govern the behaviour of structural
systems) that if a rail in any sleeper bay is subject to a longitudinal

force, then the fastenings on either side of the bay will each absorb a fixed proportion of that force, leaving the remainder to be passed into the adjacent sleeper bays. The next fastening in turn will absorb the same fixed proportion of the remaining force, and so on, so that a listing of the forces in all the fastenings will take the form of a geometrical progression whose common ratio "c" is less than unity. Although such a list would be infinitely long, the sum of all the forces in the fastenings on both sides of the force would equal the original force in the rail.

Using these principles it can be shown that the force B_x in a fastening at a distance x from the point of application of a brake force P is, to a very close approximation given by the formula:

$$B_x = \frac{P\,(l-c)\,c^{x/L}}{(c+l)} \quad \text{...\,...\,...\,...\,...\,...\,...\,...\,...\,...\,...\,...}\, 2-37$$

where c is a dimensionless constant (which we may conveniently call the "distribution ratio") whose value varies from zero (corresponding to an infinitely stiff fastening) to unity (corresponding to an infinitely flexible fastening) and derived from the expression:

$$c = \frac{(2Z+1) - \sqrt{(4Z+1)}}{2Z} \quad \text{...\,...\,...\,...\,...\,...\,...}\, 2-38$$

In expression 2-38, Z is given by:

$$Z = \frac{AE}{K_f\,L} \quad \text{...\,...\,...\,...\,...\,...\,...\,...\,...\,...\,...\,...}\, 2-39$$

where:
       A = Cross-sectional area of rail;
       E = Young's Modulus for rail steel
       K_f = Longitudinal fastening rigidity, expressed as force per unit deflection per fastening

L = Distance between fastenings.

The inverse of equation 2-38 is

$$Z = \frac{C}{(l-c)^2} \quad \ldots \ldots \ldots \ldots \ldots \ldots \ldots \ldots \ldots \ldots \ldots .2-40$$

The force in any fastener, due to a train of wheels, can be obtained by a summation. If the wheels are at x(l), x(2), x(i), x(t) from the fastener, then by application of equation 2-37 the total force in the fastener is:

$$B = P \sum_{i=l}^{i=t} \frac{(l-c) \, c^{x(i)/L}}{(l+c)} \quad \ldots \ldots \ldots \ldots .2-41$$

Expression 2-41 can be used to explore how B varies as a train passes, remembering that the force in the fastener is in the same direction whichever side of the fastener the wheel is, so that x is always positive. The value of P will be given by:

$$P = \mu . Q \ldots \ldots \ldots \ldots \ldots \ldots \ldots \ldots \ldots \ldots \ldots \ldots \ldots \ldots \ldots .2 - 42$$

where
μ = Coefficient of adhesion
Q = Wheel weight

If the dimensionless summation term for any given fastener position (n) is referred to as "F_n", then equation 2-41 becomes:

$$B = \mu Q F_n \ldots \ldots \ldots \ldots \ldots \ldots \ldots \ldots \ldots \ldots \ldots \ldots \ldots .2 - 43$$

In principle, the value of F_n has to be explored over a large number of locations relative to a train, in order to find its maximum value for a given distribution ratio "c", as a train consisting of a large number of vehicles passes. This maximum value of the summation term will be referred to as "F_{max}". In practice F_{max} will always be found (for bogie vehicles) at the wheel nearest to the coupling of one or other of the two centre vehicles of the train. The summation can be carried out quite quickly by anyone with access to a reasonable quality programmable calculator.

The movement of the rail relative to the sleeper at any fastener position *(n)* is given by R_n, where:

$$R_n = B_n/K_f$$
$$= \frac{\mu Q F_n}{K_f} \quad 2\text{-}44$$

2.6.11   Practical range of distribution ratio

The value of K_f for a fastener system may be taken to be about one third of its vertical stiffness. Thus for a pad having a vertical stiffness of 55kN/mm, K_f will be around 18.3kN/mm, whilst for an EVA pad with a vertical stiffness of 480kN/mm, K_f will be 160kN/mm. For BS113A rail, the term "AE" may be taken to be 1.486×10^6kN. Hence for a sleeper spacing of 704mm, the values of "Z" and "c" for fastening systems incorporating these two pads will be:

For the soft natural rubber pad: Z = 115.4   c = 0.911
For the EVA pad: Z = 13.19   c = 0.76

For most practical applications therefore, "c" will be in the range 0.7 to 0.95, and the shape of the curves is such that the most practical design curve is a graph of "Z" against "F(max)", with the "Z"-axis graduated as "Log Z". This has been produced as an example, for a train of 10 wagons with overall length 11.7m, bogie wheel base 1.8m, with 5.1m between inner bogie wheels, as representing a typical modern wagon suitable for the conveyance of high density bulk materials, and it forms the upper graph in Figure 2.15. The lower graph in the same figure represents "F_n" for a fastening under the midpoint of a wagon in the middle of the rake, and a comparison of the two curves shows how the force to be resisted by the fastening varies, as the train passes over it.

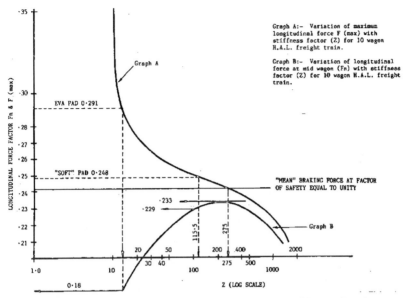

Figure 2.15 Creep resistance characteristics for 704mm Sleeper Spacing

It is emphasised that this graph is specific to the train described, and also to BS113A rail, and a sleeper spacing of 704mm.

2.6.11 Application to fastener design

For any train, the "average" brake force per rail seat will be:

$$\frac{no\ of\ braked\ axles)\ x\ (sleeper\ spacing)}{(length\ of\ train)}$$

For the train and track described, this works out to 0.241. From the upper graph in Figure 2.15, the maximum value of "Z" (and hence the minimum value of K_f) which will give this value of F_{max}, is about 275. Such a value corresponds to K_f equal to 7.68kN/mm, and in turn, to a pad with a vertical stiffness of only 23kN/mm. Such a pad would give problems with rotational stability of the rail in its seating, and also the longitudinal movement of the rail relative to the sleeper, under the wheel, assuming a coefficient of adhesion of 0.3, would be (from equation 2-44):

$$R = \frac{0.3 \times 12.5 \times 9.81 \times 0.241}{7.68} = 1.15mm$$

Such a movement would create problems of mutual wear on the contact surface of the fastener components.

However, it is not necessary to produce such a pad, in order to meet the "average brake force" requirement. Clearly, any assembly which has a "Z" value less than 275, will hold the rails, provided that the assembly is strong enough to resist the calculated creep force. This capacity is largely dependent upon the product of the coefficient of friction at the interfaces between the rail and pad, and between the pad and the sleeper or baseplate, and the pressures across those faces. There will clearly be no problem in the immediate vicinity of the wheel contact point, because here the rail seat reaction will provide plenty of reserves of strength, but a problem might be thought to arise in the uplift area of the precession wave.

Fortunately it turns out that this is not so, and it is on this point that the lower curve of Figure 2.15 is of interest, as showing how the creep resistance factor at the mid point of the vehicle falls off quite steeply with increasing pad stiffness, and that it is indeed always below the value of 0.241 corresponding to the "average brake force criterion". Confirmatory calculations, which cannot be included here for space reasons, show that in the uplift zone, which is much closer to the wheel, and where the pad relies on the clip toe load to provide the pressure to develop a frictional grip, the factor whilst above the lower line on the graph, is again, always below the 0.241 figure.

Hence, if the fastening assembly is strong enough to hold the rail stationary against a force equal to the average brake force of the train, per rail seat, with no superimposed rail seat reaction, it will restrain the track effectively, even though the longitudinal stiffness of the assembly may be very high, thus validating the specification requirement.

These calculations are the source of the assessment of the minimum coefficients of friction required to ensure freedom from creep. It should be noted that they have been done for a 704mm sleeper spacing. For the smaller spacings appropriate to the higher categories of UK track, the creep resistance coefficient for the "average brake force criterion" will be lower, and hence the demands on the fastener will be less onerous, whilst the contrary is true for the more-or-less obsolete sleeper spacing of 762mm.

The calculation for "F_n" can also be combined with the "beam on elastic foundation" calculation explained above, to estimate the combined shear and compression forces operating on the pad, and hence how the principal stresses vary with the loading conditions.

2.6.13   Effect of sleeper movement

One of the many effects tending to complicate the design process, is the probability that a sleeper, supported and restrained by ballast, will rotate slightly under the effect of traction/braking load. The effect of this will be to reduce the effective value of K_f, and hence to reduce the demand on the fastener components. It seems quite probable therefore, that one of the results of broader based concrete sleepers, and of good consolidation of ballast under and around the sleepers, will be, by tending to prevent rotation of the sleeper about its axis, to increase the demands made upon the fastening system.

Even then the analysis of pad loading and creep resistance in ballasted track can never be precise, because of the uncertainty about the detailed behaviour of the ballast bed. The analysis is of far more significance when considering non-ballasted track, and more particularly those forms of non- ballasted track (such as paved concrete track) where there is no possibility of movement of the track form relative to the track base.

2.7 INTERACTION BETWEEN VEHICLE AND TRACK

When a railway vehicle is in motion, oscillation or vibration as well as steady state forces, are experienced in its body and varying forces are applied to the track. The magnitude of these forces depends on the speed, on the vehicle design, on the track layout (i.e. the degree of curvature, cant etc.), on the design of wheel and rail, and on the size and nature of track irregularities present. The forces imposed by vehicles need to be understood by the track engineer to ensure adequate track design, and also to assist in maintaining the track in such a way as to optimise ride, and minimise track forces and undesirable environmental effects. The dynamic behaviour of railway vehicles, and the consequential loading of the track, is complex and highly interactive. The several mechanisms giving rise to lateral and vertical forces have all been the subject of much research, and in this chapter, only the main principles involved can be described.

2.7.1 Lateral interactions

Causes of Lateral forces

Five main causes can be identified. They are:

- inclination of the wheel/rail contact patch
- creepage forces resulting from microscopic movements of the wheel relative to the rail, across the contact patch
- dynamic instability
- generation of lateral curving forces other than those resulting from excess or deficiency of cant, on curved track
- inertial forces generated by wheel and/or rail irregularities.

The individual effects will be described separately, although this results in something of an over-simplification.

Force due to Inclination of Contact Patch

The profiles of the wheel tyre and the rail head are such that almost invariably, the minute area of contact between them will be sloping inwards. If this angle of slope is δ, then the wheel weight Q produces at the point of contact a reaction normal to the plane of the contact patch equal to Q/cos δ, and horizontal force acting outwards equal to Q tan δ.

Equal and opposite forces operate on the wheel. If the wheelset is displaced laterally towards one rail, the inclination of the contact normal, and hence the lateral force, increases at one end and decreases at the other. This produces an unbalanced lateral force on the wheelset which acts to restore the wheelset to its original position.

In general, for small lateral movements of the wheelset, these lateral forces are modest compared with other effects but they become significant when flange contact is approached.

Creep Forces

Careful observation of the behaviour of wheels on rails shows that even though no apparent slippage occurs, a wheel which is under traction, will rotate marginally more than expected for a given distance run. Similarly if the wheel is braked, it will rotate marginally less than expected. The fractional increase or decrease is known as creepage. This creepage is in fact, a necessary accompaniment to the generation of the traction/braking force across the contact patch, and is produced by microscopic elastic deformations in the contact patch itself.

Similar behaviour occurs if a wheelset is displaced laterally on the track, so that the two wheels have unequal rolling radii. In this case, forces are generated which are along the track in opposite directions on the two wheels, so that a resultant couple is applied to the wheelset. This couple has the effect of trying to force the wheelset to yaw.

A further possibility exists if the wheelset is constrained to run at a slight angle to its rolling direction. In this case, a lateral creepage occurs, which gives rise to lateral creepage forces.

A simple diagrammatical representation of the creep force/creepage relationship is shown in Figure 2.16. It will be seen that for small creepages (eg less than 0.5% of forward velocity), creep force is generally proportional to creepage, and at higher values the force is limited by adhesion. The magnitude of the creepages depends on the geometry of the vehicle/track system, including the geometry of the wheel/rail cross-sectional profiles, and a complex computational process is needed to derive the details.

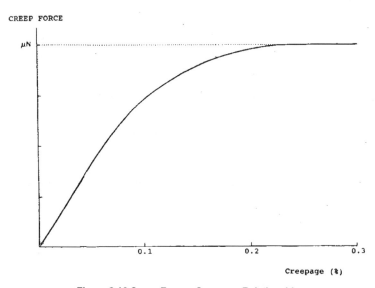

Figure 2.16 Creep Force - Creepage Relationship

VehicleInstability

The lateral behaviour of a railway vehicle is determined by the historic and fundamental feature that the wheelset consists of two coned wheels, fixed rigidly to a single axle. By this means, vehicles are guided without continuous flange contact, thus reducing flange and rail wear. As stated in the previous section, a lateral displacement causes a difference between the rolling radii of the two wheels, and consequently, creepage forces are generated which steer the wheelset back towards the centre of the track, as shown in Figure 2.17. The wheelset overshoots the centre line, a difference in rolling radii develops in the opposite sense and the wheelset again returns towards the track centre line, generating a cyclic motion with a characteristic 'kinematic wavelength' (typically in the range 5m to 15m).

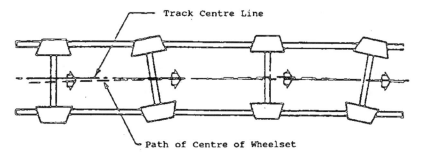

Figure 2.17 Cyclic Motion of a Railway Wheelset

The angle of the cone is critical to this oscillatory behaviour. Too shallow an angle gives inadequate guidance, but too steep an angle produces 'hunting' along the track, which is now understood to be a form of dynamic instability. Even on perfect track, energy is fed into the system, via the creepage forces, and the oscillatory motion can grow until limited by flange contact, ie, the behaviour is 'non-conservative'. For a single wheelset rolling along the track, it is relatively easy to show that such non-conservative behaviour occurs at all nonzero speeds.

When two wheelsets are brought together in the chassis of a single vehicle, the interaction between the wheelsets restrains both of them and prevents nonconservative behaviour until a specific 'critical speed' is reached. In modern vehicles, an object of design is to ensure, by a suitable combination of springs and dampers that the critical speed is above the normal operating speed of the vehicle concerned. However, hunting is certain to occur if the equivalent cone- angles of the wheels are excessive regardless of the skill with which the suspension has been designed. For new coned wheels of the angles now used, this is not a problem, but wheels with a worn profile can cause hunting in some circumstances. This is because, in general, conicity increases as the wheel and rail cross-sections become more conformal (i.e. the portions of the wheel and rail profiles which make contact, have similar shapes, the rail *convex,* the wheel *concave*). Particularly high conicities tend to arise on new rails if the head profile is too flat and the gauge too tight, or on rails which have been turned end-for-end and have subsequently worn to a very conformal profile. Conicities of 0.4 are not uncommon in the UK, although new coned wheels have a cone angle of only 0.05.

The forces generated by this hunting behaviour can be significantly greater than those produced by high cant deficiency curving. Although hunting is often considered as essentially a vehicle problem, the track gauge and rail head profile have a significant influence on how frequently it occurs, and this realisation is one of the factors leading to the revised head profile of BS 113A rail.

Equivalent Cone Angle or Conicity

Equivalent Cone Angle or Conicity is a property specific to a particular combination of wheel tyre profile, rail head profile, track gauge, and wheel spacing. If the difference between the rolling radii of the two wheels is plotted against the displacement of the wheelset from its neutral point, a graph is obtained such as Figure 2.18.

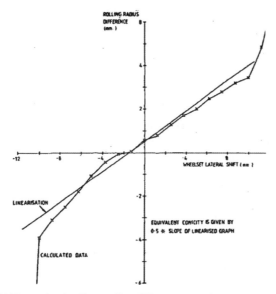

Figure 2.18 Example of rolling radius difference graph, used for calculating equivalent conicity

Typically the individual points are somewhat irregularly distributed and it is usual to obtain a "best fit" straight line through the points. The Conicity is half the slope of the line.

Vehicle and Track Interaction and Lateral Forces on Curves

The passage of a railway vehicle through a curve, and the forces and displacements developed, involves a complex interaction of a large number of factors. Explanation is simplified by considering firstly a vehicle curving at equilibrium speed and subsequently considering other factors influencing behaviour in real situations.

Curving at Equilibrium Speed

At equilibrium speed on any curve, with no traction or braking forces applied, no external forces act on the vehicle. However, the forces at the individual wheel/rail contacts can still be very large.

To see why, consider a "classic" four- wheeled vehicle or bogie, where the two axles are rigidly constrained to be parallel. It is found that the vehicle arranges itself so that its rear axle points approximately to the centre of the circle. This wheelset then behaves more or less in pure rolling and is displaced outwards to an equilibrium rolling position (i.e. the different rolling radii of the wheels compensate for the differences in rail length).

However, since the four contact patches are at the corners of a rectangle, it follows from the geometry of the system, that the leading axle is not radially disposed (the angle between the axle and the radius is the 'angle of attack'). Furthermore, the motion of the contact patches along the rail involves *both* a forward component which can be accommodated by rolling, and also a component at right angles to the direction of rolling, which can only be accommodated by creepage (or in extreme cases by sliding). This creepage generates forces which it is convenient to conceptualise from the track point of view, as a force across the leading outer contact patch (the Guiding Force), tending to push the axle towards the centre of the circle, and another force, (resisting the guiding force) across the inner leading contact patch. The two forces are more or less equal and opposite in direction and magnitude, and have the effect of trying to force the rails apart (hence the term 'gauge spreading force'). In the extreme case where gross sliding takes place, their magnitude is the product of the coefficient of friction across the INNER contact patch, and the weight on the INNER wheel.

In most vehicles there is at least some slight flexibility in the alignment of the two axles, and this helps greatly to reduce the guiding force, since the creepage forces generated by the angle of attack will tend to deflect the suspension system and turn the axle into a more nearly radial position. The degree to which this deflection is possible, depends on such factors as the suspension characteristics and the curve radius. The more nearly the two axles can be turned towards the radial position, the less will be the guiding forces. This is the principle behind the design of steerable bogies.

The whole process involves many conflicting force generation systems, and it is almost impossible to predict, without the aid of a computer, what will happen in a given situation. Paradoxically, once the calculation has been completed, it is relatively easy to understand what is happening, making use of a suitably- exaggerated 'curving diagram'.

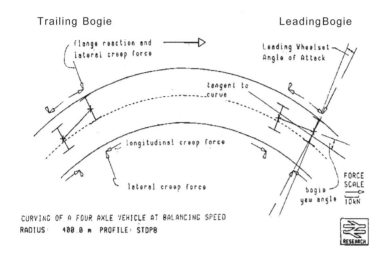

Figure 2.19 Curving diagram for a four axle passenger vehicle

Figure 2.19 shows the bogie and wheelset attitudes, and the forces which are developed by a conventional passenger vehicle on a 400m radius curve at balancing speed as computed by the VAMPIRE program. For clarity, the body of the vehicle has been omitted from the drawing. Arrows are drawn to scale to represent the forces which act on the vehicle. These forces will be accompanied by equal and opposite forces on the track.

The leading wheelset of each bogie has a significant angle of attack and large lateral creep forces are being generated at both wheels. The outer wheel is pushed into flange contact to provide a reaction force which produces 'gauge- spreading' as already described. The trailing wheelsets run with smaller angles of attack, and generate more modest lateral forces. On curves of larger radii, the angles of attack, and hence the forces, will be small, while on tighter curves the angle of attack will progressively increase. The radius at which curving behaviour becomes a problem will depend on the type of vehicle and on the value of coefficient of friction available between wheel and rail.

The main consequences of poor curving performance are an increase in flange and rail wear, an unpleasant high pitched squeal, and in extreme circumstances flange-climbing derailments.

Effects of Cant Deficiency and excess.

At speeds above or below the balancing speed on a curve, a net centrifugal (or centripetal) force, which is a function of speed, track cant, and curvature, is present. This modifies the curving behaviour of the vehicle and the magnitude and balance of the lateral forces. On curves where the leading wheelset is in flange contact, a cant deficiency force tends to cause the bogie to move outward on the curve, pivoting about the point of flange contact. This reduces the angle of attack at the leading wheelset. The cant deficiency force also changes the load distribution, increasing the wheel load on the high rail whilst decreasing the wheel load on the low rail. At considerable cant excess (ie, slow speed), these effects are reversed and in either case the effect is magnified by a high vehicle centre of gravity.

Effects of Coefficient of Friction

The magnitude of the coefficient of friction has conflicting effects on curving behaviour. A low value limits the maximum creep force, which limits the lateral forces, but also restricts the longitudinal forces and hence the available steering torques. Thus in the situation when a vehicle is able to steer its wheelsets significantly, a high value of coefficient of friction is beneficial, leading to a reduced angle of attack and hence, reduced lateral forces. However, once the vehicle is unable to steer, the opposite is true and a low friction level produces lower forces. In this situation, lubrication helps in two ways; by limiting the lateral creep forces and by minimising flange and rail wear. This latter effect is further discussed below.

Curving of Multi-axle Bogies

The above discussion relates to two-axle bogies (or 4-wheel vehicles) and a rather more extreme situation can arise if bogies with more than two wheelsets are employed. For such bogies, the leading wheelset will, as before, adopt an 'angle of attack', and the trailing wheelset will run in a relatively neutral position, or perhaps acquire a small negative angle of attack. The centre wheelset of a three-axle bogie normally also displays an angle of attack, and the resulting lateral force will be reacted by either the outer rail or the bogie frame, depending on the lateral freedom of the wheelsets in the axleboxes. The case where the load is taken by the bogie frame is potentially disastrous, since this load is then transferred to the other outer wheels (Figure 2.20).

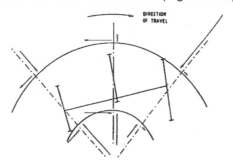

FLANGEWAY CLEARANCES PROPORTIONAL TO FULL SIZE ALL ANGLES

Figure 2.20 3-axle bogie on 95m radius curve

Very high lateral forces are generated, which cause severe wear, and distort or in extreme circumstances, destroy the rail-sleeper fastening system, or cause derailment. The results, but possibly not the mechanism of this are well appreciated by track engineers, and are, historically speaking, one of the reasons for the requirement for check-rails on curves of less than 200m radius on passenger lines.

Curving and Safety

Traditionally, curving behaviour has caused more concern at high speed than at low speeds, because of fear of overturning or track shifting. However, these problems occur very rarely, and almost always as a result of gross over-speeding, unless specific track or vehicle defects exist.

The explanation for this was shown in early work in the UK on high speed curving and is demonstrated in Figure 2.21 which shows the bogie attitude on a curve of 650m radius. At low speed, the bogie attitude is shown with a significant angle of attack on the leading wheelset, which will produce large high rail flange forces. With increasing speed, the bogie adopts a different position, with the trailing wheelset moving out under the influence of cant deficiency. This reduces the angle of attack at the leading wheelset.

At the very high curving speed, the leading wheelset has become almost radial. The underlying cause of the flange force has changed from lateral creep to cant deficiency generated, but its magnitude remains approximately the same. In addition, the angle of attack is much reduced and vertical load is transferred onto the high rail. Thus the derailment risk from flange-climbing is relatively unchanged.

The total force imposed on the track by the vehicle, increases with speed, but modern track construction, with a reasonable quality of maintenance, copes with these increased levels of force without difficulty.

Thus, a very important result is reached, that the first limit on curving speed is almost always passenger comfort, rather than safety.

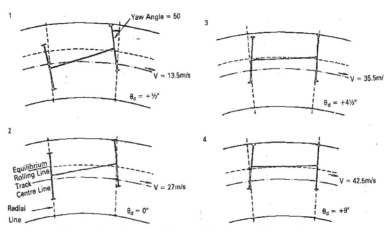

Figure 2.21 Curving diagram, high speed bogie on 650m radius curve

Wheel Flangewear and Rail Sidewear

Wheel/rail forces and creepages result in wear and so the vehicle/track parameters which improve the curving performance of a vehicle will also reduce wheel/rail wear. The geometric parameters which are under the control of the track engineer, are track curvature and track cant.

In order to reduce wheel/rail wear, track curvature should be minimised. However, because of the presence of fixed constraints such as bridges, tunnels, inclines etc., there is little scope for this on existing routes. Track cant is a more convenient parameter to change but the effects of this are far from clear cut. Apart from uncertainty on the effect on wear, changing cant affects passenger comfort, low speed flange-climbing derailments and rail head stresses. It is important that all of these are considered before major changes are made to the installed cant on a curve.

For existing railway systems, it is important to ensure at the design stage, that new vehicles have adequate curving performance, and do not produce excessive flange and rail wear on the routes over which they operate. In instances where, by virtue of other factors, vehicles are produced for a route, and unacceptable amounts of flange and/or rail wear occur where no geometric improvements to the track are practicable, then the track engineer has two possible palliatives available. These are, the use of rail and/or flange lubricators, and the installation of wear-resisting rails. Lubrication reduces available creep forces and modifies the curving behaviour of a vehicle with dramatic effect on wheel/rail wear, as shown in Figure 2.22, which has also been derived from the VAMPIRE program. Reduction factors of twenty or more have been achieved in measured wear rates in field experiments. This can also be facilitated by fitting vehicle-mounted lubricators. Premium grade rail steels have been used successfully in certain locations but their effects on wear are less marked than those of lubrication.

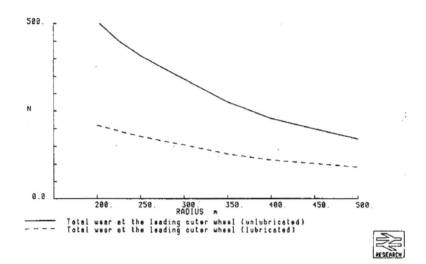

Figure 2.22 Effect of lubrication on passenger bogie wheel wear

Summary of Curving Effects

The forces and displacements developed by a vehicle on a curve, depend on a complex interaction between vehicle and track parameters. It is impossible to predict the quantitative effects of changes in one parameter without a detailed analysis of the vehicle/track system which takes account of all the relevant features. In general, any curving situation can be improved by reducing the curvature of the track, reducing the stiffness of the primary plan-view suspension, reducing the vehicle wheelbase and increasing the conicity of the wheelsets. Unfortunately however, those vehicle parameters which improve curving performance, reduce the speed at which the vehicle becomes unstable and the final choice of parameters values is often a compromise based on consideration of the particular application of the vehicle concerned.

Forces arising from Dynamic Response to Lateral Irregularities

Lateral irregularities can be divided into two groups - design cases (essentially switches and transitions) and unplanned irregularities.

Every switch or turnout, presents a very significant lateral irregularity. In the turnout direction, the train is required to change direction abruptly, without benefit of transition of either cant or curvature. It is worth adding here that whilst the historic designs used on UK track allow reasonably well for inertial effects, they fail to allow for the distance required for a wheelset to steer itself into the new direction. In general, the wheelset cannot re-adjust its direction to follow the route in a distance less than that given by its 'kinematic wavelength', and the result is that wheel/rail impacts are inevitable even on higher speed turnouts (Figure 2.23).

The forces generated can be large, and can exceed the limit for track shifting on plain track. The precise value depends on the track layout and the effective unsprung mass of the vehicle (laterally). Locomotives with heavy traction motors are normally the worst type of vehicle because the motor is unsprung laterally. Experimental work has shown that the largest forces are generated at the trailing end of a crossover, or when trailing through a sharp turnout. In these situations, low speed curving forces are being generated when the wheelset traverses the switch blades, and are added to the impact force experienced on meeting the stock-rail. Such forces are inherently present in the railway system, short of a totally different concept of vehicle design, and must be allowed for in track design.

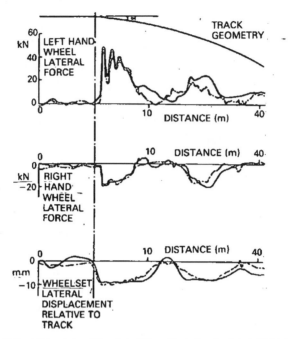

Figure 2.23 Experimental and Predicted Time Histories for 2-Axle Vehicle Traversing a Turnout

On open routes, lateral forces arise where the track alignment is imperfect. On straight track, however, misalignments need to be large before problems arise, because of the wheelset flangeway clearance of 5mm or more. By contrast, on curves, the leading wheelset runs nearer to the outer rail, and much smaller irregularities cause flange contact.

This is illustrated in Figure 2.24, which shows the increase in dynamic force produced by a lateral irregularity as the leading wheelset is displaced further over in the flangeway clearance, towards the irregularity as the curvature of the track increases. If the speed and cant deficiency remain constant, further increases in track curvature do not give rise to larger lateral forces because the lateral displacement of the wheelset equals the flangeway clearance and the wheelset is subjected to the full impact of the irregularity.

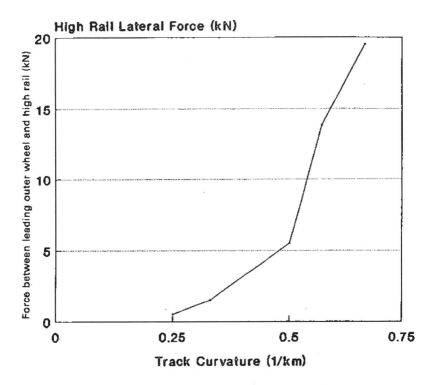

Figure 2.24 Effect of track curvature on the impact force produced at the leading axle of a locomotive with 2-axle bogies, by a lateral irregularity of 2mm, at a speed of 145 km/h and a cant deficiency of 3°

2.8 VERTICAL INTERACTIONS

This section is mainly concerned with the forces produced by the interplay of the constituent parts of the vehicle/track system under the influence of irregularities of the wheel/rail interface when the vehicle is in motion. The vehicle/track system consists of a number of masses, placed one above the other, and connected together by damped springs. In order, from top to bottom, these are:

1. The vehicle body and its contents

2. The secondary suspension*

3. The bogie frame and its attachments*

4. The primary suspension*

5. The wheel, axle attachments (the so-called 'unsprung mass')

6. The so-called "Hertzian spring", which represents the elasticity of the contact patch (this is analogous to, but much stiffer than, the compressibility of a pneumatic tyre on a roadway surface)

7. The rail

8. The rail pad

9. The sleeper

10. The ballast

11. The subgrade

*In a freight vehicle there will usually be only one layer of suspension between the vehicle body and the unsprung mass

Because the number of layers are so great, the mathematical problems involved in the formal analysis of the system are considerable, and when achieved, the essentially random nature of the input (see next paragraph), and of the properties of some of the "springs", renders the results of the analysis of limited value. Complex computational modelling can better take account of the factors involved and would today be employed in the solution of many of the problems which arise.

In general, in any system consisting of a mass connected to a support by a spring, the spring and the mass will oscillate in response (the output), when the support is caused to move (the input). The output in this case is the movement of the mass, and the variation in the tension or compression in the spring and hence the force which it exerts on both the support and the mass. There are two kinds of input, namely, the transient type where the support moves once and then returns to its original position, and the oscillatory type where the support moves repeatedly about its neutral point, in a more or less sinusoidal fashion, at a steady frequency. In the transient type of input, the reciprocal of the time taken for the movement to be completed may be taken to represent the frequency. The relationship of the output to the input depends upon the relationship between the frequency of the applied disturbance, and the natural frequency of the mass-spring system. If the two frequencies coincide, resonance is said to occur, and the result is a greatly increased amplitude of oscillation of the mass, and a corresponding amplitude of variation in the force in the spring. Depending on the degree of damping, this force may, at resonance, oscillate between zero and *twice* the weight of the mass at rest.

Where two or more mass-spring systems are superimposed on one another, the combined system will possess two or more natural frequencies, and hence if the input (ie, the shape of the wheel/rail interface, seen as a function of TIME) is a complex periodic one containing many frequencies, resonances may occur at several interfaces within the system. This happens in the vehicle/track system, and several important sets of interactions can be identified. Firstly, in some freight vehicles, fairly long vertical track irregularities may excite the vehicle body at low frequency (0.5 to 5Hz) on the primary suspension. If these are repeated in a cyclic fashion, they may produce resonance and consequently reduce the effective wheel load sufficiently to cause derailment, as discussed in more detail below. Secondly, irregularities at medium frequency (say 10 to 100 Hz), which may be caused by track or wheel defects, excite the unsprung mass to vibrate on the track spring (ie, the combination of the rail pads and ballast). If the irregularities are transient (eg, a solitary dipped joint or weld) the result is a dynamic increment in the wheel load. If the irregularities *ere* oscillatory (eg, out-of-round wheels, or the small variations in rail top level associated with the sleeper passing frequency) they not only produce a dynamic increment but also cause ground vibration with its attendant environmental problems. Finally, rail corrugations produce a high frequency excitation (say 800 Hz) which is in resonance with the natural frequency of the track mass on the Hertzian spring and produces the well-known problem of "roaring rails". This and other noise problems are mentioned later.

2.8.1 Interactions due to Long-Wave Track Irregularities - Plain Line Derailment

Nadal's Formula

A wheel flange is typically subject to a combination of a vertical wheel weight Q (kN) and a lateral force Y (kN). In the purely static situation, the direction of action of the resultant of any combination of these two forces likely to be encountered in real life, is such that when resolved into components along, and normal to the common tangent at the point of wheel and rail contact, the component along the common tangent acts down the line of contact. It thus restrains the flange from climbing. When the wheel starts to roll along the rail however, and more particularly, when it does so at an angle of attack, the surface of the wheel flange starts to move downwards relative to the point of contact. Then, due to friction between the wheel and rail, an UPWARD force develops. This upward force is related to the normal component by the coefficient of friction at the point of contact. For flange-climbing to occur, it must exceed the downward force already mentioned. This relationship is expressed by NADAL's formula as follows:

If β is the angle of slope of the common tangent at the point of contact and

μ is the coefficient of friction across the contact area, then for derailment not to occur, the ratio Y/Q must not exceed:

$$\frac{Y}{Q} = \frac{\tan\beta - \mu}{1 + \mu\tan\beta} \quad\ldots\ldots\ldots\ldots\ldots\ldots\ldots\ldots 2 - 45$$

Slow speed derailment

Flange-climbing derailment is most likely when a vehicle is starting from rest on a sharp curve, with high cant, with dry rails and an unlubricated and badly side-worn high rail. The reasons for this are:

> The value of "Q" on the high rail is minimised by wheel weight transfer due to cant excess;

"Y" is in this case the gauge spreading force, which as already seen, is related to the wheel weight on the inner rail (maximised by cant excess), and the coefficient of friction across the low rail table (maximised by the dry rail and starting conditions);

"μ" is maximised by the lack of lubrication and the starting condition;

"$\square$" is reduced by the side-worn rail condition.

For BS113A rail in good condition, the flange angle is 68°, so that for a dry rail with μ equal to 0.3, such as might occur on starting, Y/Q must exceed 1.25 for derailment to occur. Once the vehicle is moving, the value of μ will fall, whilst if the high rail is well lubricated, μ will fall still further. If its value is as low as 0.1, the critical value of Y/Q increases to 1.90. From the discussions of lateral curving forces earlier in this chapter, it will be seen that "Y" values of the requisite size cannot develop. Hence, in the steady state under normal plain line conditions, flange-climbing derailment is extremely unlikely.

Flange-climbing derailments at normal operating speeds

For a flange-climbing derailment to occur at normal operating speeds, just as much as at slow speed, a combination of high lateral force Y and low wheel weight Q must occur. As already indicated, high Y values can result if wheelset hunting occurs, whilst low Q values may arise from resonant pitching if the wavelength of cyclic defects of "top", coincides with the natural pitching frequency of the vehicle at its speed of travel. Pitching however, requires that both rails dip together, and it is more likely that one rail only will subside. If this occurs, the cant or cross level will change with distance along the track. The rate of change of cant with distance is referred to as TWIST. Clearly, if the track is twisted, the contact patches of the four wheels of a vehicle or bogie cannot all be in the same plane, but the chassis is to all intents and purposes rigid, and planar. Hence the spring deflections and thus the effective wheel weights at the four wheels are no longer equal. If the twist varies cyclically, then the vehicle body undergoes a cyclically varying set of spring forces which set it rocking, and if the natural frequency of the vehicle in this mode coincides with the periodicity of the cyclic twist, marked periodic reductions in wheel weight at the leading wheel occur. The frequency of this effect will not normally coincide with the lateral kinematic wavelength of the vehicle, but it is nonetheless possible for a minimum "Q" to coincide with a maximum "Y" if the vehicle encounters a cyclic twist when hunting.

With the more primitive suspensions which were common in freight wagons in the UK until the mid-1960's, this coincidence was regrettably frequent, and was one of the main causes of derailment. Another possibility which may supply the necessary high Y value, is a cyclic misalignment. Improvements in both design and maintenance of wagons, as well as improved means of detecting and correcting severe and cyclic track twists, have rendered this type of derailment much less common than formerly. It must be recognised however, that this is only achieved at the price of vigilance. This is the reason why, in track geometry recording, emphasis is placed on the measurement of twists on two wheelbases, 3m and 5m. It is also the reason for the level-2 exceedences discussed in Chapter 8.

Interactions due to Transient Irregularities

Transient irregularities are the ones most readily explored by analytical mathematics, and three types of such defect are discussed in this section.

Dipped Joints and P_1 and P_2 Forces

The classic transient irregularity case is the dipped rail joint. This is modelled as a simple ramp whose angular deflection is the sum of the downward and upward slopes of the running off and running on rail heads, and it gives rise to two readily distinguishable dynamic increments, the first acting over a very short time span (i.e. at high frequency), denoted the P_1 force, and the second, the P_2 force, which acts at medium frequency. Due to the time taken in compressing the various 'springs' involved, there is a phase displacement of each force peak beyond the instant of change of direction.

The P_1 force results, as does the effect of corrugation, from the track mass oscillating on the Hertzian spring of the wheel/rail contact area, and as a transient, its effects are very localised. Its peak occurs a few centimetres beyond the joint, and it contributes to rail end and fishplate damage, and to the formation of bolt hole fatigue cracks.

The P_2 force on the other hand, results from oscillation of the unsprung mass on the track spring, as already stated. It takes much longer for this force to reach its peak value. It typically peaks over the running on sleeper. This force also contributes to rail end batter, star cracks, etc., but it is also the source of various types of sleeper damage in the vicinity of rail joints.

Nomenclature and units in the well-known formulae which give P_1 and P_2 in kN are:

P_0 Steady state wheel/rail contactforce (kN)

V Train speed (m/s)

a Ramp angle (radians)

M_{T1} effective track mass for P_1 calculation(kg) (The value of this parameter is typically 98kg)

M_{T2} effective track mass for P_2 calculation (kg) (The value of this parameter for typical BR track is typically 274kg)

h effective Hertzian contact stiffness (MN/m) (say 1.244×10^6 MN/m for a 25-tonne axle load)

C_T Track damping (kN.s/m per sleeper end)

M_u Unsprung Mass (kg)

K_{T2} Stiffness of the track spring (MN/m). *NB* This parameter is identical to the Lumped Track Stiffness, K_L

Using this nomenclature, P_1 is given by 2 -46:

$$P_1 = P_0 + 2 \ ex \ V \ \sqrt{\frac{h \, M_{T1}}{1 + M_{T1}/M_u}}$$

It is noted that the Hertzian stiffness (h), is itself dependent on P_1 so that the calculation is strictly iterative, and would involve recalculating "h" after each evaluation of P_1. The formula for "h" involves knowledge of the contact radii of wheel and rail, and is not quoted here. The value of "h" quoted is however, accurate for practical purposes.

The formula for P_2 is:

$$P_2 =$$

$$\frac{P_0}{V} + 2\alpha \left\{ \frac{M_u}{M_u + M_{T2}} \right\}^l \frac{1}{2} \left[1 - \frac{C_T rr}{4 K_{T2 (Mu+M_{T2})}} \right] \sqrt{JK_{T2} M_2} \dots\dots\dots 2 \text{ - } 47$$

The term involving the track damping (C_T) is usually neglected, and since the value of the expression $M_u/(M_u + M_{T2})$ is, (for ordinary ballasted track), only very slightly less than unity, the rather cumbersome complete formula becomes approximately:

$$P_2 = P_0 + 2\alpha V \sqrt{JK_{T2}} M_2 \dots\dots\dots\dots\dots\dots\dots 2 \text{ - } 48$$

Soft Spots and Hollows

Work done in the 1970's by British Rail Research, indicated that the dynamic increment $\square P$ due to soft spots and hollows in the rail profile can be expressed in dimensionless terms as:

(1)   For soft spots:

$$\frac{\square P}{P_0} = f\left[\frac{M_u V^2}{K_{T2} L^2}\right] \text{-----------------} 2 \text{ - } 49$$

(2)   For hollows:

$$\frac{\square P}{K_{T2} d} = f\left[\frac{M_u V^2}{K_{T2} L^2}\right] \text{----------------} 2 \text{ - } 50$$

Where P_0, M_u, V, and K_{T2} have the meanings assigned above, and:

L = Length of irregularity (m);

d = Depth of hollow (m)

Equation 2-49, for a soft spot with a stiffness of one-third of that of the firm track on either side, and equation 2-50 generally, are expressed graphically in figure 2.25. For small values of (M_u V^2/K_{T2} L^2), equation 2-49 becomes:

$$\frac{\mathit{\Pi}P}{P_0} = 190 \left[\frac{M_u V_2}{K_{T2} L^2} \right] \text{------------------} 2 - 51$$

and equation 2-50 becomes:

$$\mathit{\Pi}P = 60 \left[\frac{M_u V^2}{L^2} \right] \text{------------------} 2 - 52$$

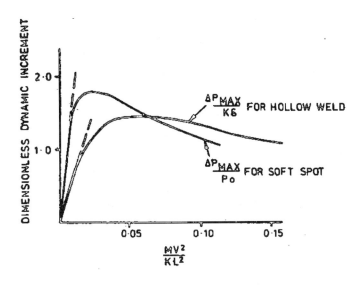

Figure 2.25 Dimensionless Dynamic Increment for a Dipped Weld and a Soft Spot

The graphs show that for soft spots, the dynamic increment peaks at a value of $\square P/P_0$ of about 1.75, i.e. P(max) is $2.75P_0$. This peak occurs when (M_u V^2/K_{T2} L^2) is approximately 0.025. For a hollow (eg a dipped weld), the corresponding peak occurs when $\Box P$ is about $1.45K_{T2}d$ and $(M_u V^2/K_{T2} L^2)$ is about 0.06.

General effects of transient defects

The shape of the curves described in the previous sub-section, is of particular interest. It reflects the fact that above a certain value of the ratio V/L, the wheel starts to 'fly' across the depression, since the spring force pushing it down is no longer sufficient to make the wheel follow precisely the curve of the depression. A similar effect would operate over dipped joints if the approach to the angular change of direction were a downward curve rather than a straight line, but the effect is not modelled in the P_1 and P_2 formulae.

From the track point of view, these four formulae are of value as enabling limits on track defect parameters such as a, L, and d, to be set for a particular track category. They are also used in vehicle design specifications. Since the peak vertical forces predicted by the formulae are large enough to produce cumulative damage, either to the rail, sleepers, or ballast, they form the basis of track deterioration models which in turn, are used to predict the effect of different types of traffic on maintenance costs.

The P_1 and P_2 formulae have been used for many years in the UK to model the interaction between vehicle and track for design purposes. The familiar limiting value of the dip angle a of 0.02 radians for jointed track, has often been quoted. Since most high-speed track is now fully CWR, the relevance of this criterion may be doubted, and some authorities quote lower values of 0.005 to 0.0125 radians.

The quotation of a radian value is also, not very helpful from the practical track engineer's point of view, and it needs to be translated into a value of 'dip' measured over a given base length. Even this can be (quite misleading unless the shape of the dip is also defined. A measure sometimes quoted is a limiting dip of 5mm over a one metre length, but this entails assuming that the rails on either side of the dip are straight in elevation for at least 0.5m, which appears unlikely, but is actually by no means unknown. An alternative possibility is that the rails could be bent into two circular curves, each of which was tangential to the general profile at a point some distance from the cusp. In that case, if the curves extended for one metre on either side of the joint, the limiting value of 0.02rad. would be reached at a dip of 3.75mm, measured over a distance of one metre, and if the curve only extended over 0.5m either side of the joint, the limiting value would be reached at a dip of only 2.5mm, again measured over a metre. These considerations are more important in relation to the P_2 than to the P_1 force, because if the dip runs out at or before the running on sleeper, as in the last mentioned case, it is possible that the P_2 force would not develop as predicted in the simple ramp model. Thus, considerable thought needs to be given to the definition of the critical defect size in order to provide a criterion which is both simple to measure, and an attainable maintenance goal, whilst also not inhibiting vehicle design and the progress towards higher speed running.

Periodic Vertical Interactions (I) - Vibration

The vertical interaction between wheel and rail normally displays a continuous pattern of undulations, of varying amplitudes and wavelengths, caused by irregularities of both the wheel and the rail profile. The movement of the train combines with these undulations to produce vibrations over a wide range of frequencies, which propagate both upwards into the vehicle, where they potentially disturb the payload, and downwards into the track, ballast, and subgrade. The track engineer is mainly concerned with the latter.

The effect of the combined spring/mass systems of the rail, rail pad, sleeper, and ballast, is to attenuate the vibrations over those frequencies which do not coincide with one or other of the natural frequencies of the system. In this respect, the track structure bears some resemblance to the floating foundation systems which are sometimes employed to isolate machinery from adjacent sensitive areas, with the significant differences that the floating part (the sleepers) is much less massive by comparison with the vibrating element (the wheelset etc) than is the case with machinery foundations, that the ballast is generally a very stiff and lightly damped 'spring' compared with what is needed to isolate the track from the ground, and also that the spectrum of vibration input is much wider. Where resonance does occur, the track will pass the vibration down into the ground more or less transparently. Such of the energy as does enter the ground is transmitted through the ground in three ways, and ultimately, even though attenuated by distance and the deadening effect of the ground itself, may excite the foundations of a building, to cause more or less nuisance to the occupants thereof.

To a considerable extent, this process is beyond the control of the practical track engineer, since the origin of part of the driving oscillation is above the wheel/rail contact point, and much of the transmissibility is a function of track design, rather than of construction or maintenance. Nevertheless, it is obvious that vibration will be kept down if the top, and the rail table, are maintained in smooth condition, and if resilient components are properly maintained.

If environmental problems do arise, the investigation of them, and the design of palliative measures, is a specialist task, but some possible means of improvement include the use of softer pads, the use of resilient baseplates, altering the spacing or even the type of sleepers, using soft pads under the sleepers, or even of using ballast mats under the ballast. The effect of this latter possibility is to convert the ballast to some extent from its usual role of spring to become in effect, part of the floating mass. Most of these prescriptions are exceedingly expensive, and must never be embarked upon or suggested without taking a second opinion.

The problem may become particularly acute if ballasted track is replaced by non-ballasted. This kind of operation is more likely to be required in a tunnel than in the open, and one would not expect nowadays that such a scheme would be embarked upon without study of the environmental consequences, possibly involving dynamics and acoustics experts. This would almost certainly be the case if a new section of railway track were being contemplated.

Periodic Vertical Interactions (II) - Corrugations and Noise

The higher frequency parts of the spectrum of oscillations described in the previous sub-section cause trouble of a different kind. Here, the web and head of the rail are excited into vibration, as well as the rim and disc of the wheel, and these vibrations are transmitted into the atmosphere as audible noise. Two well known examples of this effect are the noise associated with corrugated rails, and the violent squealing associated with some sharp curves. To a greater or less extent, the effect is inseparable from any form of wheeled transport, since it is associated partly with the effect of roughnesses of the rolling contact surfaces, and partly with the effect of the mutual deformations of two elements. The effect is magnified in the steel wheel on steel rail operation, by the greater stiffness of the rolling contact surfaces, and by the stiffness and relative lack of damping of both the upper part of the rail, and of the outer parts of the wheels.

Apart from the noise aspect, corrugations are a serious matter from the track engineer's point of view, since the consequent vibrations of the rail can seriously damage the rail fastenings, pads, and sleepers.

Where noise nuisance due to wheel/rail noise in general, arises from track on ballast in the open, whether on fill or on level terrain (noise nuisance from track in cuttings seems less likely), the palliative method most likely to succeed is shielding by noise barriers of some form. These must be placed as close to the track as possible, they must reach high enough to confine the noise (say up to about platform level), and they need to be heavy. They are also likely to be very expensive, and anyone faced with a noise nuisance problem must seek advice and authorisation before embarking upon any palliative measures. It is worth knowing that there are now a number of specialist suppliers who design and build noise barriers for motorways etc., who are available to give advice on suitable forms of construction, should such work become necessary.

Possibilities short of noise barriers might include the fixing of damping materials to the wheels and/or rails, the use of resilient wheels, or changing the shape of the rail. The experiments with the so-called

"HUSH" rail, with its squat, thick-webbed cross-section, have been directed to this end.

Bridges carrying the track present a somewhat different noise problem, since here, particularly if the track is directly fastened to the bridge deck, the bridge itself acts rather like the sound box of an acoustic gramophone, with the wheel being effectively the groove in the record, and the rail acting as the needle. The only possibilities here at track level, are to use resilient baseplates, or, if the bridge will stand the weight involved, to construct some kind of floating intermediate layer with built-in resilience.

CHAPTER 3

RAILS

3.1   THE USE OF STEEL IN TRACK ENGINEERING

Rails, rail fastenings, base plates, distance blocks, the high tensile wire used to prestress concrete sleepers, sometimes even the sleepers themselves, are all made from one form or another of iron and steel. It may be said as with many other things which are commonplace in modern life that the railway could not exist without iron. The great variety of types of article produced from iron are only possible because of certain fundamental peculiarities in the way iron behaves when heated. These peculiarities also bring with them problems if iron or its derivatives are carelessly heated or cooled. To understand and appreciate the behaviour of steel rails and other parts made of iron and steel in relation to the processes used in the laying and maintenance of railway track, we need to understand the metallurgical behaviour of these materials and how they are produced.

3.2   THE PRODUCTION OF IRON

In the traditional method of smelting iron; iron ore (which is in principle iron oxide) is mixed with carbon (either coke or in early days charcoal), and a flux (usually limestone), and the mixture is heated in a blast furnace. The carbon/coke burns to produce carbon monoxide (CO) and heat. CO is a highly reducing gas which readily reduces the iron oxides to iron. The iron melts due to the high temperatures generated as the coke burns in the blast of hot air. Other impurities present are caught up by the flux to form a slag (which being less dense than iron floats on the surface). The slag also protects the iron from further re-oxidisation. The molten iron is tapped off at appropriate intervals. The pig iron which results after solidification will contain an appreciable quantity of carbon left over from the smelting process. As the iron cools and solidifies, what happens to this carbon which is mixed with it?

Before answering this question we need to know a little about the behaviour of pure elemental iron when heated and cooled. Pure iron exists at room temperature in a crystalline form known as ALPHA IRON. In that form it is only moderately hard and ductile, and it is also magnetic. When heated magnetism is lost at 769°C, and at 910°C the crystalline structure is transformed to a form known as GAMMA IRON. The iron undergoes a volume change as a result of the rearrangement of the atoms in the crystals. It also becomes softer, and more ductile. The process is reversed if the iron is allowed to cool slowly to its original temperature.

As the liquid iron from the blast-furnace solidifies it takes up the gamma form, and in this form can take up to 1.7% carbon into solid solution to form austenite. The excess carbon can combine with iron to form the chemical compound cementite (Fe_3C). With carbon levels of 3% and above, and dependant on the cooling rate, free graphite will be precipitated from the solidifying liquid at the same time as the formation of austenite and cementite. On further cooling the austenite itself transforms to alpha iron (a low carbon content phase) called ferrite, and a lamellar phase called pearlite, which is made up of alternate laths of ferrite and cementite.

When final cooling has been achieved and dependant on cooling rates and other elements present, particularly silicon, pig iron can have a structure of pearlite and graphite (grey iron) or, pearlite and cementite (white iron) or, pearlite, graphite and cementite (mottled iron). Blast furnace pig iron is generally not suitable for the production of castings and further remelting and refining is required.

3.3 THE PRODUCTION OF STEEL

Steel is an alloy of the element iron with a very small percentage of carbon and other elements such as manganese. Steel is made from pig iron by heating it once more until it is molten, and then blowing oxygen through the melt which combines with some of the carbon to reduce the proportion of this element in the alloy. To produce steel, the carbon content must be brought below the percentage at which free graphite will form in the austenitic phase (i.e. less than 3%).

With a carbon content as high as 2%, the steel consists of granules of cementite in a matrix of pearlite. As the carbon content decreases towards 0.8%, the proportion of free cementite in the alloy reduces until at 0.8% the steel if slowly cooled through the austenitic-pearlitic transformation, consists entirely of pearlite. Below this carbon content, the steel takes the form of a mixture of granules of ferrite, and granules of pearlite.

Since cementite is a compound which is both much harder and more brittle than ferrite, the hardness and brittleness of the resulting steel will be related to the proportion of cementite, and hence to the proportion of carbon. Steel with 1.3% carbon would be used to make something like a file or a razor. The wire used in the tendons of prestressed concrete sleepers would be made from steel with from 0.8 to 0.9% carbon, while rolled steel joists are made from steel containing only about 0.10-0.20% carbon.

Normal and wear resisting rail steels of Grades A and B contain 0.45% to 0.8% by weight of carbon. Hence the rail steels usually used are referred to as pearlitic steels, with a medium to high carbon content.

3.4 RAIL MANUFACTURE

After conversion from pig iron to steel, the steel is traditionally in the form of an ingot, formed by pouring the molten steel into a mould and allowing it to cool slowly. In producing a rail, the steel passes through two further processes. In the first, the ingot is reheated until it is white hot and then rolled out into a strip having a rectangular cross-section, and about three to four times as long as the original ingot. This strip is then cut into sections called blooms, which are either further processed immediately, or allowed to cool for storage. The final process involves again heating the bloom to white heat and further rolling it through a series of specially shaped rollers to produce the final rail section. Again a bloom which starts off around four metres long will roll out to at least around 36 metres of finished rail.

3.4.1   Modern Steelmaking Processes

Current steelmaking technology takes two short cuts. The first short cut is to convey the newly smelted iron directly from the blast furnace to the converter in liquid form instead of running it into moulds and allowing it to cool to the ambient temperature. The ladle in which this is done is a large and very heavy steel cylinder with a refractory lining, usually mounted on a railway truck. The weight of this so-called "Torpedo ladle" is such as to involve axle weights of 45 tonnes making such traffic (on the internal tracks of steelworks only) the heaviest rail borne freight in the country. The main effect of this modernisation of technology is to eliminate the energy loss involved in cooling and reheating the raw iron. The second short cut is more significant from the track engineer's point of view. It is to make blooms directly from the converted steel by means of the continuous casting process.

In the continuous casting process, molten steel is fed into a tundish with openings in the bottom leading to a corresponding number of rectangular tubes into which the molten steel flows. These tubes, in which the initial solidification of the outer skin of the steel takes place, define the ultimate cross sectional shape of the bloom. The remainder of the process is concerned with the controlled cooling, straightening, and cutting of the blooms.

3.4.2   Continuous casting and piping

Clearly, the continuous casting process saves both the energy required to reheat the ingot, and that required to roll the ingot into blooms. The importance of this process change rests also on the elimination of the possibility that a defect known as PIPING will be present in the finished rail. Piping occurs because when liquid steel is poured into an ingot mould, cooling and solidification set in immediately. The parts of the ingot which cool quickest are the sides, which are in contact with the cold walls of the mould, and even more so, the free surface at the top of the mould which is in contact with the atmosphere. Quite quickly, a skin of solid steel starts to form on these surfaces, and as soon as that happens, the exterior form and dimensions of the ingot are determined. However, as the molten steel continues to cool, it shrinks. (When water freezes into ice, an expansion takes place (vide icebergs), but with steel, shrinkage continues through the solidification process.) Since the centre of the ingot cools and solidifies last, liquid steel is constantly being drawn towards the colder outward regions until at last the originally level top surface of the ingot is sucked inwards in a kind of solid vortex, and in the middle of this surface a narrow tube or pipe leads down into the upper part of the ingot. As a result of this phenomenon, the upper 10% or so of the ingot is always useless for rolling, and it is cut off and fed back into the stock of refined steel. Even so, sometimes the vortex extends further down the ingot than usual, and when this happens, it persists through the bloom and rail rolling processes to finish up as a very narrow pipe shaped void in the body of the rail. Continuous casting eliminates this type of defect completely.

3.4.3   The significance of the rolling process

Hot working by rolling ingot or concast blooms produces many benefits to the properties of steel. The grain structured is refined, the orientated cast structure is destroyed, the metal becomes more homogeneous, diffusion of segregated alloy additions is promoted, undesirable brittle films are broken up, cracks, blow holes and porosity (provided they are unoxidised) are welded up, and toughness and ductility are enhanced. Some directionality of properties is incurred, the better properties being along the rolling direction.

3.4.4   Other constituents of steel

According to BS 11, the commonly specified grades of rail steel will contain proportions by mass of elements other than iron as shown in Table 3.1.

TABLE 3.1 Chemical Composition of Rail Steels

	Normal	Wear Resisting	
	Grade	Grade A	Grade B
Element	Percentage	Percentage	Percentage
Carbon	0.45-0.60	0.65-0.80	0.55-0.75
Silicon	0.05-0.35	0.10-0.50	0.10-0.50
Manganese	0.95-1.25	0.80-1.30	1.30-1.70
Phosphorus	0.040 max	0.040 max	0.040 max
Sulphur	0.040 max	0.040 max	0.040 max

Carbon has already been discussed.

Phosphorus and sulphur are present as impurities, mainly because they form a proportion of the naturally occurring ore from which iron is smelted, and it is difficult to eliminate them altogether in the smelting and refining process.

Silicon is also present in most steels, arising from the refractory materials used in lining blast furnaces and steel converters. Siliceous materials are also generally present in most ores of iron. Generally silicon at controlled levels is beneficial to the properties of steel, and is added deliberately to liquid steel prior to ingot or continuous casting to remove excess oxygen from the steel. This process is called "killing" the steel, and is most essential.

The other common element used for 'killing' liquid steel is aluminium, but silicon is currently preferred for rail steel production, as oxides of silicon which solidify as inclusions in the solidified steel are less harmful than aluminium derivatives for subsequent service performance.

By contrast, sulphur is a highly injurious impurity. The reason for this is that at the high temperatures involved in steelmaking, sulphur combines chemically with the iron to form iron sulphide (FeS). This compound is soluble in molten steel but it is virtually incapable of forming a solid solution with it. Consequently, as the molten steel solidifies, so the FeS is rejected from the solid part of the ingot and is ultimately deposited as a thin layer along the grain boundaries. Steel so vitiated is virtually useless.

To prevent the formation of FeS, manganese (Mn) is added to the steel at the conversion stage. The sulphur then forms manganese sulphide (MnS) in preference to FeS. This has two effects, both of which are due to the fact that MnS is insoluble in molten steel. The first is that most of the MnS floats off the surface of the melt in the slag. The second is that the remaining MnS forms independent globules which are distributed throughout the steel, and solidify in this form. The MnS has none of the deleterious qualities of FeS.

Manganese combines readily in solution with austenite and ferrite, and also forms a stable carbide analogous to cementite, so that it has no ill effects on the properties of the steel. Moreover, it has other values in increasing the depth of hardening of the steel, and improves strength and toughness. These valuable properties are reflected in the comparatively high proportions of this element which are allowed in rail steel.

3.4.5   Nitrogen and hydrogen - two other undesirable impurities

Both these impurities make the steel brittle, but in different ways.

The gas nitrogen which forms four fifths of the atmosphere tends to be absorbed by molten steel during the manufacturing process. This was particularly so when steel was made in Bessemer converters because the blast used was air, and the absence of nitrogen from the blast used is an important reason for blowing pure oxygen rather than air, through the melt when making steel (the "Oxygen Process").

Nitrogen dissolves in the liquid steel and having a small atomic size the nitrogen atoms are located between the iron and carbon atoms making up the bulk of the steel, hence the nitrogen is called an interstitial atom. As such it increases the strength of the steel but there is a corresponding reduction in ductility. This phenomenon accounts for the reduced carbon level specified for Acid Bessemer rail steel (0.4—0.5% C) compared to Open Hearth rail steel (0.5- 0.6%C). The reduced carbon level in the Acid Bessemer steel is compensated for by the nitrogen content.

Water (either as vapour or chemically combined as for example in rust) can come into contact with steel at various times during the purification process. The hydrogen formed by the breakdown of water during steel making is in the form of atomic hydrogen which, being the smallest of all elemental atoms, is again an interstitial atom, which moves freely by diffusion within liquid and solidified steel. Solidified steel contains many cavities (on a microscopic and sub microscopic level) and the hydrogen atoms diffuse preferentially to those sites, and form molecular hydrogen which then, as more hydrogen molecules gather in the cavity, exert an increasing pressure on the metal surrounding the cavity. If sufficient hydrogen is present in the steel the build up of pressure causes small fractures to occur within the steel, known as hydrogen flakes or shatter cracks. These small fractures can then, under the influence of the various forces acting on the rail in service, initiate fatigue cracks which grow and finally cause a brittle fracture of the rail to occur. This type of failure with its characteristic fracture features is known as a Tache Ovale. Higher strength steels are more prone to hydrogen embrittlement than low strength steels, so that the higher strength wear resisting rail steels require more careful control procedures to reduce the risk of its occurrence.

Modern steelmaking techniques, combined with controlled cooling of the finished product, reduce the hydrogen content of modern rails to one or two parts per million, in spite of which this type of defect remains one of the main hazards which the track engineer must guard against, and the principal one which can be classed as a manufacturing defect.

3.4.6   Heating and cooling defects

One of the most important and fascinating features of the behaviour of steel, and one which has been a major contributor towards its economic and cultural importance, is the fact that if a piece of ordinary carbon steel is made red hot and then plunged into water, it becomes extremely hard. In this condition it is possible to grind a sharp edge onto the article, which will remain sharp for a much longer time in use, than would be possible with the same article cooled slowly.

The process of cooling hot metal very quickly is called "quenching". Quenching may be done, not only with cold water, but in oil, or by a blast of cold air, and in many other ways and at many different temperatures. In metallurgical terms, it has already been explained that steel above a temperature of 910°C consists of austenite based on the crystalline structure known as gamma iron, and that it changes to ferrite and/or pearlite based on alpha iron as it cools slowly to normal temperature. When it is rapidly quenched, the iron crystals still change from the gamma form to the alpha form, but the carbon atoms no longer have sufficient time to combine with iron atoms as cementite, and hence the formation of pearlite is no longer possible. The surplus carbon atoms become effectively trapped in the crystalline structure and distort it. The amount and type of distortion depends on the proportion of carbon in the steel, but in all cases the resultant steel becomes harder than either pearlite or austenite. If sufficient carbon is present in the alloy an extremely hard material is produced known as martensite. This appears under the microscope as a uniform mass of needle shaped crystals, although in fact they are more like discs.

The problem with martensite is that not only is it very hard, but it is also very brittle. In the making of steel tools etc, the finished product can be made less brittle, at some sacrifice of hardness, by the process of tempering, which involves reheating the steel slightly so as to encourage the carbon atoms to rearrange themselves. During tempering the fully martensitic structure progressively changes towards the equilibrium structure of ferrite and cementite.

Rail steel contains sufficient carbon for a fully martensitic structure to develop if it is cooled rapidly enough. However, in the manufacture of rails, the objective throughout is to achieve a fully pearlitic structure, and it is an important part of the rail maker's skill to ensure that at each stage in the processing, the product is allowed to cool slowly enough to ensure a return to the desired crystal structure.

Once in track, rails sometimes have to be made very hot (e.g. for welding), and sometimes get very hot by accident (as in wheelburns). What then happens, if the rail gets hot enough to change to the austenitic structure, and then is carelessly allowed to cool very quickly, is that martensite forms in place of pearlite/ferrite in the heat affected zone. This will soon crack under traffic, leading to a complete rail fracture. This is unavoidable with wheel burns because the volume of steel affected is so small by comparison with the rest of the rail, and the damage will be done before anyone can get to it. In alumino-thermic welding it can be prevented by suitable measures to control the rate of cooling of the weld.

3.4.7   Getting austenitic steel at normal temperatures

Plain carbon steel, whilst it has many good qualities, is by no means an ideal material. The realisation of this fact meant that, concurrently with the development of the processes for its mass production as described above, much research was devoted into iron alloys having desirable properties, such as hardness combined with toughness, or freedom from rusting, which ordinary carbon steel does not possess.

One of the results of this research was the discovery by Sir Robert Hadfield in 1882 that by adding a comparatively large dose (around 12.5%) of manganese to high carbon steel (ie steel containing about 1.2% of carbon), it was possible to make the austenitic structure possessed by the alloy at high temperatures stable through the quenching process so that for the first time austenitic steel was available at normal ambient temperatures. It has been found that in this form the steel is extremely tough and shock resistant. In the "raw" state austenitic manganese steel (AMS) is relatively soft (its Brinell Hardness is around 200, which is slightly softer than BS11 Normal Grade rail at 220 to 260), but in use it very quickly becomes extremely hard indeed (Brinell Hardness of around 400 to 550, compared with 280 to 320 for BS11 Wear resisting Grade B). This rapid development of hardness when worked makes it necessary to use special techniques if it is desired to machine AMS. AMS is extensively used by railways around the world in the form of castings, to make monobloc crossings. It is also possible to make it into rolled rails, and these were formerly extensively used in sharp curves subject to heavy side wear, and for making switches and stock rails for use in situations of exceptionally heavy traffic. However the difficulties involved in manufacture, tendencies to develop certain kinds of defects, and the availability of other hard rail steels has led to the decline of use of AMS in the rolled form.

When two pieces of rail are welded together in the track, the method commonly employed is the alumino-thermic process. In this process the weld is made by pouring molten metal around the ends of the rails to be joined. From the account of the metallurgy of steel already given it should be clear that one of the conditions for a successful weld is that after pouring, the weld must be allowed to cool slowly enough for the austenitic-pearlitic transformation to take place. Furthermore, the chemical composition and crystallographic structure of the two rail steels should not be too dissimilar. Whilst most common pearlitic rail steels can readily be welded to one another, and also to the bainitic steel (see below) used for cast crossings, it is not normally possible to weld AMS components, such as castings or rolled rail, directly to pearlitic or bainitic steel rails. However, developments in welding techniques have been made to create acceptable welds between cast crossings and rails (see Chapter 4).

3.4.8   Bainitic steel

Bainitic steel is yet another phase or arrangement of atoms of carbon and iron which can be produced by certain types of controlled quenching from the austenitic phase, or as with the case of the alloy developed for cast crossings, by the additions of the elements boron and molybdenum, which modify the transformation behaviour from the austenite phase. In this instance the austenite-pearlite transformation is suppressed and the austenite transforms to bainite under natural cooling. Additions of chromium, nickel and copper are also made to the alloy to tailor the properties to meet the requirements of the service environment. Bainite is somewhat similar to pearlite, in that it consists of alternate layers of cementite and ferrite, but the grain is very much finer, and the toughness and hardness much increased, qualities which of course are just what is wanted for a crossing, which has to withstand heavy abrasion and impacts.

It will be evident that since bainite is a phase rather than a separate chemical compound it is possible for its chemistry to be very similar to that possessed by a BS11 rail. If then a crossing leg made of bainitic steel is alumino-thermically welded to BS11 rail, weld metal of the normal composition can be used, and appropriate control of the cooling rate will result in a successful weld, and a suitable gradation of hardness between the two parent rails. Alternatively it becomes relatively simple to weld the two together by flash butt welding.

3.5   THE FUNCTIONS OF A RAIL AND HOW THESE DETERMINE ITS SHAPE

As was explained in Chapter 1, the original function of the rail was to act as a hard and unyielding surface to carry a rigid tyred wheel without rutting or abrasion, and this remains a prime function of a rail to this day. The other prime functions are to act as a beam and thereby transmit the wheel loads to the sleepers, and to act with the tread and flange of the wheel tyre in steering the vehicle in the desired direction.

The shape of the rail that has become standardised for all modern main line railways is the Flat Bottom (FB) rail (sometimes called Vignoles rail after its inventor). The bull-head rail which was once universal in the UK and now limited to slower speed, specialist locations, underground and heritage railways. A few notes on its development and the reasons for changes will be found in Chapter 1. For the sake of any reader who may be totally unfamiliar with the terminology, Figure 3.1 shows a typical cross section of a wheelset on a railway track, with the names given to the different components. The reasons for the shapes of the various parts of the rail are discussed below, with particular reference to the standard rail section, the BS113A rail, the dimensional details of which are shown in Figure 3.2.

Considering first the shape of the rail head, note the subtle combination of curves of three radii varying from quite sharp to very flat. These are designed to fit with the shape of the wheel tyre, to form a combination which will both have good riding qualities and minimise contact stresses. One of the features of a well matched combination of wheel tyre and rail head is that, when the axis of the wheel set coincides with the longitudinal axis of the track, and the rail is at its correct inclination, the point of contact between the two is very close to the centre of the rail. This is very desirable from the point of view of the rail, since it minimises the twisting effect which an eccentrically applied wheel load would have on the rail, and by keeping the wheel/rail contact area away from the gauge corner the likelihood of gauge corner shelling and head check fatigue damage is reduced. Side wear is also reduced. It also implies a practical limit to the overall width of the rail head, and even the heaviest rail sections in use have head widths very little greater than those of the BS113A rail.

The BS113A rail head has sides which slope outwards at 1:20. This is to compensate for the 1:20 inwards slope of the rail, and not only makes it simpler to control gauge, but ensures that when side wear takes place, the associated gauge widening is minimised.

Even when rails are welded into CWR, it is occasionally necessary to join them together by fishplates,, and the lower faces of the head and the inner parts of the upper faces of the foot are designed so as to maximise the efficiency of the fishplate in transmitting the longitudinal forces associated with the bending of the rail end under wheel loads, and to allow wear on the mating surfaces to be taken up by re tightening the fishbolts.

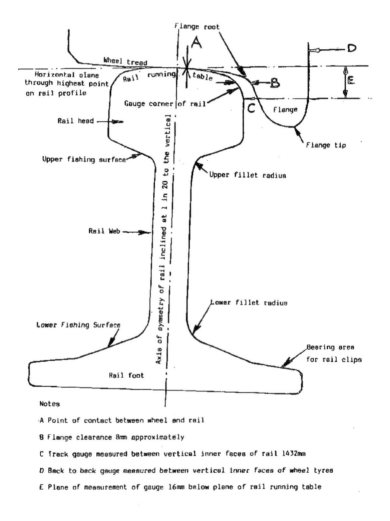

Notes

A Point of contact between wheel and rail

B Flange clearance 8mm approximately

C Track gauge measured between vertical inner faces of rail 1432mm

D Back to back gauge measured between vertical inner faces of wheel tyres

E Plane of measurement of gauge 16mm below plane of rail running table

Figure 3.1 Standard wheel profile running in its central position on 1432 mm gauge track with BS 113A rail

The design of the fishing surfaces also ensures alignment of the rail ends without any contact between the rail web and the fishplates. When it is necessary to install insulated fishplates in CWR, the fishing surfaces are also called upon to transmit the thermal loads from the rail into the fishplate and vice versa.

The thick, sturdy web of the BS113A rail is designed to give the rail adequate shear strength to guard against fatigue failures, particularly around fishbolt holes even under the impact loading imposed by the heaviest axle loads passing over dipped joints. Another feature to note here is the generous radii of the transitions between the head and web and between the web and foot. These fillet radii as they are called are necessary because the guidance forces on curves (see 5.5 below and elsewhere) may be of the order of 35% of the axle load. The obliquity of the direction of application of the resultant wheel load and the asymmetry of the point of application of the wheel loads, results in the development of secondary stresses within the rail itself, particularly at the fillets between the head, web, and foot, and the large fillet radii of the BS113A rail are designed to keep down these secondary stresses, again to guard against the possibility of fatigue failure.

Finally we come to the foot of the rail, which is broad to give stability to the rail against roll over, remembering that the steering forces mentioned above produce significant torsional forces which have to be resisted by the clips which secure the rail to the sleeper. The rail is made flat underneath to distribute the wheel reaction evenly over the surface of the baseplate. The reason for the slope of the inner portion of the upper surface of the foot has already been mentioned. The much flatter slope of the surface towards the outer edge of the foot is provided as a table upon which the rail clips may rest, and the surface is planar rather than cylindrical to enable precise control over the deflection of the clip, to simplify the shape of any insert provided between the clip and the rail, and to minimise contact stresses between clip and rail.

In addition to its primary functions the rail has secondary functions, notably that it may convey return traction current, and that it may carry electric currents for signalling purposes, usually as some form of track circuit.

3.5.1   The Rail as a Beam

In the most modern applications the rail is supported at discrete intervals by sleepers. In the case of concrete sleepers thin resilient pads, typically 5mm- 10mm thick, lie between the rail and sleeper; where timber sleepers are used a steel baseplate lies between them. On bridges, baseplates may be positioned directly onto bridge girders. Thus the rail can be seen to be acting as a beam and indeed its transverse section is similar to that of an I-beam.

As explained in Chapter 1, the loads from train wheels produce bending moments and shear forces in the rail. The bending moments cause longitudinal compressive and tensile stresses which are mainly concentrated in the head and foot of the rail whilst the shear forces produce shear stresses which occur mainly in the web. It is the need to provide adequate resistance against bending (referred to as resistance moment) which determines the required areas of the head and foot of the rail, and its overall depth.

It is to be noted at this point that even if the rail is continuously supported, as for instance in slab track, the track form almost always possesses a degree of vertical resilience (eg in the form of a continuous rail pad). In this case also, the stiffness of the rail distributes the point load from the wheel over a determinable length of support, and bending moments and shearing forces arise from this distribution of load. Even with continuously supported rail, the rail's size must be commensurate with its function as a beam.

Even in jointed track a rail will be continuous over 24 or more supports. Characteristically for a continuous beam over many supports carrying rolling loads, the rails are subject to repeated reversals of bending and shear stress. These stress reversals are a source of metal fatigue. To combat this they are limited to an allowable level which takes into account CWR thermal stress and residual stresses caused by the manufacturing process and wheel/rail contact.

A further factor to be taken into account is, that in course of time, the rail head gets worn away by wheels on its top surface, and galled by abrasive contact with baseplate or sleeper on its underside. Corrosion (especially of the rail foot) leads to loss of rail section, and surface pitting may itself reduce the fatigue resistance of the rail. In wet tunnels severe localised head loss may occur due to dripping water. This too will reduce the cross section of the rail, but more significantly the resulting running surface irregularity will cause higher than usual dynamic loads. These features have to be taken into account when determining rail removal criteria.

Once these allowances have been made, the strength and the moment of inertia and depth of rail required in any particular situation may be determined by considering the bending moments produced by the known wheel loads, (suitably factored to take account of the effects of impact forces produced by track irregularities). The calculation of the bending moment has to take account of the spacing of the sleepers and the fact that the sleepers themselves sink into the ballast as the wheel passes and recover afterwards. A method of doing this is explained in Chapter 1. It should be noted that the strength of rail steel required is also likely to be influenced by wear resistance requirements and the need to combat rolling contact fatigue problems.

3.6 USE OF RAILS IN THE UK

3.6.1 Rail Sections

There are two standards of rails in use in the UK, UK56 and UIC60. The UK56 rail is the European description given to the UK developed standard FB rail shown in Figure 3.2, and is known by the code name BS11-113A, where:

> BS11 stands for the British Standard
>
> 113 indicates the mass of rail in lbs per yard
>
> A indicates that the sides of the rail head are inclined outwards at 1:20

The geometrical properties of UIC56 rail are given in Table 3.1. Earlier FB rail used in the UK includes rail of 98lb, 109lb, and 113lb/yard, as well as BS11-110A.

Areas – Head	**2882.83 sq mm (40.12% of whole area)**
Web	**1711.93 sq mm (23.83% of whole area)**
Foot	**2589.74 sq mm (36.05% of whole area)**
Whole section	**7184.50 sq mm**
Calculated mass	**56.398 kg/m**
Moment of inertia I(xx)	**2332.4 cm ^4**
Section modulus Z(xx)	**277.95 cm ^3**
Distance from top of	
rail to neutral axis	**83.914 mm**

TABLE 3.2   Area, calculated mass and properties for BS 11-113A railway rails in metricated units. (Reproduced from Table 11 of BS 11-1985)

Only the last named of these is currently featured in the British Standard Specification for rails (BS11, see below, section 3.5). In common usage the figure 11 is usually omitted from the coding.

BS110A rail is a section of some significance, since it is identical with one of the standard rail sections used by railways in mainland Europe, where it is known as UIC 54. The letters UIC stand for Union Internationale des Chemins de fer (ie International Union of Railways). The figure 54 stands for the mass of the rail in kg per metre. Some administrations use UIC56 which is the European equivalent of BS113A.

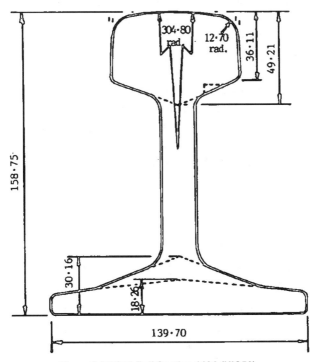

Figure 3.2 BS11 Rail Section 113A (UIC56)

At the time when CWR track became the standard on concrete sleepers the BS110A rail was adopted as the standard UK rail, but following a tragic accident in late 1967 involving much loss of life, and attributed to a rail end failure involving this rail, concern began to be expressed that the web of the BS110A rail was too thin in the middle, and after some consideration it was decided to thicken up the web, with parallel sides rather than the elegant curves of the original rail, and this became the BS113A rail. This change added about three lbs per yard to the mass of the rail, which is otherwise identical to BS110A (otherwise UIC 54) rail.

Another change in 1993 made in the UK was to make a further modification to the rail section to be supplied as nominally BS113A. The difference will be observed if Figure 5.2(a) is compared with Figure 3.2. It is that the width of the central band at the crown of the rail having a radius of 304.800mm is reduced from 19.05mm to 12mm. The effect of this is significantly to reduce the average crown radius, hence increasing the effective conicity of the wheel-rail combination. This improves the lateral dynamic stability of bogies on straight track without degrading performance in curves. Initial wheel-rail contact stresses are however slightly increased.

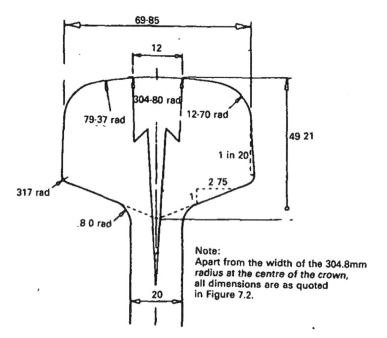

Figure 3.2(a) BS Rail Section No 113A with 12mm wide crown as modified by BR

BS11 contains descriptions of a wide range of rails varying from the 50 'O' rail with a mass of 24.83kg/m to the 113A at 56.40kg/m, but only the dimensions of BS 113A rail which is the one most likely to be found on UK tracks are given in Figure 3.2.

UK and Continental railways now use for their premier lines a heavier rail than UIC 54, namely UIC 60, which as its name indicates has a mass of 60kg/m. The leading dimensions of this rail are quoted in Figure 3.2(b).

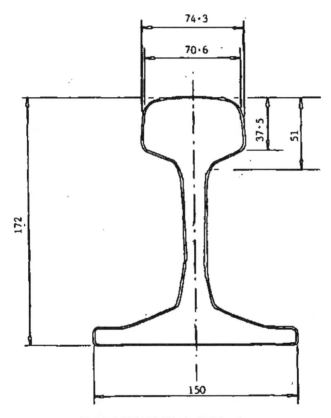

Figure 3.2(b) UIC 60kg/m Rail Section

3.6.2   Switch Crossings and Rail

For Switch and Crossing work certain variants of the basic FB rail are available and should be mentioned. The main features of these rails are that they have a much thicker web than is customary, and are much shallower and squatter in section. They come in both a symmetrical form, where the foot of the rail is the same width on both sides of the web, and asymmetrical, where the foot on one side is virtually absent. These rails are mainly used for switch rails and are illustrated here in Figures 3.3 and 3.4.

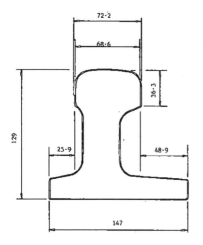

Figure 3.3 Asymmetrical Shallow Depth Rail Section for use with UIC 54 or BS 113A Rail Mass 68.5kg

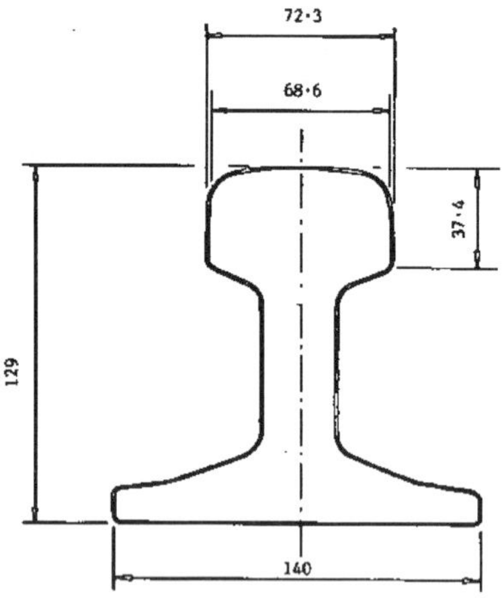

Figure 3.4 Symmetrical Shallow Depth Rail Section for use with UIC 54 or BS 113A Rail Mass 59kg/m

Another rail section sometimes encountered is the special rail used originally on the Continent for making check rails. Its code is U33, again denoting it's mass in kg/m. It is illustrated in Figure 3.5.

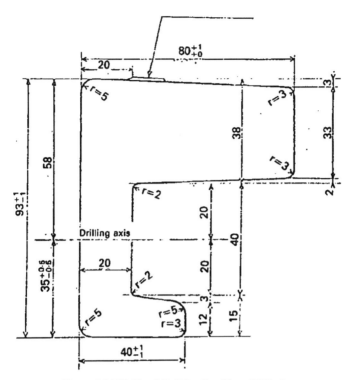

Figure 3.5 U33 Check Rail Section Mass 32.3kg/m

3.6.3   Tram Rails

In tram systems for street running grooved rail is essential. No BS is available for grooved rails, but certain designs of German origin are available, under the codes Ri 60 and Ri 59 (Ri stands for Rillenschiene, the German word for grooved rails). These are shown in Figure 5.6. It should be pointed out that the "Ri" rails have a groove which is much too narrow to take a standard railway wheel tyre flange. This is because a groove wide enough for a standard vehicle would endanger cyclists, and so where street running is anticipated, the vehicles must have an appropriately designed tyre profile with a narrower flange. This in turn creates problems with switch and crossing work when it is desired to run Light Rapid Transit (LRT) vehicles over the same metals as standard vehicles.

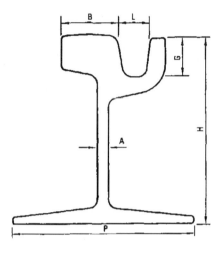

Rail Designation	Weight	H	P	B	A	L	G
	Kg/m	mm	mm	mm	mm	mm	mm
Ri59	58.96	180	180	56	12	42	47
Ri60	60.48	180	180	56	12	38	47

Figure 3.6 Tram Rails Section properties

3.7 WEARING RESISTANCE

3.7.1 Abrasive wear

Abrasive wear occurs when there is contact between the side of the flange of a wheel and the gauge face of the rail. Such contact usually takes place between the leading outer wheel of a vehicle or bogie, and the outer rail of a curve. In this case the wheel is directed towards the outside of the curve, so that as that part of the flange which is forward of the area of contact between the tyre and the rail head moves downwards, it attacks the side of the rail head (the angle which the wheel makes with the line of the rail is called "the Angle of Attack"). It is to be noted that this orientation of the wheel is caused, not by centrifugal force, or the momentum of the vehicle, but rather it is associated with the requirement that the longitudinal axis of the vehicle must change direction as the curve is traversed, and hence the outer rail must be continually pushing the outer leading wheel inwards towards the centre of the curve. These effects (known as "guidance forces") are discussed in more detail in chapters 4 and 15. The effect of the wheel flange rotating ever downwards in contact with the side of the rail head is akin to that of a grinding wheel and the result is the same, namely loss of metal from the side of the rail head by abrasion.

3.7.2   Plastic Flow

When a wheel rolls along a rail, with its flange clear of the gauge corner of the rail, the area of contact between the two is extremely small. In theory contact between the conical wheel flange, and the rounded rail head is only possible at a point, but this implies an infinitely large pressure at the point of contact, and in practice, both surfaces deform slightly with the result that a contact patch forms. The shape of this contact patch varies but its maximum area is typically about 100 sq mm. Consequently the contact pressure is very great. Under the wheels of Heavy Axle Load (HAL) wagons, pressures will average around $1200N/mm^2$ across the whole patch. Such a stress is much greater than the yield point of the steel, and causes the metal in and near the contact patch to become plastic, so that if it were not for the restraining effect of the main bulk of the metal in the rail head, the material under the wheel would squeeze out sideways like toothpaste. Indeed the metal on the surface of the rail head does in fact flow under the combination of the contact stress and tangential forces between the wheel and the rail. A frequently observed example of this effect is the flow of steel on the head of the low rail in curves, from the gauge side towards the field side of the rail. This effect is particularly marked on sharp curves where trains move slowly, and it is caused by the gauge spreading force which is a by-product of the frictional forces involved in curving, rather than by cant excess (see Chapter 4 for further discussion of the forces resulting from curving).

A further effect, not observable with the naked eye, is that even on straight track, the longitudinal forces between the wheel and rail, cause the surface metal to flow along the rail. After some time in service, examination under the microscope of a section taken along the rail, reveals changes in the shape of the crystals of steel, and eventually even, cracking commences, leading to the development of the defect known as a "squat".

3.7.3   Limits on Wear

As a result of both plastic flow and abrasion, metal is lost from the rail head and in the process of time 10% or more of the area of the rail head may get worn away. Eventually the loss of section becomes such that the rail must be removed from service, either because its strength as a beam is no longer sufficient or because the head has become so misshapen that the conditions for satisfactory (or even sometimes, safe) running can no longer be guaranteed.

3.7.4   Headwear

Loss of depth of the rail due to wear of the crown is a slow process which ultimately reduces the bending stiffness of the rail, and also increases the likelihood of a tall flange hitting track fittings in the flange way. There is, consequently, a headwear limit which is expressed as a remaining rail depth.

3.7.5   Sidewear in Plain Line

Sidewear at many locations has been regularly monitored and it is observed that the maximum flange contact angle remains virtually constant as the sidewear progresses across the rail head. The shape of the head, however, tends to become increasingly concave as more of the flange overlaps the original rail outline.

The limit now set for UK rails is 61mm of remaining head width (i.e. 9mm loss of width) at 16mm below the crown. This limit applies to plain line curves with reasonably constant sidewear. There is a subsidiary limit which dictates that the wear scar should not cut the bottom corner of the gauge face. This sets a combined headwear/sidewear limit, so that tall flanges are kept clear of fishplates, etc. The limits for intermittent sidewear on curves and cyclic sidewear on straight track are currently being reviewed.

The standard sidewear gauge now adopted for use in the UK is shown in Figure 3.7. To use the gauge it is placed on the railhead and the straight jaw lined up with the outer face of the rail as shown. A stepped wedge is inserted into the slot in the centre of the gauge face jaw. This wedge will enter the slot by a varying amount depending on the amount of sidewear. The numbers engraved on the wedge indicate the "lift-off" measured in millimetres. The rate of sidewear can be judged from the change in 'lift-off measured over a period of time. The letters 'P' (minimum point for planning rerailing) and 'I' (immediate rerailing) are engraved on the wedge to correspond with 3 and 0mm lift-off respectively. The use of this gauge on turned rails is currently under discussion.

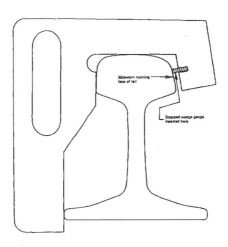

Figure 3.7 UK Sidewear gauge in use

3.7.6   Switch and Crossing Wear

A standard gauge for measuring wear of switch and stock rails is used in the UK and Europe.

The gauge is shown in Figure 3.8 and has two faces. Gauge 1 is used to check the lateral wear of the stock rail, and Gauge 2 to measure the wear of the switch rail within the planing length.

A further gauge is now in use to measure the lipping of rolled AMS switches which has featured in some recent derailments of multiple unit rolling stock.

3.7.7   Obtaining Wear Resistant Rail

Much research and practical experimentation has been devoted to the complex problems involved in understanding the mechanisms by which wear takes place, and of selecting the proportions of the alloying elements which will give the best possible wear resistance without making the rails less serviceable in other ways.

Since it is abrasion (side wear) which can cause the most rapid loss of metal from the head, the first priority has been seen as producing an abrasion resistant steel. After much experimentation it has been established that for pearlitic steel there is a correlation between abrasion resistance and hardness (measured as Brinell hardness number), so that the greater the hardness, the greater the abrasion resistance. It happens also that there is a similar positive proportional link between hardness and ultimate tensile strength (UTS). This latter property is the key parameter by which steel quality is the key parameter by which steel quality is measure, and to obtain high UTS it is necessary to ensure that the finished steel has as fine a grain structure as possible. This fine grain structure can be obtained in three main ways:

- by modifying the chemical composition of the steel, or

- by careful heat treatment, or

- by a combination of both.

The first method is used to produce what are termed "naturally hard" steels, of which the "Wear resisting Grade A" and "Wear resisting Grade B" rails of the BS11 and UIC60 specifications (qv below) are examples. The Brinell Hardness (BHN) values of these rails are about 270-290 compared with 230- 250 for BS11 Normal Grade rail.

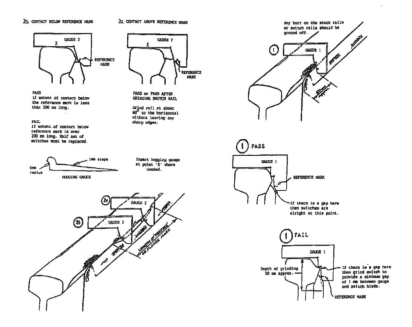

Fig. 3.8 Switch Sidewear Gauge

Still finer grained metallurgical structures are obtained by using an alloy containing up to 1.20% of chromium (Cr). The resulting steel has a UTS some 30% higher than Grade A, and correspondingly enhanced hardness (around 330-350 BHN) and abrasion resistance. Rails are manufactured from this steel and have been used for many years in industrial railway tracks, under combinations of very high axle loads (45-60 tonnes) and severe curvature. More recently such rails have been used on heavy haul railways, and very selectively on more conventional railways and metros, where service conditions are extremely harsh. Rails made from 1.2% Cr steel, are however very difficult to weld, and tend to be subject to mechanical damage during handling. They are not therefore normally recommended for use.

Clearly, the part of the rail which it is most important should be wear resistant is the head, and if a method could be devised whereby in an otherwise normal rail, the wear resistance of the head could be greatly enhanced, the result would probably be economic by comparison with a rail made entirely of wear resistant material. The first method of doing this consists of reheating the head of the rail until it is austenitic, by electromagnetic induction (this method of heating involves placing electrically conducting, water cooled coils close to the rail head; when a high frequency alternating current is passed through the coils the constantly changing magnetic field in the rail causes it to heat up), and then passing the rail under a series of hoods through which a mixture of water and cold air is blown. By controlling the force of this blast it is possible to obtain an austenitic/pearlitic transformation temperature low enough to develop the required degree of hardness in the rail head.

A more recent development has been the introduction of what is termed Mill Heat Treated (MHT) rail. In this process the rail is subjected to a computer controlled air/water quenching system as it emerges from the final shaping rolls. The cooling is controlled so that a very fine pearlitic structure is produced, and hardness levels between 300 and 390 BHN can be achieved. Three qualities of rail are offered by UK steel suppliers; 320, 350 and 370 BHN. These are all from the same composition range, namely the American Railroad Engineering Association (AREA) standard grade, which is slightly above the carbon range for BS11 Grade A.

From what has been written about the effects of flash butt and thermit welding on the crystalline structure of rail steel it will be clear that the thermal cycle induced during welding (i.e. the heating up and subsequent cooling of the rail ends making the welded joint) will modify the metallurgical structures developed by these accelerated cooling treatments. To avoid the retention of softened areas produced in the weld heat affected zones (HAZ) which could lead to excessive wear at these positions, special cooling treatments need to be applied to the weld to induce hardness levels similar to those in the unaffected parts of the rails making the joint. See Chapter 6 for details of the process involved.

Some indication of the economic value of the increases in hardness discussed above may be gained from the fact that the rate of wear per million gross tonnes (MGT) of traffic if rail of 370 BHN is used is about one fifth of that experienced when rail of 260 BHN is used. However, nothing is entirely unalloyed advantage, and it has to be pointed out that as the rate of surface wear is reduced, so the tendency towards rolling contact fatigue increases.

3.8 RAIL SPECIFICATION

3.8.1 BS11

The main features are given below.

(a) The Specification applies to rail with a mass of 24.8kg/m or more.

(b) The Purchaser's obligations to supply information to the supplier in his enquiry and order are defined.

(c) The Specification includes tables of dimensions, areas, properties and mass per unit length, with which rails ordered by BS rail section number must comply. The values for BS113A are reproduced as Table 3.1.

(d) The manufacturer is responsible for deciding the source of the steel and the steelmaking process employed, although if the purchaser asks to be told the process to be used, this cannot be changed without informing the purchaser.

(e) BS11 recognises three grades of steel for the manufacture of rails, viz, "Normal", "Wear resisting A", and "Wear resisting B". For each of these grades the

Specification quotes a permitted range of chemical compositions (see Section 3.3 above), and lays down how the required tests shall be carried out to confirm the chemical analysis. It should be noted that in UIC 860-0, the equivalent wear-resisting grades are called Grade 900A and 900B respectively.

(f)   The toughness of the rail steel is required to be checked by a Falling Weight test, which is described in detail.

(g)   The strength of the steel is required to be checked by tensile testing. The required ultimate tensile strengths (UTS) and minimum elongations at rupture for each of the three grades of steel are specified. Minimum UTS's specified are $710N/mm^2$ for Normal grade and $880N/mm^2$ for Wear-resisting grades. The mean UTS of rail as supplied is about $826N/mm^2$ for Normal and $932N/mm^2$ for Wear-resisting grade A rail steel.

(h)   Standard brand markings to be rolled in relief on the side of the rail to enable the origin of the rail and its grade of steel to be permanently recorded are laid down. Provision is also made for the identification of the purchaser. In addition, each rail is required to be identified on its web in such a way that the complete manufacturing history of the rail can be traced.

(i)   The manufacturer is required to supply master templates and a set of plus and minus limit acceptance gauges to the purchaser (subject to certain limitations). He must also provide facilities for the purchaser to observe the manufacturing process and inspect the finished product.

(j)   Dimensional tolerances and requirements for freedom from surface and internal defects are laid down.

3.8.2   Other Types of Rail Steel

Rails made from steels not described in BS11 are available. Perhaps the best known of these, as it has been available for very many years, is the rolled Austenitic Manganese Steel (AMS) rail. The process for making AMS is described in Section 3.4.7 above, from which it will be seen that in order to obtain the desired austenitic crystal structure, it is necessary, after the rail has been rolled, to quench it in cold water. With the development of the low carbon austenite manganese steel (LCAMS) it is possible to produce AMS rails without the quenching into water necessary with the higher carbon original Hadfield composition. This is now the standard alloy produced by BSC in rail form. AMS rail tends not to be of the same high geometrical quality as rails to BS11. Nevertheless the rail is very useful in localities where sharp curves and very heavy traffic are combined.

Other alloys sometimes used for making rails are those including small percentages of chromium (Cr).

3.9 FRACTURE TOUGHNESS AND FATIGUE

The most common cause of failure of structures and components made from metals is that of "fatigue", and rails are no exception. Failure by fatigue almost always occurs in three stages:

> (i)   the initiation of a crack,
>
> (ii)   the growth of the crack,
>
> (iii)   final fracture or collapse of the structure.

Stages (i) and (ii) of the fatigue process are caused by repeated loading of the structure which, in rails, is primarily due to the passage of wheels. (In rare cases, stage (i) of the process may not occur, or be very short, if the structure contains a crack-like defect which is created by some other mechanism). Stage (iii) is caused by one loading event which may be abnormally severe or, as is more likely, the final occurrence of a normal load.

The initiation of a crack (stage (i)) may require millions of repeated loadings and during this time there is no way of detecting the forthcoming crack. The growth of the crack, particularly in its early stages, may also require the application of millions more loading cycles. Stages (i) and (ii) therefore take place over a period of time, often years, and during this period there will be no perceptible change in the shape or dimensions of the rail. The final stage of fatigue failure in the common types of rail steel described in this chapter will be sudden brittle fracture during a period of increasing load. It is the *lack of warning* of fatigue failure, that has caused so many notorious incidents involving, for example, aeroplanes, ships, road vehicles, power generation equipment, pressure vessels, pipe lines, off-shore platforms and trains. The time required to develop fatigue cracks does, however, at least offer the opportunity of applying ultrasonic testing methods in an attempt to find them before final fracture occurs.

With the exception of wheel/rail contact stresses, which are a special case, the stresses in rails are only a fraction of the UTS (see Chapter 2). However, fatigue failures occur at stress levels well below the UTS. This is caused by the gradual accumulation of fatigue damage with each stress cycle. As the range of stress (i.e. the difference between the maximum and the minimum stress levels in the cycles) decreases, more stress cycles are required before failure occurs.

This phenomenon continues until a stress range is reached when, for all practical purposes, fatigue no longer occurs. This stress range is sometimes referred to as the material's fatigue strength and for new, as-rolled normal quality rail steels, its value is about 300 MPa. The fatigue strength of rail steels, in common with many other steels, is established at about 2 million stress cycles and for this stress range, and all lower ones, the fatigue life of the rail steel may be considered to be infinite. The fatigue strength may appear to be high compared to, say, nominal bending stresses in the rail; however, the rail presents many localised areas (stress concentrations) where these nominal stresses are substantially increased.

These stress concentrations are caused by many features that disturb the smooth flow of stress contours. Some are caused by the design of the rail itself such as:

- fish bolt hole
- radii between the head and web and foot and web

Other stress concentrations may be unintentionally caused by the manufacturing process:

- hydrogen shatter cracks
- non-metallic inclusions
- roller guide marks
- scores and scratches
- pits caused by indented mill scale

Some stress concentrations are caused intentionally by the user:

- drilled holes for cables and other attachments
- stamp marks

Stress concentrations are also caused unintentionally during use of the rail:

- damaged hole surfaces caused by poor drilling or using crow bars or badly fitting bolts
- rail surface damage caused by hitting the rail with unsuitable tools
- wheelburn cracks
- plastic flow of the rail head
- foot galling
- corrosion pitting

Apart from stress concentrations, other features occur in track which raise nominal stresses.

These include:

- track irregularities such as bolted and welded joints
- wheelburn depressions
- poor support conditions
- rail head depressions caused by dripping water in wet tunnels
- transitions, off and onto bridges

Residual, or locked in stresses, which are originally produced during rail manufacture (roller straightening) and subsequently modified by service loadings, also play a role in determining the fatigue performance of the rail and possibly add to the stress causing final fracture of the rail. Thermal stresses in CWR play a similar role to residual stresses but they, of course, vary according to weather conditions and time.

The final fracture of the rail will occur when the fatigue crack length and applied, residual and possibly thermal tensile stresses combine to cause a critical, unstable condition and, as mentioned above, in common rail steels this results in an instantaneous brittle fracture of the rail. This unstable condition is determined by the fracture toughness of the material, sometimes described as its resistance to fracture in the presence of a sharp fatigue or fatigue like crack. Fracture toughness is measured in tests based upon the principles of fracture mechanics, and it is found that it decreases with decreasing temperature and, in general, decreases with increasing loading rate. Consequently, fracture will occur in a fatigue cracked rail more readily at low temperatures and with impact loads (caused for example by wheel flats and badly dipped joints).

As noted above the only practical way of controlling rail fatigue failure in service is by ultrasonic inspection. Any periodic ultrasonic testing policy should:

(i)   ensure that detection techniques will detect cracks which are less than the critical size,

(ii)   the time periods between inspection should not allow a crack to grow to its critical size,

(ii)   sizing of cracks should be accurate enough to allow logical withdrawal criteria to be   determined.

Developments in steel technology have reduced and in some cases eliminated some traditional fatigue failure types, whilst improved ultrasonic test methods and a better understanding of fatigue mechanisms and fatigue life prediction, are helping to contain the problem in service. A better understanding of wheel rail interaction has also led to vehicle and rail design improvements. Nonetheless the constant drive for higher speeds and heavier axle loads will demand that a continued effort be maintained to keep rail fatigue failures under control.

3.10 CORRUGATIONS

Rail corrugations are more or less periodic undulations which form under traffic on the running table of a rail. They vary in wavelength from 50mm or so up to 300mm or even longer. In the case of the shorter wavelength corrugations the phenomenon becomes troublesome at amplitudes from peak to trough of 0.1mm or less, and amplitudes of 1mm or more occur on the longer wavelengths. The interaction between the wheel and the rail, when passing over short wave corrugations particularly, produces a very high noise level, audible both in the train and at the lineside, the latter leading to complaints of noise nuisance. As well as these environmental problems, very high frequency vibrations of the rail on its support occur. These are liable to damage the fastenings, pads and sleepers. Particularly severe damage may be caused to concrete sleepers which may render the sleepers unserviceable in a very short space of time.

The ideal solution to this problem would be to prevent the corrugations forming in the first place, but to achieve this it is necessary to know what causes the phenomenon.

3.10.1 Causes of Corrugations

In spite of much research, insufficient is known about the origins of rail corrugations to make it possible to design a corrugation proof track form. The present state of knowledge is broadly as follows:

The wheelset, and the rail, are parts of two independent mass-spring-damper systems. The two systems are connected by the contact patch, and it seems that corrugations form when the complex dynamic interactions between the two independent mass-spring-damper systems affect the adhesion conditions across the contact patch (i.e. whether at anyone instant there is relative movement across the patch, and whether the movement if any is creepage or sliding) in an unfavourable manner. Some researchers (e.g. CO Frederick) have built mathematical models which express the relationships involved, and have even been able to explain why corrugations have formed at certain locations and not at others.

In closely controlled conditions (eg on non ballasted track) there seems some prospect that in the not too distant future it should be possible to design a relatively corrugation proof track, but on ballasted track the variability of the stiffness and damping of the ballast is such that occasional corrugation is inevitable, and the only remedy is correction by grinding. (See Chapter 14)

3.10.2 Reprofiling of worn rails

In a typical grinding machine the planning heads are mounted on a roller which runs along the rail head. Pinch rollers locating on the sides of the rail head determined the lateral position of the planing heads relative to the rail.

The machine profiles both rails (simultaneously if required) by making a number of passes with the cutters set at predetermined angles to regenerate the rail head profile. The number of passes required depends on the extent of reprofiling required. Typically:

- to regenerate the gauge corner requires three passes;

- to regenerate the gauge corner and head profile requires seven passes;

- to regenerate the complete head profile requires ten passes.

Using this equipment it has been found in practice that the output in terms of metres of reprofiling per hour depends upon the hardness of the rail being treated, and the depth of cut required. The maximum amount of metal that can be removed per pass is affected by the depth of work hardening of the rail surface and the presence or absence of corrugations, whilst the excessive hardening of the steel in the vicinity of wheelbums can seriously damage the cutters. For this reason it is emphasised that the planing of secondhand rails is not a straightforward process, and it demands skilled planning and preparation if it is to be a success.

In reasonable conditions up to 0.5mm of metal per pass can be removed at 35 m/min, enabling 1000 metres of complete reprofiling of the whole rail head in a 6 hour possession. Whilst with the development of aggressive grinding technology, metal removal rates achievable by grinding have reached values comparable with those attainable by the rail planer, it is cheaper to run the planer than a grinding train, and since the process results in rails capable of being run over at the highest speeds permitted on tracks where recycled rail is allowed, the process remains, at least potentially, financially well worthwhile.

It is noted in passing that the concept of very rapid metal removal has been taken a step further forward recently by the marketing of a rail mounted milling machine, in which the profile is renewed by a set of rotating profile milling cutters.

One of the problems associated with this type of operation is the quantity of swarf to be removed from the track. A convenient way of dealing with this is to have an ancillary trailer vehicle equipped with an electromagnetic pickup which lifts the swarf and collects it in a suitably sized container.

3.10.3   Modifying the head profile of new rail

A principal reason for high rates of rail wear in curves is the aggressive attitude adopted by the leading outer wheel of a bogie. It can be predicted from the advanced curving analysis methods available today that if a means can be found of increasing the effective conicity of the wheel-rail interface, then for any given lateral displacement of the wheel the restoring force able to be exerted across the interface will also be increased. It also follows that for a given forcing level, the lateral displacement experienced by the wheel will be reduced and consequently the likelihood of flange contact will be decreased.

In curves with differing profiles on the high and low rail the point of zero rolling radius difference can also be offset radially to reduce to some extent steady state curving forces. This effect is most significant in shallow curves (say of radius greater than about 900m), and decreases with reducing radius.

The theoretical development of the concept was originally undertaken on behalf of certain heavy haul railways in Australia, and differing preferred profiles were developed for the high and low rails of curves. A series of practical tests were carried out with the rails strain gauged to measure lateral and vertical forces. It was found that by adjusting the rail profiles the forces involved in steady state curving could be reduced by up to 90% in curves of a 900m radius and greater, and by about 50% in curves down to 550 radius. The dynamic components of the wheel-rail interaction forces were also reduced, and also the effect was to redistribute the lateral forces so that lateral forces at the leading outer wheel were reduced and those at the outer trailing wheel increased.

These force reductions when applied to actual railway conditions were found to lead to significant reductions in rail side wear.

The reprofiling process requires the removal of about 1mm of metal from the rail head in the vicinity of the 79.4mm transitional radius and has the effect of making the average crown radius less than normal. Removal of this thickness of metal with earlier generations of grinding train was a very slow process and this work was the impetus towards the development of the so-called aggressive grinding which is possible with current marks of grinding trains. These methods are now being applied in the UK and Europe.

3.11 THERMAL EXPANSION AND ITS EFFECTS - RAIL END GAPS AND RAIL ADJUSTING

Most materials expand as they get warmer and contract again on cooling, and railway rails are no exception. The extension of a bar of material, expressed as a proportion of its original length, which occurs when the temperature of the bar is raised by one degree, is called the COEFFICIENT OF LINEAR EXPANSION. The "dimensions" of the Coefficient of Linear Expansion (see Chapter 2 for a definition of the term "dimension") are:

$$[LENGTH] \text{ I } [LENGTH] \text{ I } [DEGREES] \quad = \text{1I } [DEGREES]$$

Hence values of the Coefficient of Linear Expansion are always expressed as a figure per Degree C (or per Degree F). The Coefficient of Linear Expansion of rail steel is usually taken as 0.0000115 per Degree °C.

If a rail which is nominally 18.3m long is laid into the track at a temperature of 0°C, then on a warm sunny day when the rail temperature has reached 30°C its length will have increased by:

0.0000115 x 18300 x 30 = 6.31mm

The rail, even in UK, may get a good deal hotter than this, say as hot as 45°C, by which time its length will have increased by around 9mm.

3.11.1   Rail expansion and the stressing of CWR

If the rail is constrained so that it cannot expand, then in order that it shall remain at its original length, the rail must undergo a compressive strain which is equal and opposite to the thermal strain, and according to Hooke's Law, is subject to a compressive force determined by the expression:

Force = (Strain) x (Cross sectional area) x (Modulus of Elasticity)

In the above expression, let

Cross sectional area be A.   (A =7184.5 mm² for BS113A rail)

Modulus of Elasticity be E.   (E =200000 N mm² for rail steel)

Coefficient of Linear Expansion be C.

Temperature rise be t °C.

The thermal strain is equal to the Coeffficient of Linear Expansion multiplied by the temperature rise, so that can write:

Thermal Force = C . t . A . E

$$= 0.0000115 \times t \times 7184.5 \times 200000$$

$$= 16524 \ N/°C$$

$$= 1.7 \ Tonnes/°C \ approx$$

$$= 76.5 \ Tonnes \ for \ 45°C \ temperature \ rise.$$

A compressive force in each rail of this magnitude is sufficient to cause the track to buckle and it is essential for safety that it is prevented from developing such a force. This is done in CWR by artificially extending the rail at the time of laying and fixing it down when in tension. By this means the rail does not go into compression until the temperature exceeds some predetermined value (this value is called the neutral or "stress-free" temperature and it is 27°C in the UK), thus reducing the thermal force on the hottest day to less than half what it would otherwise be. The rail may be artificially extended either by warming it, or as is universally done nowadays, by stretching it with a tensor. In either case the amount by which the rail must be extended is calculated by multiplying together the difference between the actual rail temperature and the required neutral temperature, the length of rail being treated, and the coefficient of linear expansion, and the track engineer must check that the calculated extension is obtained, that it is evenly distributed down the whole length of rail, and finally that the rail at either end does not move.

The neutral temperature takes into account the average climatic conditions prevailing in UK. It should not be blindly applied elsewhere. If called upon to design CWR working instructions in a "green field" situation, the engineer should first study the climate, so as to form an estimate of the likely maximum and minimum rail temperatures, and fix a neutral temperature about midway between these values. Ideally records should be kept over as long a period as possible to validate any preliminary estimate made. The possibility of the temperature range influencing the choice of rail section is discussed in Chapter 2.

3.11.2 Rail end gaps in jointed track

In jointed track the build up of compressive force is prevented by building in expansion gaps at the rail joints when the track is laid. The size of gap varies according to the rail temperature at the time of laying. In the UK these should be:

Below 10°C   10mm

10-24°C   6mm

34-38°C   3mm

Over 38°C   nil

Special shims of the thicknesses prescribed in the table above (usually referred to as "Expansion Irons") are available and must always be used when laying jointed track. They are slipped over the ends of the rails as they are laid in, and the rails are pulled up tight against them before the fishplates are assembled and bolted up. Care must be taken to see that the irons are all collected up when the work is completed.

A rail thermometer should be used to determine the rail temperature.

3.12 CONDUCTOR RAILS

There are two main types of system:

(a) The "Third Rail System", where the power (positive) rail which supplies the traction current is located to one side of the running rails, and either one or both of the running rails is used to return the (negative) traction current to the substation. The conductor rail may have the contact surface at the top, under or inside.

(b) The "Fourth Rail System", where two top contact conductor rails of similar properties are provided, the power (positive) rail located to one side of the running rails and the return (negative) rail located between the running rails on the track centre-line. This latter is the system installed on the lines of London Underground Limited.

On the majority of electrified railways where conductor rail is installed, it is supported by the track structure. A selection of typical conductor rail installations are shown in Figure 3.10.

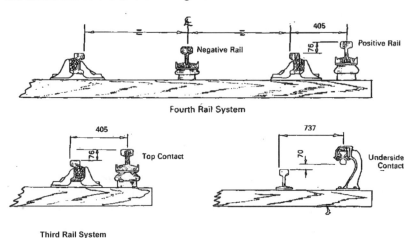

Figure 3.10 Typical Conductor Rail Systems

The following notes reflect the practice in the South East England 660/750 volt direct current (DC) electrified lines, which is the world's largest conductor rail installation. It is a third rail system with top contact conductor rails. The term "conductor rail" should in this section be understood to refer to the positive conductor rail which supplies the traction current.

3.12.1 Materials and Sections

Steel conductor rails

Conductor rails are made of steel containing 0.04% carbon and chromium respectively, 0.36% manganese, and a trace of silicon. Such steel is designed to be of high electrical conductivity, but containing as it does only a tenth of the carbon of normal rail steel, it is also extremely soft, and of low strength. Apart from this metallurgical difference, conductor rails are made to the same standards of straightness and dimensional accuracy as the running rails. Whilst such low carbon steel is not so brittle and notch sensitive as rail steel, its softness and low strength imply that it also needs to be handled with care. Conductor rails are rolled to weights of 52.6 and 74.4 kg/m. The dimensions of the sections of these rails are shown in Figure 3.11.

Conductor rails can be joined by fishplates of special design to fit in the space between the fishing surfaces, which are much closer together than those on running rails. Where fishplated joints are used it is necessary to fit laminated or stranded copper bonds to ensure good electrical continuity across the joint. In the pressed or riveted type of bond connection the terminals are pressed hydraulically into holes drilled through the bottom flange of the rail. Alternatively welded bonds can be used. These are placed under the rail, the terminal being welded to the underside of the rail foot. However, the use of conductor rail fishplates is to be avoided, the rails being welded into as long lengths as possible.

Aluminium - Stainless Steel Composite (ASC) conductor rails

Steel conductor rail has been the most economical and effective material for heavy traction current railways despite its relatively high electrical resistance (14 milliohms/km when fully worn) compared to other components of the traction current supply system. The lower resistance of aluminium and the wear resistance of stainless steel have been combined in the Aluminium- Stainless Steel (ASC) composite conductor rail. This is lighter and smaller in cross-section than conventional steel conductor rail, yet it is robust enough to resist mechanical and electrical damage during its service life. Mechanical, electrical, and accelerated corrosion tests, and service trials, indicate a service life in excess of 45 years.

Two sizes of ASC conductor rail are used in the UK:

(a)   Standard conductivity (resistance 14 milliohms/km) - a rail with a resistance similar to steel conductor rail for installation as a replacement on existing systems.

(b)   High conductivity (resistance 7 milliohms/km) - a rail with a lower resistance which enables an improved electrification system design, and a reduction in the substations required for the power supply. A section of the ASC conductor is shown in Figure 3.12.

The rails are supplied in 15 metre lengths and connected on site by aluminium fishplates secured by four pre-tensioned multi-groove locking (MGL) pins. The fishplates are sized to provide both the mechanical and electrical connection. The high conductivity section shown in Figure 5.12 weighs 16.5 kg/m, the light weight by comparison to steel making handling and installation easier, and enabling the lengths of rail to be transported to site by light weight self propelled rail vehicles. The ease with which ASC conductor rail may be removed and reinstated in association with track maintenance work is seen as making a long term contribution to reducing maintenance costs.

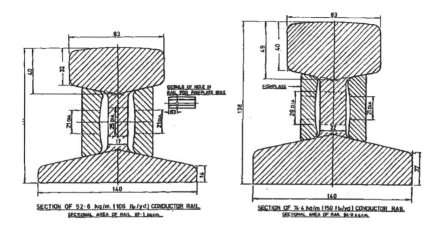

Figure 3.11 Steel Conductor Rails

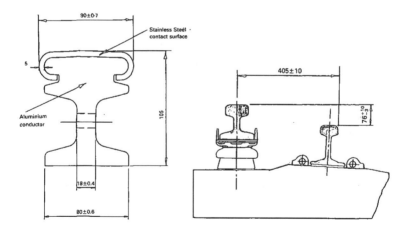

Figure 3.12 Composite Conductor Rail   Figure 3.13 Position of Conductor Rail

3.12.2 Supports

For steel conductor rails

The conductor rail is supported by insulator assemblies spaced at every eighth, sixth or fourth sleeper depending on the track curvature, location, and type of fixing. The insulator assembly consists of a porcelain pot, the top of which is cemented into a cast malleable iron cap. The purpose of the cap (sometimes termed the ears) is to locate the conductor rail transversely relative to the running rails, without restraining it longitudinally. The insulators lie on top of the sleeper, and are restrained by a pair of cast malleable iron wrap- round base clips which are fixed to the sleeper. Where timber sleepers are used the clips are fixed by coachscrews, but when concrete sleepers are used, fixing is either by coachscrew into plastic insert, or by Seetru or similar expanding bolts, the sockets for which are inserted into holes preformed in the sleeper during manufacture. For this reason, and also because a level space must be formed at the sleeper end to support the insulator, concrete sleepers for third rail electrified track are manufactured to a special design.

The rates of head wear of conductor rail and running rail are not as a rule the same, and therefore provision must be made in the supports for adjustment of the height of the conductor rail above the sleeper. This is done at two locations, viz. between the rail and the ears, and between the insulator and the sleeper. The adjustment between the rail and the ears takes the form of a number of steel or plastic shims which are inserted when the rail is first laid and can be removed if the rail has to be lowered. That between the insulator and the sleeper consists of timber or composite material packings which are added if it is required to lift the conductor rail.

For ASC conductor rails

ASC conductor rail is similarly supported by porcelain or dough moulded resin compound (DMC) insulators and is free to move longitudinally. However, because the rail is so light it has to be restrained vertically as well as laterally to avoid displacement. The top cap of the insulators has ears which are curved over the bottom flange for this purpose. The insulator is mounted on timber or concrete sleepers by a malleable iron baseplate, which has a square spigot which projects up the centre of the insulator to permit easy removal, whilst preventing the insulator from turning. The baseplate is secured to the sleeper by coachscrews or expanding bolts like the fastenings for steel conductor rail. The insulator is held down in position on the base spigot by the weight of the rail. The rail can be removed by raising the insulator clear of the baseplate spigot and turning the insulator through 45 degrees.

Anchorages

To prevent longitudinal movement or creep of the conductor rail, which arises due to variation in the neutral point from which expansion and contraction occurs as the rail temperature changes, "anchor insulator assemblies" are installed at the midpoint of all conductor rail lengths. The number of assemblies required depends on the length of conductor rail to be anchored.

3.12.3   Detailing Conductor Rail Layouts

The location of the conductor rail relative to the running rail is shown in Figure 3.13. The conductor rail may be placed on either side of the track. In most modern twin track installations however it is usually placed between the two tracks, where it is out of the way of personnel walking along the cess. In stations it will be positioned if possible away from the platform edge, and through switch and crossing work it will be so arranged as to minimise loss of contact between shoe and rail.

Wherever possible, steel conductor rails are laid in long continuously welded lengths up to a maximum of 550 metres on straight track, the length being reduced proportionately on curved track below 600 metres radius. Similar arrangements apply to ASC conductor rails, the maximum fishplated length for the high conductivity section being 310 metres.

Gaps are required at level crossings, and at substations and track paralleling huts (TP huts) where the traction feed cables are attached to the conductor rails.

Whenever the conductor rail comes to an end, ramps must be provided to ease the collector shoe on and off the rail with minimum impact. In the early days of electrification these were made by bending a section of plain conductor rail downwards with jim-crow in two places as shown in Figure 3.14. As train speeds increased this arrangement was found to be unsatisfactory, and the present practice is to use a single bend ramp, the extreme end of which is supported on a shallow insulator as shown in Figure 3.15. The ramp gradient should be quite closely controlled, to minimise the impact caused when the collector shoe strikes the ramp. Low speed ramps of steeper gradient are installed in depots and the low speed tracks of crossover roads. Prefabricated horizontal base ramps as shown in Figure 3.16 are installed at locations where the clearance to ground is inadequate with the standard ramp formation.

To provide a clear passage for the collector shoes at turnouts and junctions, the conductor rails are terminated as shown in Figure 3.17. Where this arrangement creates unacceptable "gapping" (i.e. where it is possible for *all* the collector shoes on a train to be out of contact with a conductor rail at the same time), consideration may be given to installing a continuous length of conductor rail on the higher speed track. The collector shoes of trains approaching from the low speed track are then raised on to the top contact surface of the conductor rail as they pass towards the switch and vice versa by the use of a 'side entry ramp' which is bolted to the inside of the conductor rail. The principle and construction of side entry ramps is shown in Figure 3.14. It is important that the conductor rail to which the side entry ramp is attached, is installed and maintained to a high standard, to ensure that the position of the ramp does not change, and that the top is level with the contact surface of the conductor rail. Side entry ramps are to be avoided if possible, because the impact between shoe and side ramp can cause very heavy wear, both on the ramp and on the collector shoe. The speed at which trains approach the ramp from the side must not exceed 20 mph (32km/h).

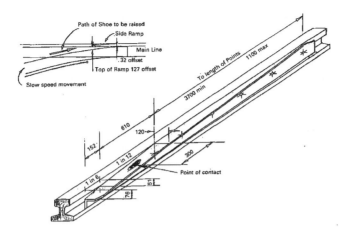

Figure 3.14 Side Entry Ramp

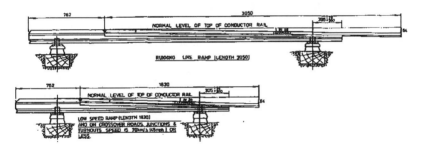

Figure 3.15 Conductor Ramp Ends (Prefabricated-Horizontal Base)

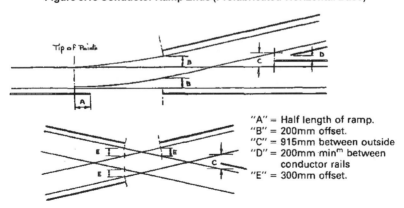

"A" = Half length of ramp.
"B" = 200mm offset.
"C" = 915mm between outside
"D" = 200mm min^m between
      conductor rails
"E" = 300mm offset.

Figure 3.16 Termination of Conductor Rail at Points and Crossings

3.13 DEFECTIVE AND BROKEN RAILS: THEIR SIGNIFICANCE, CAUSES, DETECTION AND REMEDIAL ACTIONS

Rail steel has a considerably higher carbon content, and hence in the nonheat treated state (pearlitic) is more brittle than mild steel. Consequently as described in Section 3.9 a variety of stress concentrating defects in rails, combined with the alternating loads from the passage of traffic, can produce a slowly propagating fatigue crack. On attaining a critical size, this results in an almost instantaneous brittle fracture. Broken rails are an obvious hazard to the safe passage of traffic and therefore need to be minimized as far as practically possible. The significance of broken rails is emphasised by the fact that the UK has a statutory obligation to report certain types of rail fractures.

Most wear resistant rail steels have similar fracture properties to the normal grade steel. However, austenitic manganese steel is very much tougher than normal grade rail steel and does not produce a brittle fracture provided it has been correctly heat treated. In spite of this, any slow growing fatigue crack in this material could still result in a complete, although ductile, fracture of the rail, or even collapse of the rail.

3.13.1 Causes

Initiation of fatigue cracks comes from a variety of causes, such as the passage of traffic where vehicle wheels can damage the running surface. The manufacturing process can leave inherent defects in the rail such as hydrogen shatter cracks piping and non-metallic inclusions. Welds used to join lengths of rail or for build-up repair can also contain defects which eventually result in a rail failure. Corrosion and mechanical damage can produce a sufficient stress concentration to initiate fatigue crack growth. Thermal damage in the form of accidental heating can produce a very brittle structure which easily forms cracks that can propagate further by fatigue.

3.13.2 Classification

As well as having a statutory obligation to report certain rail fractures there is also a need to make a continuous assessment of the rail defect situation, both for safety and a financial reason. In order to do this, it is necessary to classify the types of defects so that trends can be detected. The original classification scheme was produced in conjunction with the International Union of Railways (UIC). The classification scheme deals with all defects, cracks and fractures found in rails, most of which require removal from track soon after discovery. The permanent way engineers also need to monitor defects and cracks which have been detected in track but are not sufficiently serious at that point in time to warrant removal. In order to manage these, a database is needed with the key features of monitoring minimum actions. Rail authorities now have quality systems for classifying and recording rail failures.

There are approximately 50 different types of defect in plain rails (Switches and crossings are dealt with separately). However, when the rail failure statistics are examined, it is found that about 75% of these defects are accounted for by just six or seven types. These are squats, taches ovales, alumino-thermic welds, bolt holes, vertical splits of the head and web, rolling contact fatigue and wheel burns. The following sections give a brief description of these major types of rail defects.

Squats

These are fatigue cracks initiated at the running surface by wheel/rail contact forces. They are visible on the running surface as cracks, usually with a black central area, where the wheel loses contact with the rail. The crack propagates at a shallow angle to the horizontal until it reaches 3 to 5 mm below the surface. At this depth, it usually branches downwards to form a defect which is very similar in appearance to a Tache Ovale (see below). If left in track, this would eventually produce a brittle fracture and a complete transverse break of the rail.

There are two types, random and periodic. The former can occur anywhere on the rail with multiple defects common at spacings from a few millimetres upwards. The high rails in high speed curves are particularly susceptible to this form of the defect. This common occurrence of multiple defects increases the possibility of the loss of a section of running rail thereby representing a considerable safety hazard.

The periodic type occur at spacings of about 2.9 m, resulting from some hard object becoming embedded in a vehicle tyre and indenting the rail. As the rail wears down, the indentation forms a crack, which propagates as described above. In a similar manner to the random type, the fracture of two adjacent defects could result in the loss of a section of running rail.

Another form of rolling contact fatigue defect is the shell and detail fracture. These usually result from large non-metallic oxide inclusions just below the running surface at the gauge corner. This form of the defect is very rare mainly because of the use of rail steel which has been killed with silicon rather than aluminium.

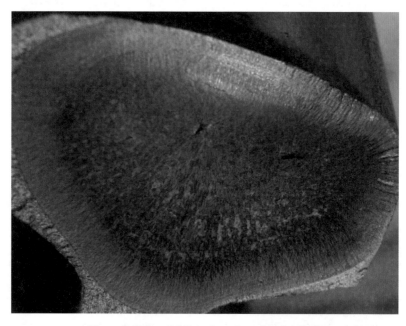

Figure 3.17 Squat failure (courtesy Jay Jaiswal)

Taches ovales

This defect name comes from the French meaning an oval spot, which describes the shape of the fatigue crack in the rail head, (Figure 3.18). In ingot produced rail steel (pre 1974), these defects initiated from shatter cracks, which result from excessive hydrogen in the metal during or soon after the production stage. It was hoped that with the changeover to continuous casting the annual number of these defects would decline. To date, the rail failure statistics show little evidence of this. The main problem with this type of defect is that it is only rarely visible on the rail surface before it produces a brittle fracture. The only satisfactory method of detecting it is by ultrasonics. The other important aspect is that when it results from hydrogen embrittlement in ingot produced steel, the whole rail and possibly all other rails from the same cast are likely to contain multiple defects increasing the potential hazard to safety.

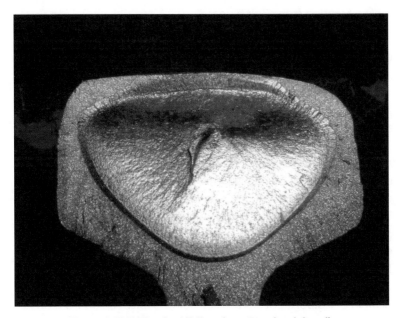

Figure 3.18 Tache Oval Failure (courtesy Jay Jaiswal)

Defective Alumino-thermic welds

Alumino-thermic welds are used to join lengths of rails together in track. Therefore, they are made under more adverse conditions than the depot produced flash welds and as a result are subject to more defects, see chapter 4. In addition, the process itself is more prone to produce defects than flash welding. The most common of these are lack of fusion, where the weld metal has not completely bonded to the ends of the rail; slag or sand (from the mould) inclusions; gas porosity; and hot tears resulting from the hydraulic tensioners being removed too soon or slipping. Although some of these defects manifest themselves on the surface, most are internal and therefore are only detectable using ultrasonics. Since the complex geometry of an alumino-thermic weld makes a full ultrasonic examination very time consuming and expensive, the avoidance of failures relies mainly on improvements to the welding consumables and practice. Radiography is also a possible method of examining the whole weld.

Cracks radiating from Bolt holes.

These are often referred to as star cracks because of the characteristic manner in which the fractures occur at 45° to the vertical, (Figure 3.19). In this case, the stress concentration is due to the bolt hole itself, which under higher than normal service stresses can initiate a fatigue crack. In this situation, the critical fatigue crack size to initiate brittle fracture is only a few millimetres. The actual size at fracture is determined by the magnitude of the impact forces, which are a function of the traffic speed and dip angle in the running table at the joint. The likelihood of this type of defect developing is increased by damage to the bolt hole such as the use of a crowbar to turn the rail, or bad drilling (or reaming) producing a burr. Due to the small size of the initiating fatigue cracks and the fact that most bolt holes are covered by fishplates, the only satisfactory method of detecting them is by ultrasonics. The current method of inhibiting crack initiation is bolt hole expansion. Pulling a mandrel through the hole fitted with a split stainless steel sleeve produces compressive residual stresses in the surrounding material. This resists the formation and propagation of fatigue cracks.

Figure 3.19 Bolt Hole Failure (courtesy AREMA)

Vertical splits of head and web.

These defects, (See figure 3.20), are mainly a problem in old rails produced by the ingot process. Insufficient cropping of the ingot leads to fissures or bands of non-metallic inclusions in the rail, which act as fatigue crack initiators. Although splits in the rail head can exist for several metres, they fortunately appear to have little tendency to form transverse fractures.

These types of defect can be detected visually as a black line or crack on the running surface, or cracking in the upper fishing radius in the case of head splits. Visual detection of web splits is usually by observation of the bulging effect. Both types of defect can be confirmed by the use of callipers to determine the existence of head or web bulging. However, the most common method of detection is by ultrasonics using a recently developed procedure. As the old rail in track is gradually being replaced by rails made from continuously cast steel, these types of defects should decrease in number.

Figure 3.20 Vertical Split Failure (courtesy Jay Jaiswal)

Wheel burns

Wheel burns result from spinning of a driving wheel on the rail. The resulting heat input from friction is often sufficient to produce a metallurgical transformation in the steel to an extremely hard and brittle structure, which easily breaks up under traffic. Fortunately, most of these defects simply shell out and rarely form a transverse fracture. They are found in two forms, isolated and continuous. The former are produced by a stationary vehicle and often show the bogie wheelbase pattern on the track. The latter are formed when the vehicle is moving slowly whilst the wheels are spinning. These defects are easily detected visually and are dealt with either by replacing the rail or weld repair depending on the economics of the situation.

Rolling Contact Fatigue

The phenomenon known as rolling contact fatigue or "gauge corner cracking" has been around for many years and was often so some extent ignored. This was because the small cracks (See figure 3,21) did appear to get worn away by varying profiles of wheels. The cause is related to creep and plastic flow due to rolling stresses that eventually cause fatigue failure. Two key processes govern RCF - crack initiation and crack propagation. These processes are governed by a number of factors including environmental conditions, rail and wheel profiles, track curvatures, grades, lubrication practices, rail metallurgy, vehicle characteristics, track geometry errors, and rail grinding practices. They all play a role in the formation of RCF and – universally – can be used to control and minimize RCF.

The amplitude and position of the crack initiating stresses varies depending on thecontact geometry, load, and friction conditions. Under high friction conditions shear stresses are large but very shallow. Under low friction conditions, the peak shear stress decreases but extends deeper into the railhead. (See figure 3.22). The result is that some RCFdefects are initiated at the surface and others below the surface.

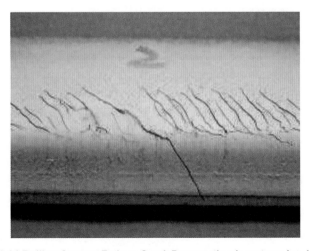

Figure 3.21 Rolling Contact Fatigue Crack Propagation (courtesy Jay Jaiswal)

**Figure 3.22 Rolling Contact Fatigue Cracking leading to Failure
(courtesy Jay Jaiswal)**

3.14   DETECTION

The first source of defects likely to cause rail failures is the manufacturing process. However, over the years, the manufacturers have made considerable improvements to reduce the number of defects or make their detection much easier. In 1974 rail production by British Steel was changed from the use of ingots to continuous casting, reducing the likelihood of piping and related defects. On-line non-destructive testing using visual inspection and a variety of ultrasonic probes on the head, web, and foot, guards against most of the surface and embedded defects commonly associated with the manufacturing process. After production, the rails are inspected by appointed quality assurance personnel but this is mainly aimed at geometrical tolerances.

Once the rails are in track, they are subject to regular and frequent (depending on track category) visual inspection for gross defects. However, most defects are hidden and therefore not detectable by visual inspection. Hence various types of non-destructive testing techniques such as ultrasonic inspection are employed, as described below.

3.14.1 Ultrasonic Detection

The ultrasonic (sound at a frequency higher than that detectable by the human ear) examination of rails for internal defects is carried out on a regular basis by pedestrian operators or a rail borne ultrasonic testing unit.

The principles of operation of ultrasonic testing are similar for both these methods and are as follows. The flaw detector equipment produces a short electrical pulse which is applied to a piezo-electric transducer (the ultrasonic probe). The piezo-electric transducer has the ability of converting an electrical pulse into an acoustic pulse and conversely of converting an acoustic pulse into an electrical signal. The probes are manufactured to produce a beam of ultrasound which propagates into the rail at a given angle. The commonly used angles (measured from a perpendicular line to the rail head) are 0°, 40° and 70°.

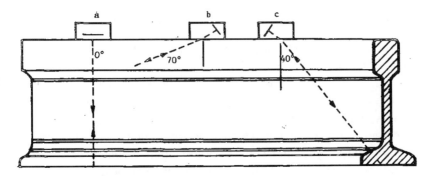

Fig. 3.23 Probe Arrangements

Figure 3.23 shows the typical probe arrangement. On the left it shows how, when a 0° probe is placed on a rail head with a coupling medium of water or oil, the pulses of sound will pass from the probe down through the rail section to the rail foot, where they will be reflected back (like the way in which light is reflected from a mirror) to the rail head. At the rail head some of the sound passes back to the probe and the remainder is reflected back to the rail foot as a repeat echo but at a lower amplitude. The sound pulse that returns to the probe is reconverted to an electrical signal. The time difference between the two pulses indicates the distance the pulse of sound has travelled and the amplitude of the signal represents the strength of the returned pulse. If there is a discontinuity, such as a bolt hole or a defect, the pulse will return at a shorter time, and hence at a shorter range than that from the rail foot. Figure 3.23 shows the 70° probe which scans the rail head for taches ovales. And shows the 40° probe which scans the web of the rail for bolt holes and associated defects.

The 40° probe will locate either a bolt hole or a bolt-hole crack. The detection of the crack rests on the fact that a sound bolt hole will produce one reflection, whereas a bolt hole with a crack will produce either two or three reflections, and the time intervals between sending and receiving pulses or echoes will be different.

Two tests are required to detect cracks in the "S.W-N.E. direction" and the "N.W.-S.E. direction".

3.14.2 Pedestrian Operators

The pedestrian operator's equipment, (Figure 3.24) is portable and the pulse echo display is presented on a cathode ray tube which allows the operator to interpret the signals to find any abnormalities within the rail. The operator has two methods of testing, the first is with a small trolley which is pushed along the rail head and the second is with hand held probes for bolt hole testing or assessing defects such as squats. The trolley is fitted with a 0° probe for the detection of horizontal defects and a 70° probe for the detection of tache ovale defects. Hand held probes are provided at angles of 0°, 40° and 70°. The main limitation of using pedestrian operators to detect defects is the limited amount of track they can cover in a given time. However, they are able to test awkward areas such as points and crossings and assess the severity of a defect much better than any vehicle borne equipment could.

Figure 3.24 Sperry Walking stick in use. (Courtesy of Nexus)

3.14.3   Ultrasonic Test Unit

Ultrasonic test units, (Figure 3.25), enables the rails to be tested at speeds up to 70 km/h on continuously welded track and good quality jointed track. The unit consists of a three car self propelled unit with the end vehicles being the power/accommodation vehicles and the centre vehicle housing the instrumentation. One bogie of the centre vehicle has had the brake gear removed and a frame fitted to the axle boxes to give a stable platform to fit the ultrasonic probes and associated control gear. This vehicle uses a number of small wheels with ultrasonic transducers fitted to their axles which are enclosed by polypropylene tyres. The tyres are filled with a water-glycerol mixture and a water couplant is used between the tyre and rail. The ultrasound travels from the transducers through the internal fluid, the tyre, tyre/water couplant and into the rail to the rail foot and then returns back to the transducer. The probe is guided down the centre line of the rail head by means of a servo system and maintains contact with the rail head by its own weight. At fishplated joints the resilience of the tyre allows contact with the rail to be maintained. However, if the rail joint is in poor condition with an excessive amount of movement between the rail ends then probe rail contact could be lost and the testing speed must be reduced to allow the rail to be tested satisfactorily. The three wheel probes on each side of the vehicle contain a total of 14 transducers. These produce signals, each of which is processed by its own microprocessor. Further processing by microprocessors allows a defect or anomaly to be identified in real time, so that its position can be immediately marked on the rail. The type of defect is printed out with a record of the mileage and route being tested. The pattern of the signals in the vicinity of the defect is stored for further processing and confirmation of the defect. Non-contacting eddy current probes are also employed for the detection of surface defects. These allow squats and wheel burns to be detected.

Figure 3.25 Ultrasonic equipment on UTU 2 (Courtesy of Network Rail)

3.14.4   Other Methods

There are several methods of non-destructive testing, other than ultrasonics, which may be employed to detect defects in rail. Those in current use are: Visual examination Magnetic particle inspection Dye penetrant testing Eddy current testing

A brief description of each method and its application to rail testing follows.

(a)   *Visual inspection*

This is the primary and most easily applied method for detecting and evaluating defects. However, there are cases where other methods of testing are required to show the severity of the visual defect. An example is the squat defect where ultrasonic testing is required to indicate the severity of the sub-surface defect. Conversely, there are cases where visual indications can lead to an incorrect conclusion, such as interpreting the fusion lines at a weld boundary as a defect. In cases of doubt other methods of non destructive testing must be used as a means of confirmation.

(b)　　*Magnetic particle inspection*

This technique can only be applied to ferro-magnetic materials and therefore cannot be used on austenitic manganese steel (AMS) rails, which are non-magnetic. When a magnet, either a permanent or electromagnet, is applied to a ferro-magnetic material magnetic lines of force flow through the material. If a surface defect, or one very close to the surface, is present within the magnetic field, the lines of force are interrupted and a leakage field occurs at the defect. If iron particles are applied in the vicinity of this leakage field they will be concentrated at the point of leakage and the crack will be indicated by the presence of the particles. The particles are highlighted by first coating the component to be tested with a white background paint.

(c)　　*Dye penetrant testing*

This method can only be used on surface breaking cracks and should only be used on non ferro-magnetic materials, such as AMS rails, as it is considered to be a less sensitive test than magnetic particle inspection. This technique relies on a dye which has the properties of being highly mobile. The component under test must be clean so that when the dye is applied it will penetrate into any defects open at the surface. Surplus dye must be removed from the surface and a developer, consisting of a white powder, sprayed on the surface. Any defect will be filled with dye and the developer acts as a blotter which will soak up the dye showing the position and extent of the defect.

(d)　　*Eddy current testing*

This technique is used for the detection of surface breaking defects in both ferrous and non ferrous metals. This is an electro-magnetic method of detecting defects that uses eddy currents induced into the component being tested by means of a small coil which is fed with an alternating current. The currents in the component interact with the current in the coil and change its characteristics depending upon the nature of the material being tested. When the component contains a surface breaking defect, the eddy currents around the defect are altered to the extent that they can be detected by the coil. This allows an instrument to be calibrated to detect these changes in the coil which can measure the presence, and under certain circumstances the depth of a defect.

Other possible sources of detection of broken and defective rails are visual and ride observations by train crew, and signalmen reporting loss of track circuit continuity.

3.15   CAST CROSSINGS

The majority of the methods of detecting defects in rolled rails listed above are applicable to cast crossings. However, there are a few exceptions. The magnetic particle method cannot be applied to non-magnetic materials such as AMS or stainless steel. In addition, due to its large grain size, cast AMS cannot be inspected using ultrasonic equipment. In contrast, both methods can be used on cast bainitic crossings.

3.16   REMEDIAL ACTIONS

Visual and ultrasonic (manual and vehicular) testing of track for defects are carried out at specified frequencies, which depend on track classification. When a defect is detected, it is reported and entered into a computer system by track engineers. From the details entered, the programs are designed to immediately indicate the minimum action which must be taken to deal with the defect. These minimum actions, which depend on the type and severity of the defect, range from imposing a 5 mph temporary speed restriction, fitting emergency clamped fishplates and removing the section of rail within 36 hours, to simply retesting at the specified frequency. In between these two extremes, defects can either be subject to removal of the rail within a set time period or repaired by welding. Another technique currently under development is the use of the corrugation grinding train to remove fine running surface cracking to prevent the formation of squats.

CHAPTER 4

TRACK WELDING

4.1   INTRODUCTION

Track welding comprises three main areas:

- Welding of plain rail into long strings for use in relaying or new track construction
- Manufacture of switch and crossing components from plain rail
- Repair and resurfacing of rails and crossings

A range of processes are used in these activities. These are summarised in Figure 4.1, and they can be divided into two main areas of operation, shop or depot welding, and site welding. In the UK, fixed installations are used for the manufacture of Long Welded Rail (LWR) by the Flash Butt Welding process (FBW). Welded common crossings are made by an approved supplier using arc welding or the electroslag process, working to procedures and standards set by infrastructure owners. Crossings may be repaired in temporary or permanent workshops by a supplier's welding staff.

The most important site welding process is the aluminothermic welding of LWR strings on site to form Continuous Welded Rail (CWR). Mobile FBW welding machines are also in use in appropriate locations. Apart from this work, site welding is used in the maintenance and repair of plain line and S&C work, using the processes identified in Figure 4.1.

4.2   MANUFACTURE OF LONG WELDED RAIL BY FLASH BUTT WELDING

The manufacture of LWR as a step in the production of CWR may be seen as a continuation (and perhaps even as the culmination), of the process of elimination of rail joints which has continued almost throughout the history of railways, as described in Chapter 1.

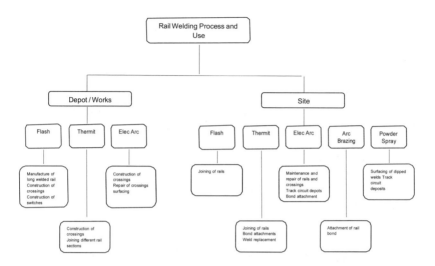

Figure 4.1 Summary of Historical Rail Welding Processes and Applications

The lash butt welding process production consists of rail preparation, rail end cleaning, welding, trimming, enhanced cooling, pressing and grinding. As shown in Figure 4.2, the plant is designed to enable the production activities to operate concurrently as each work station is located at intervals corresponding to rail length.

4.2.1   Rail Preparation

New rails are checked and quality assured prior to welding and secondhand rail ends are cleaned to remove rust and scale from the electrode contact areas of the rail by passing through a degreasing unit, and shot blast cabinet, which cleans the top, bottom and end faces of each rail.

Second hand rail is examined for the presence of wheelburns, damaged running edge, and alumino thermic welds. Should a rail contain any of these features that cannot be removed during the sawing operation, the rail is rejected. Flash butt welds are permitted provided they comply with the standard for straightness, and are not located closer than 3.5m to a rail end. Acceptable rails are passed through an ultrasonic law detection

**Figure 4.2 All Terrain Mobile Flash Butt welding machine
(courtesy Thermit Welding (GB) Ltd)**

unit which scans the rail head for defects. The rails are then cropped to remove fishbolt holes, alumino thermic welds and rail damage. The minimum length accepted for welding is usually 9.14m.

After cleaning, the rails are fed into the welding machine, which automatically aligns the two rails, which are then clamped in position, ready for welding.

Flash welding is an automatic process, and welding parameters are normally selected by means of procedural trials. The process is a solid phase pressure welding process where the two rails are clamped and a high current is passed across the components at low voltage, during which one of the work pieces is projected in a controlled manner toward the other. Electrical resistance heating first causes contacting surface irregularities to melt and subsequently raises the temperature of the whole interface to near melting point. Once the components are sufficiently heated they are forged together, and excess molten steel at the interface is forced out of the weld area.

The welding procedure currently applied to BS11 113A FB section rail results in a welding time of less than 1 minute. When the weld has solidified integral shears remove the excess upset from the periphery of the weld. The upset is left approximately 1mm proud all around the weld.

On completion of trimming the welded joint is conveyed to water spray units for enhanced cooling.

The major problem with a rapid cooling rate is that if the rail cools too rapidly martensite or bainite transformations will occur, resulting in embrittlement and possibly cracking. However, once the pearlite transformation is complete, the steel can be cooled rapidly, without danger of embrittlement. Extensive testing has shown that the best cooling rate can be achieved by using an air/water mist spray, followed by a direct water spray once the weld temperature has fallen to a safe level. When the weld has cooled to the ambient temperature, the welded joint is pressed vertically and laterally using a Geismar press with a maximum loading capacity of 200 tonnes vertically, and 100 tonnes laterally.

The current UK tolerances for weld alignment are measured over a 1 metre span and are vertically - 0.4mm and horizontally + 0.4mm.

Grinding equipment is used to smooth the straightened weld to profile and to remove deviations from straight local to the weld. Only the rail head is ground, ensuring that the grinding operation does no metallurgical or physical damage to the welded joint. The UK specification allows a maximum deviation from lat on the smoothed surface of 0.2mm.

4.3   SITE WELDING FOR CONTINUOUSLY WELDED RAIL

On arrival at site, the rails are welded to form continuous welded rail using the alumino thermic (or 'thermit') welding process. While a small amount of rail is welded alongside the track, either supported on temporary stillages or clamped in alignment beams, the majority of rail is site welded in its final position. All thermit welding is carried out under a complete possession, and traffic is not permitted to pass until the weld has been trimmed and ground, and all clips have been replaced.

It is estimated that there are over 1.5 million thermit welds in track in the UK, with an annual installation rate of approximately 55,000. Currently all thermit welds are manufactured by welding staff operating in teams of two. Each welder is trained in the appropriate techniques and is subjected to rigid inspection by both welding supervisors, ultrasonic examination, and a two yearly practical and written examination.

4.3.1   The Alumino Thermic Welding Process

The alumino thermic reaction, discovered in 1896 by Professor Hans Goldsmidt, is based on the reduction of heavy metal oxides by aluminium. The process is commonly called "Thermit". It was found that once the reaction had been started using a suitable heat source the aluminium reacts with the metal oxide to form Al2O3, liberating the metal from the oxide, and generating sufficient heat to raise the temperature to approximately 2500°C, so that both the metal and the Al2O3 are in liquid form. The heavy metal separates from the less dense Al2O3, which floats to the top of the reaction vessel. The process can be described by the general chemical reaction:

$$3MeO + 2Al = Al_2O_3 + 3Me + heat$$

When applied to the reduction of iron oxides the following reactions can occur:

$$Fe_2O_3 + 2Al = 2Fe + Al_2O_3 + 760 \text{ kJ};$$

$$3FeO + 2Al = 3Fe + Al_2O_3 + 780 \text{ kJ}$$

The generation of a liquid metal and a large amount of heat means that it can be used as a method of welding materials. For making welds, the workpieces are clamped and held with a gap at the joint. A refractory mould is placed around the joint and the thermit portion is ignited in a refractory crucible positioned above the moulds. The portion is a combination of powders which after reaction will produce a weld metal which matches the chemistry and metallurgy of the rails which are being welded. When the reaction is complete the crucible is tapped and steel pours into the moulds to form the weld. (Figure 4.3).

Thermit welding was first used to join rails in Hungary in 1904. Most of Europe had adopted the process by 1927, but the UK did not accept the process for general use until 1938, when the Boutet thermit welding process was introduced. In the 1950s the SmW process, marketed by Elektro-thermit GmBH was adopted in the UK. The process used pre-formed CO_2 hardened silica sand moulds and an oxy/gas preheat with a 7.5 kg thermit portion. The process was used as standard in the UK until the mid 1970s with over half a million welds installed.

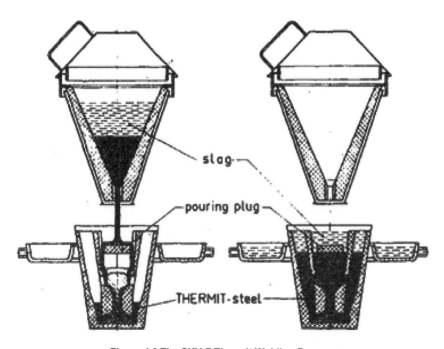

Figure 4.3 The SKV-F Thermit Welding Process

In 1974 the Elektro-thermit SKV (Schweissverfahren mit Kurz Vorwarmung) was available (see Figure 4.3). With this technique a timed 1.5 minute oxy/LPG preheat was used in conjunction with a 12.5 kg thermit portion, and the claimed advantages were:

- Fixed preheating parameters reduced the incidence of preheat faults.

- Installation time was slightly reduced.

- A narrow heat affected zone was produced in the rail.

- The mould design incorporated a centre pour (rather than a side pour) which aided uniform fusion of the rail ends.

Following a number of trials and improvements the SKV-F method (marketed in the UK by Thermit Welding (GB) Ltd.) was adopted as the standard in 1978 and used for over 30 years. A lat collar SKV-F version produced a weld with a small collar formation which improved fatigue properties. This used either oxy/propane preheating or oxy/acetylene (for use in confined spaces) and the process used to weld rails with similar amounts of head wear with moulds being available for all the sections of rail installed in the UK and also for combinations of rail sections.

At the turn of the century, the Thermit® SkV-E Welding Process, with a single use crucible (SUC) became the standard on main lines in the UK (Figure 4.4) and its development in conjunction with Network Rail focussed upon the ease of use with maximum lexibility and reliability.

The Welding process is approved in accordance with the European Specification Pr EN 14730 pt 1 and the key advantages are:

(1) Immune to common weld faults

(2) The finished weld can be easily cleaned for final inspection

(3) Factory prepared crucible, charged with the portion and tapping system included

(4) Preheater position is pre-set.

(5) Moulds are designed for precision fit to the new rail profile, with adjustment for worn rail

(6) No hazardous materials - No unusual fume or smell

(7) No special transport or disposal requirements

Figure 4.4 The SKV-E Thermit Welding Process with Single Use Crucible

There are other specialist welding systems that complement the above:

Welding rails on concrete slab track - A three piece version of the mould has been developed for in situ welding in slab track where the clearance beneath the rail foot and the supporting slab is restricted.

A three piece mould process for installing crossings - Version developed to enable crossings to be inserted into the gap left after a crossing has been removed by cutting through the thermit welds. The separate bottom mould aids fitting of the moulds around any remnants of the old thermit weld.

Stepped Moulds - A stepped mould system has been developed to aid in producing a more satisfactory weld between rails of similar section but of different depth - a situation generally arising when re-railing, relaying or replacing defective or worn rails. The stepped moulds, which are "handed", are designed so that the operator does not have to reshape the under-foot collar formation in any way. Rails with up to 9mm difference of head wear can be joined using stepped moulds.

Wear Resisting grade B and BSC 110 kg/mm2 Rail Steel - Although carried out, the welding of 110 kg/mm2 Cr Steel is not common. The SKV process can be used in this situation, but because of the high hardenability of 110 kg/mm2 Cr steel, cooling must be retarded. A limited amount of BS11 grade B rail has been installed, and this also requires the post weld cooling rate to be slowed down.

Figure 4.5 Typical SKV-E Weld (courtesy Thermit Welding (GB) Ltd)

Wide Gap Welds - If SKV welds are used, replacement of a broken weld requires a new piece of rail at least 4.5m long and two thermit welds. To replace any broken weld directly requires a system capable of producing satisfactory welds with a rail gap of 75 mm. This dimension was determined by the 65mm collar width of old SMW welds. (Figure 4.5).Wide gap welding is estimated to give savings of 75% and 35% in money and time respectively compared with the SKV method.

Figure 4. 6 SKV- L80 composite weld between 56E1 rail and BS90A rail
(courtesy Thermit Welding (GB) Ltd)

Welding of Low Carbon Austenitic Manganese (LCAMS) - As explained in Chapters 6 and 13, normal AMS containing 1.2% C cannot be welded to other rails, but this problem has to some extent been overcome by the development of LCAMS. This steel, being more thermally stable, can be welded to adjacent rails of similar composition. To achieve this it has been necessary to develop welding portions and moulds which are compatible with the chemistry of the LCAMS. The preformed moulds are of a non siliceous material to minimise the problem of silicon contamination of the weld metal and the formation of low melting point compounds. UK rail authorities have installed a number of welds in rolled LCAMS rail in the London Underground lines between Drayton Park and Moorgate, and between St Pancras and Moorgate.

Figure 4.7 Railtech PLA Weld (coutesy Railtech UK Ltd.)

4.3.2   Process of Site Installation of Alumino Thermic Welds (Thermit)

All welds installed are manufactured to working instructions issued by the infrastructure owners. The stages of manufacture are briefly described as follows:

1) Preparation of the Rail Gap
The SKV-E process requires a 22-26mm gap and this can be achieved by lame or mechanical cutting using approved devices. Each rail end must be square and vertical and the end faces must be cleaned to remove loose oxide and burns.

2) Joint Alignment
Unless welding at lineside all welds are manufactured with the rails located in the baseplates or chairs. Rail alignment is achieved with the aid of wedges between the rails and baseplate or sleeper, and checked by means of a 1 metre straight edge. The running surface

must be aligned so that the rail gap is 1.5mm high relative to datum points 500mm either side of the gap, and this is achieved by means of a straight edge with nibs on each end.

3) Attachment of Moulds

After the rails have been accurately aligned the preformed moulds are fitted, adjusting as necessary to accommodate rail wear. The gap between the rail head and the moulds is blocked using one or more corrugated card inserts provided with the moulds. Once the moulds are located about the rail joint, luting sand or putty is used to seal the gap between mould and rail. The type of sand and moisture content is strictly controlled to avoid porosity in the weld.

4) Preheating

The preheating procedure consists of adjusting and maintaining gas pressures at specified levels, setting the lame conditions, and correct location of the preheat appliance within the moulds. The preheater is located into the mould and fitted on its stand. Preheating may be by either oxy-propane or oxy-acetylene.

5) Preparation of the Crucible

On completion of the preheat and removal of the preheater, the crucible is fitted in position and the central plastic seal is removed. The igniter is lit and inserted into the hole in the lid and pushed into the portion through the refractory cover. On completion of the reaction the Thermit steel automatically discharges into the mould. After a couple of minutes the crucible can be lifted off the moulds and placed in a safe position. When cold the steel lid may be removed to enable the container to be filled with the welding debris and removed from site. The crucible is located on top of the moulds either by using the SUC mould shoes or by using a crucible locating frame, which fits on top of the mould.

6) Thermic Reaction and Pour

On completion of the preheat, the crucible is positioned over the moulds and the Portion is ignited. On completion of the Thermit reaction, the Thermit steel is automatically discharged into the mould. The time between ignition and commencement of the pour is in the range 20-35 seconds, by which time the more dense steel has collected in the bottom of the crucible, with the slag loating on top. The molten steel pours into the mould, filling the gap between the rail ends together with the riser. The slag, being less dense than the steel remains in the top of the mould, with excess running off into slag bowls attached to each side of the moulds. After a couple of minutes the crucible can be lifted off the moulds and placed in a safe position When cold the steel lid may be removed to enable the container to be filled with the welding debris and removed from site.

7) Trimming

The weld is allowed to solidify undisturbed for a minimum of five minutes, after which the excess steel around the rail head is trimmed using hydraulic shears. After 20 minutes all the mould material and excess steel and slag, is removed from the weld, the welded joint is clipped down, and the joint is ground to profile. When cooled the weld is trimmed with the aid of a hydraulic shear, after which the rail head is ground to profile

8) Inspection

The joint is cleaned of loose sand and mould material to facilitate inspection of the weld. The welder must check the vertically and squareness of the weld, the extent of weld metal lashing and the ground surface must be inspected for surface imperfections, and the straightness must be checked with the aid of a one metre straight edge. Finally the weld must be stamped with the welders identification number.

4.3.3   Quality Assurance of Site Welds

Although quality procedures have vastly improved over the last 20 years, there are still a number of defective or substandard welds. All site welds should be inspected within 1 month of installation and some will have to be replaced. These will have been removed either because they have fractured, or because they contain visible defects, or after detection by ultrasonics. Weld failures are mostly due to brittle fracture from welding defects, the most common being lack of fusion.

The maintenance of product quality is the responsibility of the supplier, who must be accredited. Thermit welding products are based around weld requirements which include composition, strength, and hardness, and only products which comply with the specification are permitted to be used on track. The welding consumables must be carefully stored as they are flammable and explosive.

Thermit welds should be subject to regular quality audits which include analysis of failures, quality tests on sample welds, and feedback from welding personnel.

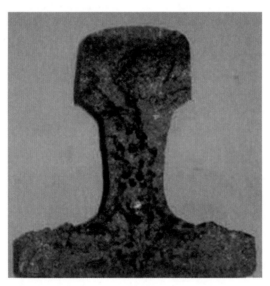

Figure 4.8 Failed Thermit weld caused by porosity
(courtesy Thermit Welding (GB) Ltd)

4.4   THE MANUFACTURE OF PART WELDED CROSSINGS

The constructional details of built up crossings, and their problems, are described in Chapter 9.

There was an increasing awareness of a need for a more rigid, lower maintenance type of crossing and the welding of the fabricated type of crossing was thought to be an alternative answer. In the UK it was decided to manufacture crossings where the point and splice rails were welded together and the wing rails welded to these. Considering the rail profile, it may be seen that its complex cross-section does not lend itself readily to being welded to another rail lying parallel. In the technique adopted, the machining of the point and splice rails was changed to give symmetrical proportions, the rails now being referred to as the vee rails. Welding preparations were machined in the top and bottom and a semi-automatic CO_2 process was used. Welding was finished on the running surface by two layers of manual metal arc welding using alloyed electrodes for batter resistance. The wing rails were attached by what was known as "plug-welding". In this technique, mild steel spacer blocks were welded to the component rails through oval holes in the rail web using the CO_2 welding process in the downhand position. Baseplates, again made from mild steel, were originally plug-welded to the crossing but these were replaced by Multi-groove Locking (MGL) pins, and the method of welding of the vee rails was changed to the electroslag process. Crossings of this basic design have been manufactured since 1966 and there are now over 6000 in service.

One of the great advantages of part welded crossings is that being made from pearlitic rail they can be thermit welded to the adjacent plain rails, with all the benefits of elimination of fishplated joints. As a consequence, it is desirable that crossings should survive the same number of years as the surrounding rails. Part welded crossing manufactured using different steels and welding products have therefore been evaluated in both the laboratory and extended trials.

4.4.1   Manufacturing Process for part Welded Crossings

Part welded crossings are manufactured by approved suppliers working to approved specifications. The wing rails are fabricated by bending two rails at the "knuckle" of the crossing. The two vee rails are bent and planed so that the weld preparation is obtained in the head and foot. An insert bar (manufactured from BS 970 080 M40 steel) is fitted between the machined vee rails and is held in place with through bolts. The complete assembly is then offered up for electroslag welding.

The electroslag welding process is a resistance melting process in which a continuous wire is fed through a molten lux pool into the weld preparation (Figure 4.9). The slag is generated by a granular or powered lux which is initially melted with an electric arc. The heat input is maintained by electric current passing through the molten slag via the continuous wire. As well as providing heat input, the slag protects the weld pool against contamination.

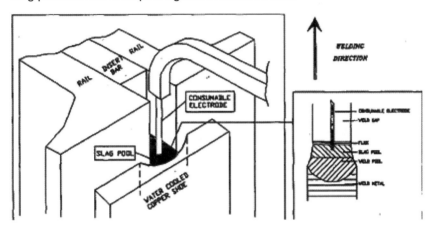

Figure 4.9 Schematic of the Electroslag Welding of rails during the manufacture of part-welded crossings

The crossing vees are welded vertically with both the foot and head being welded simultaneously. This ensures that distortion is minimised. Welding commences at the nose of the vee, which includes an extension, simultaneously removed, in which the welding process can stabilise. As the preparation is welded, the welding heads move upwards to the rear of the vee. Water cooled copper shoes are clamped to the head and foot to contain the molten weld pool during welding, and in addition, heat sink plates are placed in the web region to provide additional cooling.

The factors to be considered when selecting the optimum combination of rail steel and welding material include material and manufacturing costs, repair weldability and maintenance costs, and expected service life.

In carrying out an overall assessment on these different combinations, weldability is particularly important because crossings are normally repaired by manual metal arc welding several times in their life. Low hydrogen electrodes are used for this repair and the risk of heat affected zone cracking is largely controlled by the preheating of the material prior to repair. Determination of the required preheating temperature was carried out in laboratory trials, but the problems of obtaining and maintaining high preheat levels under track conditions are considerable and thus the higher the preheat required, the more probable it becomes that the requirement will not be met under real conditions, increasing the risk of cracking.

A controlled field trial to measure resistance to batter and deformation, and a programmed introduction to assess long term structural deterioration has been used to determine the best type of part welded crossing for use in service. The use of BS 11 wear resisting grade A rail, and BS 2901 type A33 welding wire has been found to offer the best combination of properties and part welded crossings are now manufactured as standard using these two materials.

4.5   CAST CENTRE CROSSINGS

In high speed or very heavily used track in the UK, Austenitic Manganese Steel (AMS) cast crossings have been used in many locations and this has made a vast contribution to safety, performance and efficiency for a number of important railway junctions. Originally of a monobloc construction, most were bolted and plated into track. A unique problem with AMS is that the steel is considered thermally unstable and AMS components cannot be directly welded into track using conventional processes or brittle joints would be formed. With the adoption of precise chemistry control of the manufacture of AMS and the use of stainless steel insert rails, this now allows the manufacture of cast centre crossings. These crossings contain a cast centre including wheel transfer area with rolled check rails hook pinned to the cast centre. The crossing leg extensions are machined from rolled rails and attached to the cast centre by flash butt welding. This method of construction has the advantage that the castings are smaller and less complex than conventional monobloc types, as the leg end extensions are of rolled steel they can be alumino-thermic welded or even flash butt welded (FBW) in track.

Significant improvements have been made to casting processes over the last 20-30 years. These crossings are exceptionally tough and batter resistant, but sometimes have contained internal casting defects and on occasion crossing failures have been associated with these, However, the biggest cause for service failure is fatigue, both from rolling contact type delamination of the running surface (poor maintenance of top being a primary cause) and uni-directional bending fatigue from the foot of the crossing (poor support). Edgar Allen (now Progress Rail) have been involved in design, manufacture and failure investigation since 1986. It is important to note that developments known as Low Carbon AMS and TITAN Bainitic centreblocks were made up to the 1990's, to be superseded   by High Carbon AMS centreblocks which are fully weldable into track. Explosive Depth Hardening (EDH) High carbon AMS has been available since 2012, where EDH consists of applying plastic explosive sheet (Semtex) to the running surface to create a shock wave on detonation, that work hardens the AMS to a minimum 320BHN making them more suited to resist wear from heavy axle loads and high speeds as soon as they are installed and negates the need for early maintenance grinding welding.

4.6   REPAIR WELDING

Maintenance welding in the UK covers two distinct areas:
For Crossings it is used to remove batter damage, surface cracking and restore nose topping to minimise further impact damage, using austenitic NiCrMo or Mn electrodes and is often used in the last 25% of the crossings service life.
Plain line is repaired to make good damage caused by wheelslip and to remove rolling contact fatigue defects or in some cases correct manufacturing faults.

4.6.1   Repair of Crossings

Maintenance repair involves the removal of deformed and defective material caused by the impact loading of passing traffic on the running surface, and its replacement with sound material. This type of repair is carried out in track, generally before the amount of wear on the crossing exceeds 6 mm in depth. Structural repairs to crossings are executed out of track and can involve large weld repairs to fatigue cracks or fabrication defects anywhere within the body of the crossing; the object of repair is to enhance the life of the crossing at minimal cost. However, with the introduction of flash butt welded crossings into the UK rail network, this is very limited due to the need to "cut out" a cracked crossing.

For its subsequent re-installation post repair, new FBW leg ends would be required to be put on the crossing or 4 new closure rails would have to be used.

In theory, provided welding is carried out in accordance with standard procedures, there is no limit to the number of times a component may be repaired. However in practice economic considerations (extent of damage) and general structural deterioration (loosening of fastenings, etc) will impose a limit. The decision as to whether to repair or replace a component can only be made after a detailed inspection of the crossing and its location.

Figure 4.10 The Completed Weld Repair Of A Manganese Crossing Nose And Wings
(courtesy Rail-Technology.com)

The majority of maintenance welding of crossings is to correct deformation of the running surface at the crossing vee, (especially to restore the nose topping height) or wing rails in the wheel transfer area. This deformation, or plastic flow of the crossing material results in lipping of the vee or wing rails. This can be removed by grinding and the crossing rebuilt by welding. If this lipping is not rectified, fatigue cracks propagate beneath the lip which ultimately results in the rail surface shelling or spalling and transverse cracks forming.Transverse cracking of this type is more difficult to correct and may necessitate the removal of the crossing, due to vertical crack formation. Therefore it can be seen that remedial action, at the correct time, can extend the life of the crossing, and reduce the cost and time for the maintenance work.

EDH of AMS as the benefit of reducing the incidence and / or extent of lipping during the initial period of service and helps maintain the nose topping, thus reducing impact loading on the Vee/ nose during facing movements.

4.6.2   Repair Welding of Plain Line

The adoption of manual metal arc (MMA) welding for the rectification of surface damage and defects in plain line has been stimulated by the widespread use of CWR with its attendant high cost of rail replacement. Initially repairs were restricted to inherent rail defects (rolling defects), or wheel burns. However, the identification of a large number of rolling contact fatigue (RCF) defects in the mid-1970s enhanced the cost-effectiveness of repairs by welding as an alternative to rail replacement. Welding of plain line is classed as either maintenance or repair. Maintenance welding is carried out between trains and consequently the continuity of the rail surface must be maintained, thus limiting the size of the defect. Defects requiring more extensive rectification work therefore are classed as repairs and the work must be carried out within a possession.

4.6.3   Track Repair Welding Procedures

The stress environments encountered in plain line and crossings are quite different. Crossings are mainly subject to impact loadings from the transfer of the axleload from the wing rail to the crossing nose (or vice versa), while plain line is subject to cyclic loading from the smooth passage of the wheel along the rail. The properties required of any weld metal for these repairs would appear to be quite different, but this is not necessarily the case. For crossings, resistance to impact is of prime importance, other properties required being resistance to fatigue crack initiation and growth, and a degree of ductility. A hard, tough weld metal is therefore required. Plain line repairs need to be very resistant to fatigue crack initiation and growth, and should wear at approximately the same rate as the rail steel. A high integrity weld metal that is both hard and tough is necessary.

The properties required of the two types of weld are thus very similar, and hence the procedures to be followed are also similar and will be described together.

Preheat requirements

Studies of rail steel weldability have been reported for many years and the clear need for preheating was identified, the criterion for determining preheat levels being a maximum HAZ hardness of 350HV. A preheat of 200°C was selected for manual metal arc welding (MMA) with austenitic stainless steel or carbon-manganese steel electrodes.

In 1967, low alloy, 2.25% Cr 1% Mo, hydrogen controlled electrodes were introduced for the repair of the standard grade of rail, with a preheat of 200°C. Problems were encountered with cold cracking in the weld metal and further work resulted in the preheat being increased to 350°C for both the standard grade and, more recently, wear resisting grade A rails, as reported by T L Brooks in a BR Research Report "MMA Repair Weldability - Appraisal of grades A, B and 110 Cr Rail" in 1980. Weldability studies on the grade B rail showed the need for a minimum preheat of 400°C, while the minimum preheat for fully pearlitic transformation in the BSC 110 kg/mm2 Cr rail was found to be in excess of 550°C. For practical reasons this level of preheat has been considered to be unacceptable and the repair welding of this grade of rail would not be feasible.

Welding Equipment

Until recently all track repairs were carried out exclusively using the MMA process with small portable site welding alternators capable of producing welding currents up to 250 amps. Development programmes have resulted in welding procedures and preheat requirements being developed for a range of rail steels, and a range of approved electrodes are available for use with this type of equipment. Whilst the resulting quality improvement has resulted in a relatively small risk of failure, it has been recognised that further development of the MMA process for improved weld quality and increased productivity is limited due to the characteristics of the process and the poor weldability of rail steels. By the 1970s it was recognised that the use of a continuous wire "semi automatic" welding process may offer benefits in terms of productivity and deposit quality, and self shielding lux cored wires, and wire feed units were developed for use with the conventional MMA welding machine.

With either process, the broad principles remain the same, and welding procedures are derived from:

(1) Rail/weld metal metallurgy

(2) Size of repair

(3) Welding technique

(4) Finishing requirements (smoothness).

Welding Procedures

The welding procedures for components are included in UK standards and contain the following key features:

- **Welder Competency and approved equipment**
  As with thermit welding, only welding staff holding valid certificates of competency are permitted to undertake repairs to track. Welder training takes the form of a written and practical test at a Training school, in the processes/techniques appropriate to the region or area within which the welding staff operate. It is essential that welding equipment is approved.

- **Preparation of the Component**
  Prior to any repair of track being undertaken the type of rail steel must be identified from brand markings to enable the appropriate procedure to be selected. (In the case of the 110 kg/mm2 Cr rail, repair welding is not permitted and the rail must be replaced). The extent of the damage must then be assessed, with the aid of non destructive testing (NDT) techniques, to determine the stages of the repair. This is of particular importance if the repair is being carried out between trains.

When working between trains, any excavation is limited to 15mm maximum depth, and 150mm maximum length, the width being limited by the need to maintain sufficient running surface for the safe passage of trains. Repairs out of track do not require the size of excavation to be limited, the size of repair usually being governed by economic considerations.

Defects are removed using grindstones, abrasive discs, or rotary burrs - currently arc air or gouging electrodes are prohibited on track. When the defective material has been removed, the excavated area must be subjected to a NDT inspection using either magnetic particle or dye penetrant techniques.

- **Preheating**
  The aim of preheating is to ensure that the post weld cooling rate of the component is sufficiently slow to avoid the formation of brittle structures. The procedures have evolved after considerable development work, and the preheat temperature and method of application is strictly controlled.

  Currently all preheating is carried out using air/gas radiant heaters applied directly to the component. The excavated area, and material within 75mm of the excavation is preheated to the appropriate temperature for the type of rail steel. The preheat temperature is checked with the aid of a temperature indicating crayon. Once achieved, the specified preheat must be maintained throughout the welding operation, which will necessitate repeated checks on the temperature and reheating of the component

- **Welding Technique**
  Welding techniques for both MM A and Semi-automatic are taught during the training programmes, which also include safety instruction, and checking of equipment.

  Weld repairs are subject to a fatigue environment, and so the procedures have been developed to maximise weld integrity and minimise metal lurgical damage to the parent material. Rail steels and high strength weld metals are susceptible to hydrogen embrittlement. Therefore low hydrogen electrodes are specified and stringent electrode drying procedures have been included in the working instructions.

The weld metal is always deposited in longitudinal runs (i.e. down the length of the excavation) and a small weave is permitted on all grades of rail steel apart from AMS. On completion of each weld bead, the slag must be removed, and the rail preheat temperature checked. If necessary the rail is re-heated to maintain the specified preheat. On completion of each layer, the deposit is visually inspected for defects, and finish craters lightly ground.

- **Finishing**
  When the repaired area has cooled to ambient rail temperature it can be ground to the correct running profile with the aid of a suitable straight¬edge.

- **Post Weld Inspection**
  The area of the repair is inspected visually and with the aid of dye penetrant or magnetic particle NDT techniques. Any defects must be completely removed by grinding, and the area re-welded. Any deteriora¬tion in a crossing due to the welding operation (failure of glues or bolts etc) must be reported to the local engineer.

The repair of cast manganese crossings requires special procedures to minimise heat input in order to avoid embrittlement of the casting. No preheat is required, and the temperature of the workpiece must be kept as low as possible. Heat input during welding is kept to a minimum by restricting the size of the welding electrode to 4mm diameter or less, with a maximum weld run of 125mm. No weaving is allowed.

The workpiece temperature is checked between runs using a 204°C temperature indicating crayon and if the temperature is too high the crossing must be allowed to cool before welding can continue. In some cases it is possible to aid natural cooling by introducing water into a reservoir manufactured by plugging the flange way with putty, clay or cloth. The water level is kept below the area for welding.

In the case of large defects the bulk of the repair is welded using one of the "Stainless" group of electrodes. The repair is then "surfaced" using work hardening manganese nickel electrodes on all wheel contact areas.

Bainitic (TITAN) Cast Centre Crossings were developed as a readily weldable track component, and the principal advantage of this material is that as a result of the lower hardenability, preheating is not, in theory, required. However, as these components are usually repaired outdoors, a low temperature preheat (approximately 100°C) is applied to remove moisture and prevent porosity or hydrogen embrittlement within the weld deposits.

The quality of track repairs is maintained by ensuring products are supplied to agreed quality standards and all welding materials are adequately protected against damage and moisture prior to use. The use of NDT techniques throughout the welding operation aids the production of defect free repairs, and a stringent post weld inspection ensure the repairs are of an acceptable standard in terms of weld integrity and finish.

These track repair welding methods have proved highly successful with large repairs being capable of withstanding the stresses imposed by passenger trains travelling at 200 km/h and freight vehicles with axle loads of 25 tonnes.

CHAPTER 5

RAIL JOINTS

5.1 INTRODUCTION

Rails are manufactured in relatively short lengths, typically in the UK, of 18 or 36 metres. When installed in the track, they usually require to be attached to one another end to end, and for the first hundred years or so of railway history the universal method of doing this was by the fishplated joint. This remains the fall-back method, whenever, for historical, engineering, economic or operational reasons, rails are not welded together into long strings. As a general rule, fishplates are made from rolled steel but when one rail is required to be electrically separated from its neighbour, insulated fishplates of various designs have to be used. The picture below in figure 5.1 shows a bull head standard fishplate whilst the standard plain rolled steel fishplate used with FB rail is shown in Figure 5.2. The features of insulated joints are described later in this chapter. This chapter also considers the case where rails are not actually attached to one another, and deals with the maintenance of both fishplated and welded joints. The actual processes of welding have already been discussed in Chapter 4.

5.2 HOW FISHPLATED JOINTS ARE DESIGNED

The term "fishplate" is of common application as a device for joining two beams or stanchions end to end. The structural objective of a fishplated joint is to transmit the bending moment and shearing force developed in the member by the action of external loads from one member to the other. If the combined moment of inertia of the fishplates is equal to that of the members on either side, and if the connections between the fishplate and the members are 100% efficient, then the bending moment and shearing force in the fishplates will be the same as if the beam were continuous.

For obvious reasons, when railway rails are butt jointed it is impracticable to attach plates to the head or foot of the rails, and the size and shape of the plates to be attached to the web is limited by the needto keep them out of the way of the wheel flanges. Consequently, no one has yet succeeded in designing a fishplated joint which has more than a small fraction of the moment of inertia of the rails (see Table 5.1).

TABLE 5.1   **Moments of inertia of FB Rails and Fishplates**

Moment of inertia of Rail (xx)	2332.2 cm^4
Moment of inertia of Fishplate (xx)	682.6 cm^4
Section Modulus of Rail (xx)	
Top	277.95 cm^3
Bottom	311.67 cm^3
Section Modulus of Fishplate	
Top	121.43 cm^3
Bottom	134.21 cm^3
Ratio I(xx) Fishplate/I(xx) Rail	29.3%

Figure 5.1   **Fishplated joint on bull head track showing bonding wires**
(courtesy P Ransom)

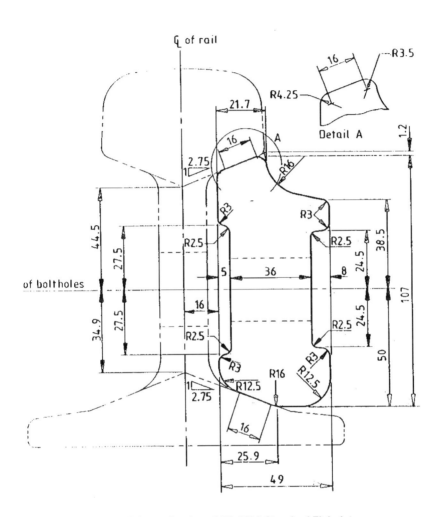

Figure 5.2 Cross Section of BS 113A Standard Fishplate

Nevertheless, the bending moment actually transmitted across a standard fishplated joint, assuming that the connections are 100% efficient, is of the, order of 90% of what it would be if the rails were continuous. The bending moment in the fishplate in the gap between the rails is produced by a couple (for definition of this and other technical terms in this paragraph, see Chapter 2) exerted on the fishing surfaces of the fishplate by the rail, and vice versa, as shown in the sketch, Figure 5.3.

In practice, the forces across the fishing surfaces are not concentrated at a point but are distributed along the fishing surface, and this distributed force varies in intensity in a more or less triangular fashion, as shown in the sketch Figure 5.3.

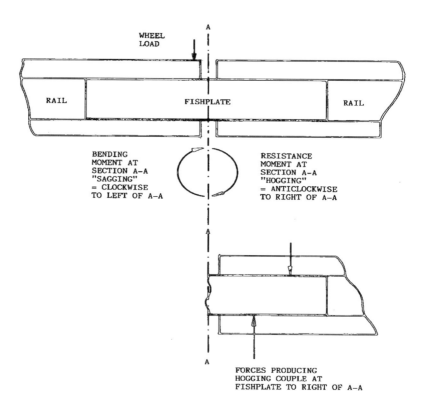

Figure 5.3 Couple Exerted on the Fishing Surfaces by Wheel Load

These forces are applied across the fishing surfaces, and are balanced by an equal and opposite force provided by the fishbolts. It is important to realise that the total force in the fishbolts is established when the bolts are torqued up, and it is the change in *distribution* from uniform (Figure 5.4(b)) to triangular (Figure 5.4(c)), and *not* a change in the tension in the bolts, which provides the resistance moment to balance the bending moment produced by the wheel load. The peak value of the pressure depends upon the general condition of the joint, and it reverses with the passage of each wheel. It is this feature which leads to wear and crushing of the fishing surfaces, and which rail joint shims are designed to correct. Provided the bolts are properly torqued up to their full working load, the 1" diameter (note that fishbolts are one of the few items still to be described in Imperial units) bolts in a standard four-hole FB fishplated rail joint will provide sufficient clamping force to develop the bending moment required to be transmitted across the joint.

It can also be shown that if the number of bolts is reduced to two, and the fishplates halved in length to match, the capacity of the fishplate will be reduced to a quarter of the strength of the four bolt plate, and the reason why such plates, used in UK before 1945, were not adequate and had to be abandoned, is obvious. Similarly, if the joint is adequate with four bolts, there is no reason to increase the number of bolts to six. The above remark does not apply to glued insulated fishplates, as is indicated below.

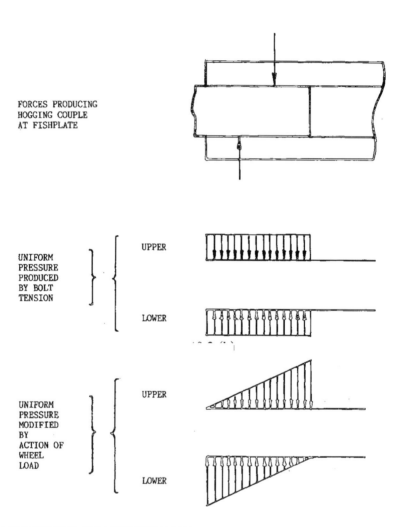

FORCES PRODUCING
HOGGING COUPLE
AT FISHPLATE

UNIFORM
PRESSURE
PRODUCED
BY BOLT
TENSION

UPPER

LOWER

UNIFORM
PRESSURE
MODIFIED
BY
ACTION OF
WHEEL
LOAD

UPPER

LOWER

N.B. THE FORCES AND PRESSURE DISTRIBUTION REVERSE WHEN THE WHEEL CROSSES -THE GAP.

Figure 5.4 Pressure Distribution on Fishing Surfaces under Wheel Load

5.3   FEATURES OF FISHPLATE DESIGN

It will have been noted that the ability of the fishplate to develop the necessary bending moment at the gap between the rails depends on the clamping force exerted by the bolts. It is emphasised that it is NOT intended that the couple should be developed by interaction between the bolts and the holes in the rail web. Both for this reason, and to give room for thermal expansion and contraction to take place, the holes in the rail web are made 30.7mm diameter, around 5mm larger in diameter than the bolts.

For similar reasons, the fishbolt holes are also made larger than the bolts. In the current standard design they are drilled to 27mm dia.

The degree of slope of the fishing surface is an important parameter. From the calculations referred to above it is clear that the shallower the slope, the larger the force across the fishing surface for the same bolt tension, and therefore, the more efficient the joint. On the other hand, the shallower the angle, the greater is the pull required to maintain the clamping force when the fishing surfaces wear. In most modern fishplates, the fishing angles slope at about nominally 1:2.75. This angle has been found by experience to be the best compromise between good initial performance and maintainability. The top and bottom surfaces slope at the same angle, as it was found that if they are different, the fishplates pull in irregularly as they wear.

The torque applied to standard fishbolts during tightening is not usually monitored or controlled (e.g. by the use of a torque limiting spanner). It has been found that when the bolt is tightened to refusal by a man of average strength using an ordinary fishbolt spanner, a torque of about 50KNm can be applied, and this is sufficient to develop the required bolt tension as indicated above.

High strength bolts should be tightened with a torque limiting spanner set at 9KNm, unless otherwise specified.

5.4   MANUFACTURING ASPECTS OF FISHPLATED JOINTS

5.4.1 Fishplates

Fishplates may be cast, forged or rolled. Ordinary fishplates in the UK are normally rolled. Rolled fishplates are of course of uniform cross section along their length. As shown in Figure 5.1, they incorporate longitudinal ribs, which have the dual function of making the plate stiffer, and of forming a groove in which the square heads of the fishbolts sit, thus preventing them from turning when being tightened. The fishing surfaces must be straight and free from twist after all drilling and machining processes have been carried out, to within tolerances laid down in the Specification. The holes are circular, and may be formed by either drilling, hot punching or machining, provided the quality of the fishing surfaces is not impaired by the process used. The positions of the fishbolt holes are as shown in Figure 5.4. Rolled fishplates are made from steel of the following composition to achieve a minimum tensile strength of $600N/mm^2$.

Carbon	0.27% - 0.48% by weight
Silicon	0.38% max by weight
Manganese	0.94% max by weight
Phosphorus	0.058% max by weight
Sulphur	0.058% max by weight

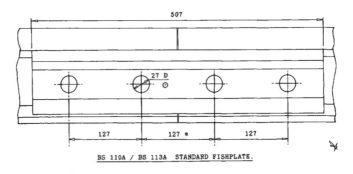

BS 110A / BS 113A   STANDARD FISHPLATE.

*   This dimension 119 for Tight Joint Fishplate.
⊙   This dimension 31 for Tight Joint Fishplate.

Figure 5.5 Holing for BS 110A/113A Standard Fishplate

5.4.2   Fishbolts and Nuts

Standard fishbolts and nuts are manufactured in accordance with exacting standards. Fishbolts are required to be forged hot from steel of a tensile strength in the range 540 to 650 N/mm² whilst nuts are forged from steel in the range 460 to 620 N/mm². Fishbolts for BS 113 A are nominally 1" (25.4mm) diameter and of uniform section throughout their length with heads and nuts 42 mm square. BSF threads are used with a pitch of 2.54 mm per turn (ten threads per inch).

Fishbolts for 95BH rail are 15/16" diameter with 11 BSF threads per inch, whilst BS 113A fishbolts for tight joints, heavy duty insulated joints, and the back joints in 1 in 24 and 1 in 28 cast crossings are 1 1/8" diameter with 9 BSF threads per inch. High strength fishbolts have been standard for all UK fishplates since 1990. They are identified by the letter "V" stamped on the head, to diameters as specified according to the type of joint in which they are used. Square headed or hexagonal nuts are specified according to the type of joint.

5.4.3   Bolt holes in rails

Since CWR became the normal primary use of new rail, it has become standard policy in the UK to order new rails with undrilled ends except where specially needed. Where drilled ends are specified, the usual drillings for BS 113A rail for four holed fishplates are 30mm nominal diameter holes with the centre line of the holes 65mm above the foot of the rail. The centreline of the first hole is 60mm from the end of the rail and the holes are at 127mm pitch.

One of the principal causes of rail failure is the development of fatigue cracks around bolt holes (the so called star crack, see Chapter 3). Much can be done to reduce the probability of development of star cracks by proper care and maintenance of the joint. One other way in which star cracking can be rendered less likely is by setting up a residual compressive stress field in the steel immediately surrounding the bolt hole.

In current practice in the UK this is done by a process known as COLD EXPANSION, or COLDEX, in which an oversized mandrel is pulled through a disposable stainless steel liner inserted in the hole. It is also important to ensure that the hole is truly circular, and correctly formed with an internal finish of very high quality. This is achieved by two stage drilling. The finished diameter of the holes after cold expanding is 30.7mm.

The holes in rails as supplied from the rolling mill with drilled ends are processed in this way as routine, but it is important that any holes drilled in the field are also correctly drilled, and cold expanded using the equipment provided.

5.5 JOINT LOCATION

5.5.1 Supported versus suspended joints

To form a supported joint, the rail ends are placed over a sleeper (which can be an extra wide sleeper), or over two sleepers placed side by side and bolted together. By contrast, to form a suspended joint, the sleepers are placed at more or less normal spacing, with the joint at the middle of the space between the sleepers. At one time it was customary to use an abnormally wide timber sleeper either side of rail joints, and a special prestressed concrete sleeper was designed for the same purpose. There is also an intermediate form of joint known as the semi-supported joint, in which the sleepers around the joint are placed much closer together than standard, and the fishplates then span the gap. Most BH track and many early lengths of FB track had semi-supported joints. These have virtually disappeared, as it was found that the joint sleepers were difficult to pack uniformly, so that the sleepers tended to tilt towards one another. It was also necessary to remove the rail fastenings in order to dismantle the fishplates.

There are arguments in favour of the supported joint, notably that the extra area of support provided by the closely spaced joint sleepers will compensate for the greater impact loadings to which the joint is subject, thus giving a better retention of longitudinal level.

The contrary argument is that a supported joint of whatever design is more difficult to pack than a suspended one. In this present age of universal mechanical maintenance by automatic machines, the convenience of regular sleeper spacing outweighs any possible theoretical considerations, and it is now universal practice to use suspended joints, and sleeper spacings are constant over the whole rail length.

5.5.2   Relative positions of joints on opposite rails

The standard positioning of rail joints is opposite one another, and out of squareness is limited normally to 64mm. To maintain this standard of squareness on curves, short rails must be inserted at intervals on the inner rail.

On some railways it is standard practice on sharp curves to stagger rail joints by up to half a rail length, on the grounds that it is easier to control alignment if the joints are not opposite one another. However, staggered joints are not recommended, partly because it is considered that if staggered joints become dipped, a cyclic twist situation will be set up.

5.6   JOINTS FOR SPECIAL SITUATIONS
Under this heading are joints in cast crossing work, joints in continuous check railed track, lift and junction fishplates, clamp plates, spliced joints, and adjustment or expansion switches. Insulated joints are given a section on their own.

5.6.1   Cast Austenitic Manganese Steel Crossings in CWR

It is customary to make the "through" side of turnouts in Continuous Welded Track (CWR) continuous with the CWR on either side. This means that the S&C components must be able to withstand the thermal longitudinal loads, and as many as possible of the joints in turnouts are therefore welded. At the same time, the joints between cast AMS crossings and BS 11 rail have to be fishplated. Specially designed joints are used for this purpose. Such joints do not have any provision for expansion, and use 28.5mm diameter high tensile steel bolts. To accommodate the larger bolts the fishplate holes are 31mm diameter. Such joints are occasionally "frozen" by coating the fishing surfaces with resin glue.

In order to bring the ends of the rails close together, tight joint fishplates have a reduced spacing between the central holes of 119mm (see Figure 5.5).

5.6.2   Joints in continuous check railed track

Where continuous check rails are used (mainly on track with a radius of less than 200 metres), special thin fishplates may have to be supplied.

5.6.3   Lift and Junction Fishplates

'Lift' fishplates are used to join rails of the same section but with different amounts of headwear, whilst 'junction' fishplates are used to join rails (new or worn) of different sections.

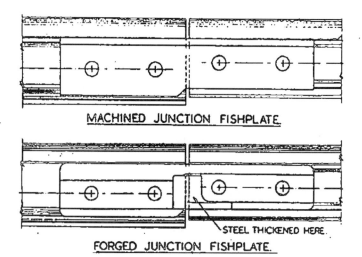

MACHINED JUNCTION FISHPLATE.

FORGED JUNCTION FISHPLATE.

STEEL THICKENED HERE.

Figure 5.6 Junction Fishplates

Lift fishplates can be forged, or machined to shape, whilst junction fishplates are usually machined to shape, especially when the rail sections to be joined are noticeably different. As it is necessary for the tops and the running edges of the two rails to be aligned, junction fishplates can be of quite complex shapes when the two rails have different depths, head widths and fishing angles. A left hand and a right hand fishplate of different shapes are required at each joint. Much as it is desirable to avoid it, it is nevertheless necessary on occasions to employ insulated lift or junction fishplates at the ends of switches and/or crossings. Lift insulated fishplates can be nylon lined forged steel plates, whilst junction insulated fishplates can be machined to shape from compressed laminated resin impregnated wood. It has been common practice to have a range of fishplates available for rails of varying depths. The conceptual form of machined and forged junction fishplates is illustrated in Figure 5.6.

5.6.4 Clamp Fishplates

Clamp fishplates are designed for temporary or emergency use. Instead of employing fishbolts, clamps around and under the foot of the rail are used to hold the fishplates in position, no holes in the rail webs being necessary. These fishplates can be used during track laying on rail ends awaiting thermit welding, or on closure rails awaiting drilling. Another purpose is to strengthen a suspect or cracked rail weld and in such cases the fishplates are joggled at their centres to clear the weld collar.

5.6.5 Adjustment Switches (overlapping Joints)

Joints with specially shaped overlapping rail ends have the advantage that the wheel tread is continuously supported across the joint. These are used in locations where CWR abuts jointed track or on each leg of unstrengthened S&C. The main purpose of this is rail expansion. The most common form of overlapping joint in present day track is the CWR adjustment switch (see Figure 5.7).

**Figure 5.7 CWR Adjustment Switch – note concrete sleeper on left
(courtesy Network Rail)**

In the adjustment switch, the two tongues are of identical design. Each tongue is joggled so that the web will lie alongside the web of the adjoining tongue, and then both sides are machined to the shape shown in the illustration. An indication of the amount of the joggle can also be obtained from the illustration by observing the shape of the rail foot through the joint. As is seen, the tongues are supported by two special baseplates with clamp brackets to hold the tongues in position whilst allowing longitudinal movement.

Overlapping joints are sometimes called for where it is required to accommodate substantial relative movement between two strings of rails, notably at bridge expansion joints. The classic example is the Forth Bridge in Scotland, but they occur on many large steel railway bridges in other parts of the world, and cases are known where up to 300 millimetres of movement has to be catered for. Such joints may call for specially designed rail sections which will be strong enough to take wheel loads even when half of the section is machined away, and check rails which themselves slide, to control wheel alignment over the substantial lengths where the track may be up to a rail head wide to gauge.

5.7 INSULATED JOINTS

5.7.1 General

Insulated rail joints are required on track circuited routes to enable rails to be physically connected whilst remaining electrically separate. It may be noted that signalling requirements may well require this electrical separation where no rail joint would otherwise be required, so that a long length of CWR may have to be broken up into several track circuit sections, and many insulated joints are required in S&C layouts. Track circuiting currents may be either AC or DC depending on circumstances, and sometimes only one rail is used for track circuiting purposes, the other typically being used for traction current return. Recent developments in electrical technology have produced the jointless track circuit, in which the two rails are connected at either end of the section by a "black box" to form a tuned circuit the electrical resonance of which is disturbed by the presence of a train. It is not always possible to use the jointless track circuit however and very many of the conventional type of track circuits will continue to be required for the foreseeable future.

Before the introduction of CWR, insulated joints were fitted with fishplates made from compressed laminated resin-impregnated beech wood, with an endpost of insulating material between the ends of the two rails (end posts, 6mm or occasionally 9mm thick are required in all types of joints and are currently made from resin bonded glass cloth or similar; nylon is not regarded as being suitable). Such fishplates never failed electrically, but their physical lives were short, firstly because they lacked the necessary structural strength (see section 5.2 above) and secondly because they weathered badly. The early development of metal fishplates with insulating sleeves and an endpost usually made of nylon. (see Figure 5.8)

Figure 5.8 Bull Head Insulated Fishplate
(courtesy London Underground Ltd.)

5.7.2 Development of Insulated joints

The type of fishplate described in the previous paragraph was found to be unsuitable for use in CWR as it did not have the necessary strength to withstand thermal stresses, and much effort has been expended over the years to produce designs of insulated fishplate which will withstand both the tensile loads produced by CWR, and the considerable increases in service loading consequent upon changes in speeds and traffic patterns over the period.

The next sections summarise present practice in relation to insulated joints, up to and including the current thinking on six-holed plates.

5.7.3 Dry Joints

Three types of dry joint survive. These are made from reinforced laminated wood, and from steel, the steel fishplates being described as plastic bonded and nylon lined respectively.

As already mentioned, compressed laminated resin-impregnated beech wood is a good material from an electrical insulation point of view, and it is easy to shape. For these reasons, attempts were made early on in the history of CWR to use insulated fishplates made from this material, using steel reinforcement on the outside of the plates to resist the thermal loads. Both four bolt and six bolt varieties were tried. Application of this type today is restricted to complex shapes such as junction fishplates in S&C.

Plastic bonded fishplates consist of a forged steel core completely surrounded by a synthetic resin bonded fabric insulating layer. Two steel plates with hardened fishing surfaces are bonded to the inner faces. Four high tensile steel fishbolts 25.4mm diameter are used with load spreading washer plates under the heads and nuts. This type of plate is expensive to manufacture, and it is possible for the steel fishing services to become debonded from the remainder of the plate. Nevertheless, the fishplate is considered robust and electrically reliable.

The nylon lined fishplate is similar in principle to the plastic bonded fishplate but comprises rolled steel fishplates which have a one piece nylon liner which clips onto their inner face. The liner is 3mm thick and has integrally moulded bushes on it which enter the fishbolt holes in the fishplate. Four high tensile steel bolts are used. They were originally 25.4mm diameter but were later increased to 28.5mm diameter. Insulating and load spreading washers are used under the bolt heads and nuts. Both types of steel cored fishplate are used extensively in jointed track, and in S&C.

5.7.4   Glued Joints

Glued joints tend to last longer than dry joints both in terms of their general performance as track components, and as regards their electrical reliability, and factory made glued joints are standard for plain line CWR wherever their installation is practical. There are a number of types in the UK.

The Permali-Edilon joint consists of a steel billet with no pre-coating. The fasteners are studs threaded for nuts at each end, and these are pre-coated with insulating material over the centre portion where they pass through the web of the rail. The fishing surfaces at the rail ends are prepared by shot-blasting, and a 3mm thick layer of polyester resin is applied between the fishing surfaces of rail and plate. The thickness of this layer is controlled by nylon spacers. The nuts are torqued up in stages to the specified torque.

A later version uses a similar fishplate to the Permali-Edilon, but uses 29mm diameter HTS bolts, insulated from the fishplates by bushes and washers but without a sleeve through the rail web. In this joint the glue is epoxy resin.

The BR mark II joint also uses forged steel billet fishplates with a thickened centre portion. In this type however an impregnated, preformed liner is epoxy resin bonded between the rail and the fishplate. The fasteners are 28.5mm diameter multi-groove locking (MGL) pins. Insulating washers and bushes separate the pins from the fishplate, but as in the previous type, they are not sleeved through the web.

Glued joints for inserting into plain line are best made in a special rail, cut and drilled accurately, rather than by juxtaposing the ends of two different rails. When made in the workshop, a rail about 18m (60ft) long, or sometimes less, is used, and this rail, with the joint in it, is then thermit welded into the CWR.

5.7.5   Six hole insulated rail joints

Experience in CWR with all the types of glued joints mentioned above has shown that none are completely trouble-free, and that their service life is typically only between a third and a half of a rail life in arduous conditions, because most resin adhesives are subject to slight deterioration of physical properties such as shrinkage or embrittlement in the course of time. Failures are associated with either "pull aparts" where the adhesive bond has failed and the joint gap opens up resulting in broken/bent bolts, or less frequently the fishplate itself cracks or breaks. Failure of the rail end, or of electrical insulation is also a risk when the joint deteriorates. Replacement of an insulated joint in CWR is not easy in cold weather when the rail is in tension, yet it is under such conditions that a failure is most likely to occur.

Analysis of the rail joint suggested that the forces developed along the fishing surfaces of a four-holed fishplated joint by the clamping force of even high strength bolts would be nowhere near enough on their own, to balance the tensile thermal forces in cold weather, and that the glue was being required to carry a substantial load for which it was not really designed. The same analysis suggested that if the fishplate were extended to take six bolts instead of four, the need for the glue to act in shear to resist the thermal loads would be, if not totally eliminated, at least reduced to a level which would be tolerable. Hence, from about 1983 onwards, the UK authorities commenced to install six-holed shop made insulated fishplates in selected arduous locations which proved to extend substantially the trouble free life of an insulated joint.

These joints are based on the principles of the BR Mk II joints, but have a very much increased area of resin bonding between the rails and fishplates compared with most previous types. The specified requirement at manufacture is, that the joint shall withstand a tensile load of 150 tons, compared with typically 100 tons for earlier types. Additionally the opening of the rail ends under such a load shall not exceed 0.25 mm.

The BR Mark III joint has forged fishplates about 900mm long with thickened centre sections and is secured with 29mm diameter HTS bolts or equivalent. Accuracy of the fishing surfaces and of bolt hole drilling are very important. Recesses have to be provided in the outer lower parts of the fishplates to accommodate the joint sleeper fastenings. Not all types of established fastenings can be fitted, and correct spacing of the joint sleepers is more important with these bolt joints than with some of the earlier types.

**Figure 5.9   Six hole insulated rail joint showing adjacent sleepers specially designed to accommodate 3rd rail power cables
(Courtesy of Cemex Rail Solutions)**

5.8 FORCES IN RAIL JOINTS

Even if the connection between the rails is fully efficient, as described above, the fact that the fishplate itself is much less stiff than the rail causes two effects:

(1) greater deflection of the running surface under the wheel than elsewhere

(2) a more angular deflected shape.

In consequence even when in perfect condition a rail joint is subject to dynamic increments in the wheel/rail forces which are absent from the plain rail track structure.

These dynamic increments, which are quantified as the P_1 and P_2 forces (see Chapter 2) in turn cause both degradation of the joint elements themselves, and a rapid deterioration of the track structure in the vicinity of the joint, than occurs in plain line. If these effects are to be minimised, joints must be inspected regularly, and the operations described in the following sections must be carried out when found necessary.

CHAPTER 6

TRACK FASTENINGS, BASEPLATES AND PADS

6.1   HISTORY OF DEVELOPMENT OF FASTENINGS

Modern FB rail fastening design originated in the late 1940's with work
which showed the need for a resilient connection between rail and
sleeper, if it is to resist the forces induced by traffic and temperature
changes over a long period of time. This statement applies both to the
resistance of the system to creep, and also to its resistance to buckling
forces in CWR at high temperatures.

It was shown experimentally that vibration caused rapid loosening of
rigid fastenings, but in addition to this the unsuitability of fastenings
without some degree of elasticity to resist both creep and the buckling
forces associated with CWR was also shown, the maintenance of clip
clamping force on the rail foot (or toe load) being very important in this
respect.

There ensued forty years of development in FB rail fastener design,
during which many different designs evolved. One well known company
alone has records of more than 700 patents connected with rail
fastening designs having been taken out over this period.

6.2   FASTENING TYPES

The many designs of resilient fastenings for FB rails developed to
date can be grouped broadly into three general types depending upon
whether the resilient element secures the rails directly to the sleeper
or to the baseplate, and in the latter case, whether or not the resilient
element is preloaded by a screw. These three types are illustrated and
described in more detail in the next three subsections.

6.2.1   Type 1 Fastenings

The first and earliest type of elastic fastening is the direct spring spike (eg Elastic Rail Spike (ERS) as shown in Figure 6.1 or Macbeth Spike as shown in Figure 6.2), in which the spike is driven directly into one or more holes drilled in advance in the (almost invariably timber) sleeper. They can be used with or without baseplates. Where baseplates are used, these can be either of rolled steel as shown in Figure 6.1 or grey cast iron as shown in Figure 6.2. Usually this type of fastening is used to hold down both rail and baseplate at the same time, the bent over part of the spike bearing on the rail foot while the shank passes through a hole in the baseplate to penetrate into the sleeper. Applications are known (eg the BR3 baseplate) where the ERS on the field side (outer edge) of the rail is replaced by a chairscrew and a lip formed in the casting, which fits over the foot of the rail to hold it down. This type of baseplate was extensively used on 3rd rail electrified track since insertion and removal of the ERS between the running and conductor rails was, to say the least, inconvenient. The absence of a pad in most applications of this type of fastening, and the lack of control over toe load, mean that this type of fastening has, long term, a very low creep resistance, and is unsuitable for use in CWR track.

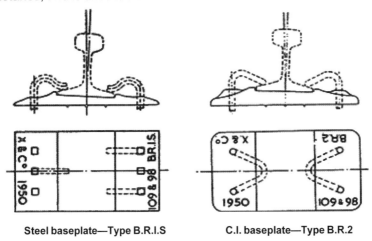

Steel baseplate—Type B.R.I.S   C.I. baseplate—Type B.R.2

**Figure 6.1 Elastic Rail Spikes and   Figure 6.2 Macbeth Spikes and
Rolled Steel Baseplate   Cast Iron Baseplate**

6.2.2   Type 2 Fastenings

With the advent of concrete sleepers the Type 1 fastening was succeeded by the Type 2 range of designs (eg, the BJB and AD fastenings used at one time on in the UK, the RN fastening used and also widely in France, and its derivative the NABLA). The RN brand rail fastening is shown in Figure 6.3 and the NAB LA brand rail fastening in Figure 6.4. These fastenings are mainly used with concrete sleepers and their common features are:

1)   a threaded element which is used to apply a force to a spring steel element, this latter being removable from the sleeper;

2)   the spring steel element, which can be of bar or plate section.

Toe load is generated by tightening down a nut to a predetermined torque (this is often done mechanically nowadays). The nut can be retightened periodically as needed to maintain toe load over the life of the fastening and it can be undone to allow component replacement. The type can incorporate features enabling adjustment of rail height relative to the sleeper, and to modify the gauge if necessary.

6.2.3   Type 3 Fastenings

In this type the bolt or other threaded element which holds down the spring in Type 2 fastenings is replaced by some kind of fixed element incorporating a hole, slot or recess into which one part of a spring clip is inserted. The spring clip when in position bears against the rail foot and against some other part of either the sleeper or the fixed element, whilst the hole etc acts as a fulcrum. The positions of the three points of contact are so arranged in relation to the shape of the clip, the hole into which the centre leg of the clip is inserted, and the ledge on which the heel of the clip bears, that the clip is delected on insertion and generates a predetermined clamping force on the rail.

Figure 6.3 (a) RN type fastening for use with FB rail and concrete

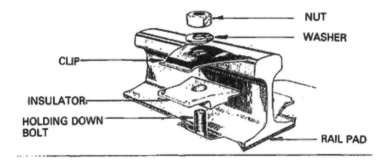

Figure 6.3 (b) RN type fastening for use with FB rail and concrete sleepers

Type 3 fastenings are well adapted to use with concrete sleepers but some types can be used on either steel or timber sleepers as well.

Where used on concrete sleepers, Types 2 and 3 have common features of a pad between rail and sleeper to absorb vibration and impact, to provide a conforming layer between rail and sleeper, to improve the coefficient of friction between rail and sleeper, and to provide electrical insulation and insulating elements to isolate the rail electrically from any metallic path into the sleeper.

Makes of Type 3 include DE, FIST HAMBO, PANDROL, SAFELOK, SHC, and SIDEWINDER.

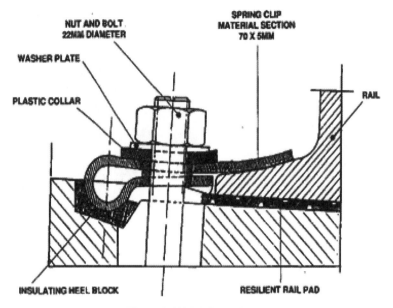

Figure 6.4 Nabla fastening

6.3 UK Practice

The standard fastening introduced in the 1970's on all new plain line track construction in the UK conforms to Type 3, and was the PANDROL brand Type "e" 1809 Rail Clip, as shown in Figure 6.5(b). This is made from 18mm diameter spring steel bar by a process which involves heating the bar into the Austenitic range, hot pressing it to shape, and then quenching and annealing. Older plain line installations, and most current S&C installations use the type PR 401A clip, made by the same process from 20mm diameter bar (some baseplate designs use the PR 303 clip, which is similar to the PR 401 A, but made from 19mm diameter bar). This shape, which is the original shape developed by Pandrol Ltd, is shown in Figure 6.5(a).

The majority of plain line track in the UK is laid on concrete sleepers without baseplates and in this case anchorage is provided by a shoulder which is cast in to the sleeper during manufacture, as discussed in Section 6.7.1 below. The non-baseplated installation is completed by a pad and a set of insulators, as described in Sections 6.6 and 6.8 below.

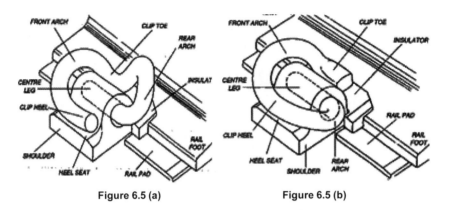

Figure 6.5 (a)   **Figure 6.5 (b)**

Where PANDROL brand rail clips are used in conjunction with baseplates, the latter are secured to the timber or sleeper by chair screws (alternative terms are coach screws, sleeper screws, or screw spikes) or LOCKSPIKES. Screw spikes are installed in conjunction with either ferrules or spring washers. Traditionally ferrules have been used on BR but the additional resilience associated with spring washers is now accepted as giving better retention of track gauge and reduced damage of the sleeper by the baseplate. The range of baseplates manufactured for use with PANDROL brand rail clips is discussed and illustrated in Section 6.7.2 below for plain line and in Chapter 13 for S&C.

LOCKSPIKES are hairpin-shaped spikes intended to hold the baseplate down on the sleeper. For this purpose the folded over end of the spike is enlarged to perform a function similar to that of the head of a nail or screw, whilst at a level designed to be just within the hole in the baseplate, there is another bulge which is compressed as the spike is driven into the plate hole. This has the effect of taking up any 'play' between the spike and the plate (thus preventing relative movement of plate and sleeper under train loading). It also spreads the legs of the spike against the hole in the sleeper, improving retention of the spike in the hole.

The latest clip is called "Fastclip" and is manufactured by Pandrol and has been widely introduced on mainlines in the UK.

Figure 6.6 Pandrol Fast Clip

6.4   MAINTENANCE OF FASTENINGS

6.4.1   Baseplated Timber Sleeper Fastenings

The current standard UK provision wherever it is necessary to lay timber sleepers (mostly Jarrah or similar these days) in running lines, is cast iron baseplates held down by coach screw or LOCKSPIKES, with PANDROL brand rail clips securing the rail to the baseplate. Designs of baseplate are available both with and without pads insulators do not normally feature in designs for use on timber sleepers. Maintenance of such assemblies is mostly confined to ensuring that the baseplate remains securely fixed to the sleeper. For attention to PANDROL brand rail clips, and pads, see below.

Coach Screws

These should be periodically tightened to ensure that they are firmly in contact with the top of the ferrule and the ferrule should be 5mm above the top surface of the chair or baseplate housing.

Where the coach screw is loose in its hole it should be removed together with its ferrule and a VORTOK coil inserted into the timber hole. A new ferrule should be used and the coach screw replaced and tightened down by a box spanner or powered nut wrench. If the coach screw is corroded a new screw should be used.

When inserting coachscrews (either replacement of existing or the insertion of new ones) in locations subject to corrosive action the screw shank should be well greased.

Lockspikes

As mentioned above, LOCKSPIKES are an accepted alternative to coach•screws as a means of fixing baseplates to timber sleepers. They should not require routine attention, once properly driven, but clearly if the sleeper becomes decayed around the spike hole, the security of the fixing will be lost. If this is found to occur, remedial measures as for ERS (see under Type 1 Fastenings, below) can be applied.

Type 1 Fastenings

Hundreds of millions of Type 1 spring spikes were installed in the 1950's and 1960's and they generally performed well in the traffic conditions of the time, (ie with jointed rail, train speeds up to 130km per hr and axle loads of 18-20 tonnes).

A crucial factor in the successful performance of such fastenings is the condition of the timber sleeper. Deterioration around the spike holes, whether by decay or mechanical damage, severely affects the ability of the fastenings to retain their hold.

Where ERS become loose, distorted, corroded or broken, additional spikes can be inserted into freshly bored holes in the spare baseplate housings provided. Spike locking devices may also be inserted into the timber hole to improve spike hold. When Macbeth Spikes become loose they can be withdrawn and a Mac-insert placed into the hole before redriving the spike. Alternatively the sleeper can be repositioned by drawing it through approximately 70mm and by boring fresh holes into the timber.

6.4.2   Concrete Sleeper Fastenings

PANDROL brand spring rail clips have been for many years the principal fastening used on running lines in the UK. From a maintenance point of view their main characteristic is that once installed it is neither necessary nor possible to adjust the clip to restore toe load. The assembly must however be inspected periodically for wear and damage. The most significant part of the assembly in this respect is the insulator.

The insulating materials of any fastening system require periodical inspection and changing where necessary. Worn or missing rail pads must be replaced and insulators between clip and rail foot must be changed before there is any likelihood of steel to steel contact which will affect the security of signalling electrical circuits. The use of the rail insulation tester for checking and locating worn insulators in all tracks is recommended.

Normally all the insulating components on a sleeper are changed at the same time. This can be done one sleeper at a time by first removing the rail clips, and then raising the rail with a jack to release and replace the insulators and rail pad. If unduly rapid deterioration of standard components is observed (eg on track used by heavy freight vehicles, or on a sharp curve), consideration should be given to using heavy duty insulators, or if necessary using harder pads or clips which exert a larger toe load.

PANDROL brand rail clips must only be removed or inserted using the approved Panpuller or the rail mounted Permaclipper machine. The correct placement of the clip can be assisted by using the Pansetter. Where a corrosion bond has developed making it difficult to release clips in the normal manner use can be made of penetrating oil and special portable hydraulic declipping equipment.

SHC clips are removed using a special swan necked tool bearing against the side of the rail. When inserting SHC clips it is important to ensure that the rail pad and insulator are correctly fitted and that the toe of the clip is engaged onto the insulator before attempting to drive the clip home. During both removal and insertion of all clips staff are advised to wear safety spectacles.

Many fastenings of older Type 2 designs exist in the UK and they present various maintenance problems. Their common feature is the need to check and retighten the all-important holding-down screw at regular intervals. To ensure that this can be done the threaded surfaces must be kept lubricated. Where due to lack of spare parts or damage to a controlling surface of the concrete sleeper further maintenance is impracticable, individual sleepers may have to be replaced with sleepers of another type or the entire length may have to be considered for renewal.

6.4.3   Fastenings in Switch and Crossing Work

PANDROL brand rail clips are the fastening most commonly used in recently laid S&C layouts and require the same kind of attention as similar fixings on plain line.

Other FB fastening types existing in S&C include the spring clip and 'T' headed bolt. These involve the use of a box spanner in order to maintain the correct tension. As with all other threaded bolt fastenings periodic oiling and tightening should be carried out.

6.5   PERFORMANCE REQUIREMENTS OF FASTENING SYSTEMS

This section reviews the general requirements of fastening systems. The factors involved are then discussed in more detail in the later sections of this chapter.

6.5.1   Vertical support

The fastening system must support the vertical loads imposed by the wheels and transmit them to the sleeper or other underlying support with a minimum of impact, vibration or damage by abrasion, remembering that the vertical steady state wheel load is augmented by dynamic increments related to speed, track, and vehicle condition.

6.5.2   Lateral support

The fastening system must hold the rails to gauge and at the correct inclination, within the tolerances set out by the railway administration, in face of horizontal forces generated by vehicle curving, by alignment irregularities, or by wheelset hunting, and by the requirement to restrain the rail against buckling in hot weather.

At the same time it is highly desirable that the fastening should be able to accommodate some degree of deliberate gauge widening on curves.

6.5.3   Longitudinal stability

If the trackform is to be able to resist creep at all, it must allow small longitudinal movement of the rail under braking or traction forces and also it must draw it back to its original position when the load is removed. In many modern fastening systems some elasticity is provided in the connection between rail and baseplate. If this is not done (as for example in baseplated FB systems without pads, or chaired BH track) then the required elasticity must be provided by rotation and/ or delection of the sleepers. Even then, the rail fastening assembly must resist the substantial longitudinal (creep) forces generated by train braking or acceleration, and by temperature changes in the rail, otherwise the rail will creep relative to the sleeper.

6.5.4   Electrical Insulation

The system must have good insulating properties, both to enable the track circuiting to work and (particularly in non-ballasted or direct fastened track in high voltage DC operated areas) to prevent stray current leakage.

6.5.5   Universality

It must be possible to use with all kinds of sleeper with little or no adaptation, and with S&C as well as plain line.

6.5.6   Ease of installation

It should be quick and easy to install without the need for complicated tests or inspection.

6.5.7   Maintenance

Day to day maintenance should not be needed. The system as a whole should last the life of the rest of the trackform, but any sacrificial components should be easily renewable. Clips etc must be able to be removed and reinserted without their performance being downgraded, and they should be resistant to unauthorised removal.

6.5.8   Economy

Taken over the life of the trackform as a whole, the system should represent good value for money.

6.6   SPRING CLIPS

The functions of the rail clip are;
* to hold the rail upright against the outward forces applied by the wheel langes;
* to fix the rail longitudinally.

Both these functions demand that the clip should exert a substantial and controlled toe load on the foot of the rail.

Type 3 fasteners typically exert a toe load of from 700 to 1400kgf. The delection at the working toe load is normally between 10 and 14mm. This minimises the effect of manufacturing inaccuracies, and ensures that the toe load remains acceptably high even after considerable wear of the contact surfaces of the components has taken place.

Spring clips are made from a variety of carbon, chrome manganese or silicon manganese steels, generally dependent upon the spring section and local economics. They are normally produced by a process of heating, forming, quenching and tempering to a specified hardness range and dimensional limits, in order to ensure the correct load/ delection characteristics.

Fatigue tests on sample clips are generally required by the user to be carried out by the supplier as part of the Quality Assurance process, a pulsating dynamic delection being superimposed on the designed static delection of the clip.

Protective coatings are applied by many manufacturers as standard. For open track use these are generally paint films of various grades but specialised coatings for exceptionally severely corrosive environments are sometimes specified by user undertakings. These include oils, bitumen, and metallic coatings of various types.

The way in which the track assembly acts as a kind of longitudinal spring to resist traction and braking forces is discussed in Chapter 2. From that treatment it is evident that for ballasted, cross-sleeper track the magnitude of the force which has to be transmitted across the pad depends not so much on the characteristics of the pad, as on the elasticity of the ballast in relation to longitudinal movement of the sleeper relative to the ground, and is very variable.

Once the size of the force has been defined, it is necessary to consider how relative movement of rail and sleeper under that force is prevented. Clearly, in the usual types of modern fastening system, it is friction which performs this function. As always, friction is a function of the force acting perpendicular to the sliding surfaces, and the coefficient of friction between them. As long as the wheel is over or close to the baseplate the rail seat reaction is such that sufficient restraining force will be developed with a very low coefficient of friction. However the rail seat reaction is absent at the mid-point of the space between the inner axles of bogie vehicles, whilst if as is commonly the case, the trackform is longitudinally fairly springy, the longitudinal force at that point will be only slightly less than it is under the wheel, and here the restraining force depends on the toe load of the clips.

It can be shown from the analysis in Chapter 2 that for a train of 25 tonne axle load wagons, and a rail/wheel coefficient of friction of 0.3, the force exerted across the rail pad during heavy braking may reach 8.65kN. (This is quite close to the value arrived at if it is merely assumed that the brake force is distributed uniformly over the length of the train.) If as is the case for a modern clip assembly, the total toe load exerted by the two clips at each rail support point is say 20kN, the system will be free from creep provided that the effective coefficient of friction between rail and pad and/or between the pad and its support (ie sleeper or baseplate) is 0.433 or more.

This analysis only works if the sleepers are bearing solidly on the ballast at all times. If there are voids present, or if the dead weight of the rails and sleepers combined is less than the uplift force associated with the precession wave, then at some points between the wheels there will be an air gap between the sleeper soffit and the ballast, so that the ability of the sleeper soffit to transmit longitudinal force into the ballast will be nil, and the sleepers will be forced against the crib ballast. Since the cribs can only resist the longitudinal force by mobilising passive earth pressure, which implies some slight movement of the sleeper, creep is inevitable in these circumstances. These very high braking forces only occur when trains are moving very slowly, but the analysis indicates clearly the reasons why creep commonly occurs near places where trains regularly come to a stand. The problem appears limited to braking but similar effects have been observed under the very high tractive efforts produced by multiple locomotives on some high duty freight lines elsewhere in the world.

6.7   RAIL PADS

Although it is one of the less obtrusive parts of the fastening assembly, the pad is without doubt the heart of the system, and an understanding of its functions and behaviour are the key to any understanding of how rail fastenings work. The essential functions of a rail pad are:

(1)   To cushion effects of vertical loading and impact loading.

This has two aspects:

   (a) by providing a conforming layer between the rail and sleeper the pad ensures even pressure between on the rail seat area;

   (b) by acting as a spring the pad reduces the transmission of vibration and impact from the rail into the sleeper.

(2)   To act as a longitudinal spring to assist in the distribution of traction and braking forces along the rail. This feature is as has been seen not essential for ballasted track, but is vital for non ballasted track forms, particularly paved concrete track.

The pad must at the same time be stiff enough to prevent excessive rail rollover. It must also have other properties if the fastening system is to perform satisfactorily, including:

- Good electrical insulation
- Long life
- Consistency of properties (notably its hardness) over a wide range of operating temperatures
- Good resistance to abrasion, and to fatigue deterioration under reversals of principal stress

- Resistance to deformation and/or any tendency to walk out from under the rail seat under load

- Resistance to the effects of moisture, ozone, ultra-violet light, hydro¬carbons or other railway related chemicals.

The most common materials from which pads are made include:

- Natural and synthetic rubbers, either solid, or with some means of aeration, such as an admixture of cork granules;

- Thermoplastics, eg Ethyl Vinyl Acetate (EVA), High Density Polyethy¬lene (HDPE), or Polyurethane.

The surfaces in contact with the rail and either the sleeper or baseplate may be either plain and smooth, or configured in some way (typically either grooved or dimpled).

6.7.1   Pad design to reduce vibration and the effect of impact

EVA, HDPE and Polyurethane have proved very satisfactory materials where axle loads are high and speeds low, as on many heavy haul railways. Where speeds are high (say above ll0km/h) then impact and vibratory forces generated by wheel and rail head irregularities become damagingly severe if EVA or HDPE pads are used. To counter these effects pads of much greater resilience and lower vertical stiffness have been developed. The difference in stiffness is illustrated in Figure 6.7. The difference in the behaviour of the system under impact loading, as between a plain EVA pad 5mm thick (material A in Figure 6.7) and a configured rubber pad 10mm thick (material C in Figure 6.7 is shown in Figure 6.8. This result was obtained on a laboratory impact rig developed by the Battelle Institute in USA, in which a mass is dropped freely onto a rail seat assembly. Its effect on the sleeper is measured and different pads can be compared.

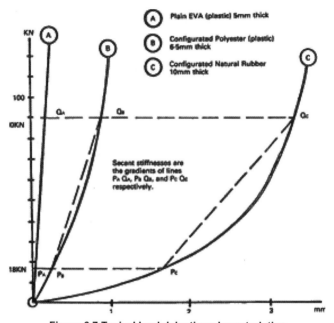

Figure 6.7 Typical load delection characteristics

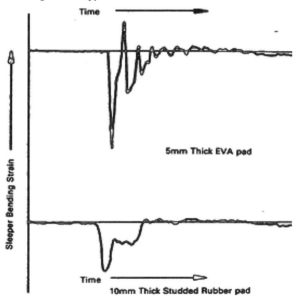

Figure 6.8 Impact effects with stiff and soft pads

The equipment is tuned to produce a first half cycle bending strain which compares with typical signals measured in track. The amplitude of the cycle is controlled by the height from which the mass is dropped, and the time history by varying the resilience of an isolating resilient pad interposed between the body of the mass and its striking head. To simulate the effect of heavy axle loads, a new development of this equipment involves the introduction of a means of applying a dynamically isolated static load of up to six tonnes upon which the impact is superimposed.

The effect of increasing the resilience of the rail pad is clearly shown. With the less-resilient EVA pad, the strain signal shows the sleeper being subjected to a sharp reversal, and resonating. With the more highly resilient pad, the amplitude of the initial strain is sharply reduced, and no reversal occurs.

Tests on materials and shape confirm the general statement that impact attenuation is primarily a function of dynamic resilience. Natural rubber provides the most resilience with the least damping. However, as with all elastomers, it is necessary to shape the surface of the pad to produce dynamic resilience and in practice, for railway use, this cannot be achieved adequately at less than about 10mm of thickness. To allow the very rapid deformation and recovery necessary to respond to high frequency forces, a maximum of force free area has to be provided while the rate of recovery depends upon the resilience of the material. The objective is to relect the energy back into the rail with a minimum of absorption or damping because absorbed energy is largely converted to heat at the rail/pad and pad/concrete interfaces where it causes depolymerisation and break-down of the pad material.

As a consequence of such work as this the use of soft pads has become increasingly widespread on high speed railways. Rail pads used on the dedicated high speed passenger train routes of the TGV in France, and the Shin Kan Sen in Japan have pad stiffnesses of the order of 90kN/mm. British Railways are now installing pads with a stiffness of 55kN/mm and improved impact attenuation properties. Combining such 'soft' pads into fastenings which may also need to carry heavy (30 tonne) axle loads is an area of continuing research, as is the optimising of pad properties.

6.7.2   Relationship between vertical and longitudinal stiffness of rail pads

Most elastomers do not strictly obey Hooke's Law, and they tend to get stiffer with increasing stress. It is however possible to apply the principles of simple elastic theory, as a reasonable approximation of the behaviour of such a material, and as a first step it is usual to express the modulus of elasticity (E) as derived from the gradient of the straight line QP in Figure 6.6 where Q and P correspond to the lower and upper limits of the working loading range of the pad. On this basis pad elastomers have E values from say 370N/mm2 down to around 40N/mm2. These values are extremely low compared with normal structural metals (E for rail steel for example is around 200,000N/mm2). At the same time the Bulk Modulus of an elastomer is abnormal when compared with that of a material such as steel. (Bulk Modulus is defined as the volumetric strain resulting from three mutually perpendicular and equal direct stresses and is equal to the reduction in bulk suffered by a solid body when immersed in a liquid under pressure).

The Bulk Modulus of an elastomeric solid is broadly comparable with that of other less "springy" solids such as steel, so that when stretching or compression takes place, the resulting volumetric change is much less than would be expected from the amount of stretch etc observed, and it is often said that the volumetric compressibility of elastomers is virtually zero.

Two consequences follow. The first may be expressed in simple terms and has already been mentioned. It is that when a rail pad is compressed, since the volume remains unchanged, the rubber has got to go somewhere. If the pad is smooth surfaced, then the only place for the rubber to go is outwards at the edges, and this is extremely difficult for the rubber since it is closely confined above and below. The necessary movements damage the rubber by abrasion and heating, and the rate at which the material can move being limited, the apparent stiffness of the pad increases. If the surfaces of the pad are configured, either by grooves or by pimples the distances which the rubber displaced by the compression has to travel are much less, so that abrasion and heating will be less, and the dynamic stiffness of the pad will be lower. A further advantage of configured pads is that because the actual area in contact with the rail and/or sleeper is less, the apparent stiffness of the pad will be lower than the stiffness of a smooth surfaced pad made of material having the same modulus of elasticity.

The second consequence is to do with an aspect of elastic theory known as Poisson's Ratio. This is the ratio of lateral strain to longitudinal strain in a body subjected to an axial load. The value of Poisson's Ratio depends upon the relationship between the Modulus of Elasticity and the Bulk Modulus, and whereas for steel or similar materials it is around 0.25, for elastomers it approaches very closely to the value 0.5 (NB it cannot quite reach 0.5 because that would imply that the Bulk Modulus was actually infinity, which is of course impossible). For a Poisson's Ratio of about 0.5, it can be determined from the laws governing the behaviour of elastic bodies that the Shearing Modulus (N) will be about one third of the Modulus of Elasticity (E).

Given this knowledge, it can then be shown that the longitudinal stiffness of the pad, expressed as force per unit longitudinal delection of the pad, will be about one third of its vertical stiffness, expressed as force per unit compression.

The vertical load delection relationship for a given pad can be used to estimate its probable longitudinal stiffness.

It is emphasised at this point that when considering ballasted track, the dominant inluence controlling the force across the rail pad will be the sleeper/ ballast interface, rather than the properties of the pad, particularly if the pad is stiff (eg EVA or HDPE) and the ballast soft. On the other hand, if the trackform is non-ballasted, and particularly with Paved Concrete Technology (PACT), the only element providing longitudinal elasticity will be the pad, and then the relationship quoted above becomes all-important.

6.7.3   Stresses in Pads and Strength Requirements for Pad Materials

Rail pads are subject to a combination of shear forces due to traction or braking, and compressive forces due to clip toe loads and rail seat reactions. It is important to draw attention to the stresses in the pad resulting from these force combinations, and to draw conclusions about the requirements which a good quality pad must fulfil in terms of the strength of the material from which it is made.

It is first necessary to understand something about the effect of the shear loads on the pads. It is noted that the result is to produce tensile and compressive stresses in directions oblique to the lines of action of the shearing forces, called principal stresses. The planes across which these stresses operate are called the principal planes.

If the direction of shear reverses, clearly the stress across each of the principal planes will also reverse, and it can be seen how the potential for fatigue arises from luctuating shear loads.

By applying a pressure to the top and bottom faces of the pad at the same time as the shear along the same surfaces, the principal stresses will still be there, but their value will change, and the orientation of the principal planes will change. If in fact the compressive stress is large enough, both principal stresses can become compressive.

The values of the two principal stresses are given by formulae which can be found in any standard textbook on Strength of Materials. These when applied to the information about longitudinal forces developed at rail fastenings, and rail seat reactions, contained in Chapter 2, enable the principal stresses in the rail pad to be worked out for any particular loading case.

There are a number of possible loading cases, eg:

(1) Toe load only (this is the base case with no train present)

(2) Toe load plus rail seat reaction (RSR) (ie under traffic but

without traction/braking)

(3) Toe load plus RSR plus traction

(4) Toe load plus RSR plus braking

(5) Toe load plus traction only

(6) Toe load plus braking only

These cases need to be considered for both Heavy Axle Load (HAL) and passenger stock, and for a range of pad stiffnesses in order to gain a picture of the range of variation in principal stress. They show that under extreme braking loads (those which occur just before the train comes to a complete stand under a full brake application under good adhesion conditions) one of the principal stresses generally becomes tensile. For any particular train loading, the tensile stress increases with decreasing pad stiffness, so that a soft, configured rubber pad suffers a higher tensile stress than an EVA or HD PE pad. It is not surprising that the HAL train produces much higher stresses than the passenger stock.

The results indicate that one of the characteristics which should be studied when specifying pad materials, is resistance to fatigue under repeated reversals of loading at the levels shown.

The features described in this and the preceding sections of this chapter go some way towards explaining the reasons why early formulations of rubber bonded cork pad ultimately proved unsatisfactory, after having apparently been adequate for many years. It would appear that what was strong enough under the types of traffic operating in the 1950's and early 60's, was unable to resist the larger stresses induced by the combinations of heavier axle loading and continuously braked freight trains of the later era.

6.7.4 Normal Provision of Pads for Rail Fastenings

There are a wide range of pads associated with the continued development and maintenance of concrete sleepers. It is usual to specify thicker pads for high tonnage and speeds up to 10.5mm in depth. Pads of different shapes are also available with lips on 'H' shapes.

6.8   SHOULDERS AND BASEPLATES

The clip or other means of holding the rail must be securely anchored to the sleeper or other supporting element. The means of doing this is either a baseplate or else, in the case of concrete sleepers particularly, some form of direct anchorage.

Shoulders and other anchorages for concrete sleepers Type 2 anchorages are normally of mild steel, and threaded at one, or both, ends. They are intended to be replaceable in the event of damage such as might be caused by a derailment, although such occurrences are unpredictable in their effect on concrete sleepers, and are at least as likely to damage the sleeper itself, as to be restricted to the fastening area and its components.

Having screw threads in the fastening assembly, they must be oiled regularly and protected; this also applies to any metal (such as a Tee-bolt head) which might be buried within the sleeper, if it is accessible to moisture.

 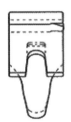 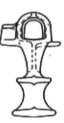 

Figure 6.9 Cast in shoulders

The fixed element referred to in the description of the Type 3 Fastening system may be removable (eg as in the FIST system) but more commonly takes the form of an anchorage or shoulder, which is cast into the sleeper at the time of manufacture. Since cast in anchorages cannot be renewed, it is important that they should last as long as the sleeper into which they are cast, and that they remain securely fixed in the sleeper throughout that life. Anchorages have generally been made from one of the ductile cast irons (typically malleable or SG iron), since these materials are resistant to corrosion and can be formed into complex geometric shapes. Experience now available, however, shows that steel pressings can provide technically acceptable alternatives to castings. The choice between malleable iron, SG iron, or steel pressing is consequently nowadays more one of economics than technology. Two modern configurations for cast in shoulders are shown in Figure 6.9.

6.8.1   Baseplates

For many years it has been the case that for any railway carrying more than minimal tonnages at the slowest speeds, baseplates are essential on timber sleepers, in order to prevent the concentrated loading produced by the rail seat reaction causing damage to the top surface of the sleeper.

A large range of baseplates manufactured in plain line with PANDROL brand rail clips is available. A typical baseplate is illustrated as Figure 6.10. Baseplates can be secured to sleepers by chairscrews and plastic ferrules in tapered holes, chairscrews and spring washers in parallel holes or with LOCKSPIKE baseplate fastenings in square holes. Baseplates are available for vertical rails or inclined at 1 in 20 and for application with or without the inclusion of a rail seat pads.

The soffit area of the baseplate must be calculated to provide sufficient distribution of loading not to overstress the surface of the timber and if several species of timber are in use the weakest of these is assumed for design purposes. The width of the baseplate is controlled by that of the sleeper on which it is placed. Plain line baseplates should normally be about 40mm narrower than the sleepers on which they rest. In S&C work narrower baseplates may be required to ensure that an obliquely placed plate does not overhang the edges of the bearer on which it sits.

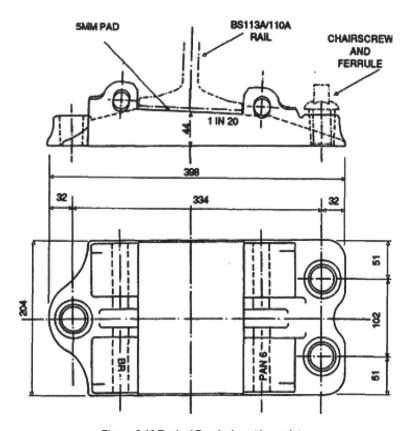

Figure 6.10 Typical Pandrol cast baseplate

Baseplates are sometimes used on concrete sleepers or S&C bearers, notably as slide baseplates for S&C with concrete bearers, and historically chairs were necessary if BH rail was carried on concrete sleepers.

Taking a world view, the vast majority of baseplates to BSS, UIC, or AREMA standards are made of rolled steel. UK engineers have found by experience that for most purposes a properly designed baseplate made from the relatively cheap 220 N/mm2 grey cast iron gives perfectly satisfactory results. Occasionally the extra strength provided by malleable or SG iron may warrant manufacture in such material.

6.8.2   Use of Rail Pads with Baseplates

Baseplate designs are available for use both with and without rail pads. Provision of a pad might be expected to give advantages in terms of improved resistance to rail creep and reduction of impact and vibration forces transmitted to the sleeper. There is however no evidence that baseplate assemblies on timber sleepers without rail pads are particularly prone to rail creep or that they are adversely affected by impact forces. In both cases in fact there are good analytical reasons (see Sections 6.5 and 6.6 above) why this should be so.

The absence of a pad eliminates a component, which can be seen as an economic advantage, but this policy results in the elimination also of a set of relatively cheap sacrificial components. If protection against traction current leakage is important (see below, Section 6.8), doing away with the pad also does away with a potentially important insulation stage.

Hence there can be no clear recommendation one way or another, and the designer must weigh the advantages and disadvantages against the prevailing circumstances of the particular case.

6.8.3   Baseplating for check-rails

When two FB rails each having a foot width of 140mm are placed close enough together to give a langeway clearance of 44mm, it is necessary to cut away part of the foot of the check-rail. This is a costly operation on continuous check-rails, and is a waste of rail. One method of overcoming this difficulty is to use a serviceable BH rail as a check with a FB running rail, the two rails being held in a special combined baseplate. Where it is required to use concrete sleepers, a special sleeper with holes for the baseplate fastenings must be used.

The ends of check-rails must be lared to terminate with an opening of 90mm to avoid impact with wheel langes entering the lange way. This is usually achieved by providing special baseplates for the first and last baseplate positions in the check. Obviously these will be of opposite hand at either end, and for this reason will be coded 'CCR' or 'CCL'.

A person requiring to ascertain whether a 'CCR' or 'CCL' baseplate is needed should stand so that the running rail is between him/her and the check-rail. The splayed end on the right needs a 'CCR' and vice versa. To identify a splay baseplate, place the baseplate so that the wider space between the running rail seating and check-rail seating is nearest to the observer. Then if the check-rail seating is on the right the baseplate will be 'CCR'.

6.9   ELECTRICAL INSULATION OF THE TRACKFORM

6.9.1   General

Where the rails form part of the track circuiting system, electrical insulation must be provided between the two strings of rail, and between the rails and the ground or other electrical earth.

Timber sleepers provide the necessary insulation of themselves, but special precautions are required where steel or concrete sleepers, or non-ballasted trackforms are used.

These take the form of:

a.   placing a pad of insulating material between the rail and the immediately underlying support. Usually the pad provided pursuant to the considerations discussed above will suffice.

b.   in addition to the above it is usually necessary to provide a separate small piece of insulating material to prevent electrical contact between the foot of the rail and any clip which holds it down, and/or between the rail and any metallic components embedded in, or otherwise attached to the sleeper.

6.9.2   Insulators

This small piece of material is referred to as the insulator. In many cases the insulator also serves as the sacrificial wear component to prevent mutual abrasion of the outer edge of the rail foot and the shoulder or other component which retains gauge in the system.

Insulators vary greatly in size and shape but are generally made from either nylon (plain or reinforced with glass fibre, referred to as GRN) or acetal resin, although many other materials have been tested. Their environment is invariably harsh, involving exposure to moisture, ultra-violet light, oil (both fuel and lubricating), detritus from trains, and any pollutants from the local environment, as well as the track forces discussed above. Hence much effort has gone into developing optimum designs, both in terms of shape and material.

The insulators used in connection with concrete sleepers are of three generic types, viz, Nylon, Glass Reinforced Nylon (GRN), and Composite (the last named incorporating a plate of malleable iron to protect the contact face which supports the clip, the remainder being made of nylon). The general shape of each of these three types is fairly distinctive, as will be seen from the illustration in Figure 6.11.

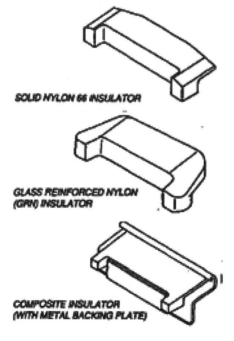

SOLID NYLON 66 INSULATOR

GLASS REINFORCED NYLON
(GRN) INSULATOR

COMPOSITE INSULATOR
(WITH METAL BACKING PLATE)

Figure 6.11 Insulators for PANDROL Brand rail clips

However, the differences in dimensions which distinguish one type of say, nylon insulator from another are quite difficult to distinguish and therefore the practice has grown up of distinguishing between different members of the nylon and composite families by means of different pigmentations. Different makes of composite insulator are distinguished by a number branded onto the metal part of the insulator. There is also a colour coded arrangement.

6.9.3   Earth Resistance of the Trackform

The effective resistance of an individual fastening assembly, dry, in factory condition, is extremely high, values of 109 Ohms being reasonably easy to obtain. However, when wet and dirty, the resistance drops dramatically. Electric current then travels through the layer of moisture (or moisture plus dirt) adhering to the surface of the insulators and pads, and the factor which is important under these circumstances is the length of the path along which the leaking electric current has to travel between the rail and the nearest "earth".

Excessive current leakage degrades the effectiveness of any track circuit which may be installed, and also, if electric traction is in use, it sets up alternative routes through which the return current may low to the busbars at the supply point. The former leads to false signal indications (eg the track circuit may fail to detect the presence of a train) whilst the latter leads to electrolytic corrosion either of track components or of lineside metallic structures or indeed the reinforcement of tunnel lining segments.

Traction return current leakage is a worse problem with DC than with AC traction, and it is worse on high voltage DC (1500 to 3000 volts) than with the low voltage (750v) used. Even on such lines however, corrosion of metallic track components can occur. It must be suspected wherever rapid loss of metal occurs in a wet and/or dirty location. For example what looks like abrasion around the points of contact between the rail and the clip on a baseplated timber sleeper in a wet area of a tunnel (this will usually not have an insulator), or perhaps also between the clip and the housing on the baseplate, may well in truth be electrolytic corrosion

Because electrolytic corrosion can very rapidly weaken a rail to the point at which it fails, and may also seriously damage any metallic structure involved, expert advice should be sought wherever it is suspected, and those responsible for track inspections in electrified areas must always be on the lookout for it.

Because high rail-to-earth resistance in polluted conditions depends on the length of the leakage path the, improved resistance is to be sought by such means as:|

- adding insulating layers

- enlarging pads to protrude beyond the edges of the rail seat

- special pads with moulded edges to form "drips" similar in shape and purpose to power line insulators

- baseplates made from insulating material

The signalling engineers, and if the track is electrified, the traction engineers also, are obliged to test track installations to check that its earth resistance is adequate for their purposes. The test usually involves several hundred metres of track, and the resulting resistance may be quite low. A conventional signalling track circuit of an older type will work even if the earth resistance of a kilometre of rail plus baseplates is as low as two Ohms. Assuming baseplates at 700mm centres, this implies an average resistance across individual baseplates of about 2900 Ohms, a figure which it is instructive to compare with the resistance obtainable under factory conditions.

If, however, the route were electrified, such a low resistance would encourage traction current leakage, and for higher voltage DC traction installations, and/ or where the railway runs in a tunnel, particularly one with reinforced concrete linings, much higher resistances are required, and a kilometre of track should then have a resistance of say at least 30 Ohms, requiring an average resistance of each individual baseplate of 43,000 Ohms. In bad environments such a value may be quite difficult to maintain.

The parameter used by electrical engineers in the measurement of earth resistance of the track has the units Ohms-kilometres (NB NOT Ohms per kilometre). This is because the average resistance of a length of track is determined by the resistance of a large number of resistances in parallel. It should be noted that if the track resistance over one kilometre is found to be say 20 Ohms so that the track resistance may be said to be 20 Ohms-kilometres, then the resistance of ½ km will be 40 Ohms, of 2km 10 Ohms, and so on, which is the inverse of what may be expected.

CHAPTER 7

SLEEPERS

7.1   INTRODUCTION

Successful sleepers must fulfil the following functions in the track:

a) They must spread the wheel loads over a large enough area of ballast to ensure that neither the ballast nor the underlying formation material are overstressed;

b) They must hold the rail to the correct gauge and inclination, within the specified tolerances;

c) The rails must not roll over sideways under traffic loading;

d) The track must not move sideways under either centrifugal or thermal forces and must resist buckling;

e) The sleepers must not move longitudinally under traction, braking, or thermal forces;

f)   The sleepers must electrically insulate the two rails from one another.

This chapter describes how concrete, timber and steel sleepers are designed and manufactured and maintained in order to meet these criteria.

The term "sleeper" is extended in this chapter to include switch and crossing bearers. Timber bearers (deeper and wider material, usually hardwood) have been widely used to support switch and crossing layouts but the obvious advantages of concrete in stability and durability led to the experimental introduction of concrete S&C bearers in the early 1970s. The performance of these layouts and continuing development work has led to the recent adoption of concrete bearers for standard use in the UK.

Steel channels can also be used to support S&C work, in a manner analogous to steel sleepers.

7.2   TYPES OF SLEEPERS AND THE MATERIALS USED IN THEIR MANUFACTURE

a)   Sleeper Types

The rails can be supported either longitudinally or transversely. Longitudinally arranged supports are usually called "rail bearers" or sometimes "wheel timbers" or "waybeams". They are not used in ballast nowadays, but on bridges or in other situations where non-ballasted track is to be found, they have a place. Since they do not of themselves resist overturning, lateral distortion, or gauge spreading forces very well, the detailed design of the track form has to take account of these forces.

The term "sleeper" is usually applied where the two rails are supported transversely. The supports under the two rails are usually joined together across the track, either by making the member of the same material right across (usually called "monobloc") or by having separate blocks under each rail, usually joined by a tiebar (twin block sleepers – sometimes known as duo-block).

b)   Materials

The traditional timber sleeper was generally acceptable to most railway administrations up to the Second World War, although its limitations in terms of quality and durability were recognised.

Even today it has advantages in some situations where such factors as good resilience, ease of handling, adaptability to non-standard situations, or electrical insulation are overriding factors, and both hardwood and softwood sleepers, almost invariably fitted with chairs or baseplates, are still used on UK mainlines and London Undergound

Steel sleepers have been used for many years, particularly in countries where termite attack of timber is a problem. Developments have taken place to improve aspects such as shape, resilience and electrical insulation including a verity of housings.

The development of concrete sleepers dates back to before the 1939-45 war. At that time, prestressed concrete was in its infancy and the early trials were with plain reinforced concrete. These were mostly unsuccessful, the exception being the French design of twin block sleeper, whose development started in the 1920's.

In Britain, the difficulties of importing timber during the war, led to the development of the prestressed monobloc concrete sleeper, initially for BH rail. The introduction of FB rail and CWR, promoted further developments to suit present day traffic requirements.

The type of concrete sleeper mainly used in UK is the prestressed, pretensioned, monobloc type, made by the long-line process, and either fitted with chairs or baseplates or with bolted or cast-in fastening systems, as described in Chapter 6. A typical prestressed, pretensioned monobloc concrete sleeper is shown in Figure 7.1.

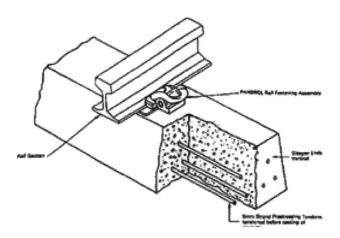

Figure 7.1 Pretensioned prestressed concrete monobloc sleeper

In France, the twin block reinforced sleeper was developed by stages to the RS type E which, by 1952, was in full production on SNCF. Further development of the design has continued since, and they are in virtually exclusive use on SNCF, including the TGV routes.

Their use has also been extensively promoted worldwide. The design is shown diagrammatically in Figure 7.2.

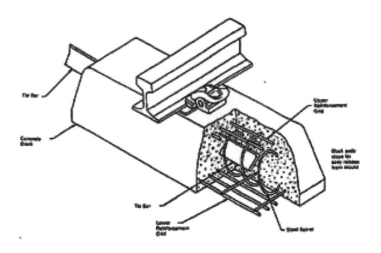

Figure 7.2 Twin-block sleeper in plain reinforced concrete

In Germany a number of prestressed concrete sleeper designs had been developed by 1955 and used two methods of prestressing, viz, pretensioning against the mould with anchors on tendon ends, and post-tensioning where the sleeper is demoulded when freshly made, and the prestressing applied after the concrete is fully matured. These methods will be discussed later.

The principle of construction of a prestressed, post-tensioned monobloc concrete sleeper is shown diagrammatically in Figure 7.3.

A feature of concrete sleepers is the need for electrical insulation of the rail fastenings. This is because although concrete of itself is not a good electrical conductor, the sleeper cannot always be relied upon to maintain electrical isolation of one rail from the other owing to the possibility of insulation breakdown between the rail clip anchorages and the reinforcement.

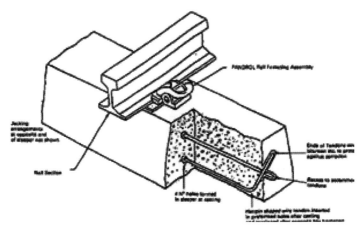

Figure 7.3 Post-tensioned prestressed monobloc concrete sleeper

c)   Sleeper Type and Spacing

As a result of close scrutiny of engineering standards in pursuit of cost effectiveness, allied with increasing knowledge about the performance of material under present day traffics, most railway authorities adopt a track category matrix. The matrixes adopted normally relate to speed, axle weights and gross tonnage over a route. The specification usually prescribes the type and material of sleepers and their contour line spacing which ranges from 600mm to 750mm. Also where these are curves sharper than 800m radius and speeds greater than 200km/h the lower values of sleeper spacing are used.

d)   Track gauge

Gauge in the UK is measured between the vertical faces of the running edges which, in the case of new rails, are nominally 14mm below the crown of the rail, and in vertical S&C work the gauge will be taken at that level. Due to the combination of the crown radii and inward inclination of the rail axes, the depth below the plane of the running table at which the gauge measurement is taken on plain line track, is 16mm as shown in Figure 3.1. Standard gauge is 1432mm through vertical S&C and in some UK CWR plain line laid in 1970 - 1990 where modern concrete or wooden sleepers and fastenings were used.

Most modern forms of CWR, (FB rail on concrete sleepers), on most types of jointed plain line track, and in S&C of inclined design, gauge as measured above, is 1435mm.

e)   Gauge Widening

On curves below 200m radius, it may be necessary to increase the track gauge to ease the movement of vehicles around the curves. The amounts of gauge widening prescribed, are shown in Table 7.1. Gauge widening is not normally employed through S&C.

TABLE 7.1

Gauge widening and langeway gaps on sharp curves

Type of Track	Curve Radius (m)	Gauge widening (mm)	Flangeway (mm)
New	200-100	6	45
	99-70	19	57
Old BH and FB	200-140	6	50
	140-110	12	56
	below 110	19	62

f)   Continuous check rails

On passenger lines where the radius is 200m or less, the fitting of continuous check-rails is a mandatory requirement of Regulatory Authorities in the UK. Where gauge widening is employed, the lange way gap must be adjusted as indicated in Table 7.1. Gauge widening should be achieved by moving the inner or low rail of the curve. The check-rail must be continued for at least 10m beyond the limit of the curve being protected, and must not finish closer than 5m from a running rail joint. Details of baseplating and lares for check-rails are given in Chapter 6.

7.3 DESIGN AND STRESS ANALYSIS OF SLEEPERS

The designer of a completely new variety of sleeper will normally commence with a brief, provided by the client railway, which will probably specify such parameters as weight and speed of trains, curve design limits, sleeper length, type of fastening system to be employed, and perhaps even maximum weight of sleeper. From this data, the designer will first work out the likely maximum stresses at the critical sections of the sleeper. This process is defined as stress analysis.

This process is more or less the same, whatever material the sleepers are to be made from, and the calculations are carried out on a "steady state" basis; that is, it is assumed that both the moving loads and the sleepers themselves, behave as if the loads were actually static. The effects of track and wheel irregularities on the actual loads are allowed for by using appropriate factors. The following sections summarise this analytical process.

It may be observed here that the effects of vibration and suchlike, are specific to the type of sleeper being considered, and as such will be described in the sections concerned, below.

7.3.1 The loading on the sleeper

The source of the load on the sleeper is wheels of the train. As explained in Chapter 2, the combination of lexural rigidity of the rail and the springiness of the ballast under the sleeper, leads to a distribution of the wheel loads between several sleepers, so that even if the wheel is directly over a sleeper, only about half of the wheel load is actually transmitted to that sleeper. The force acting between the underside of the rail and the seating area of the baseplate or sleeper is called the rail seat reaction (RSR). Thus if the wheel and rail are perfectly smooth and even, one would expect for an axle load of 25 tonnes, an RSR of about 6 tonnes. However, irregularities of which the worst are hammer- blow from locomotives, wheel lats, and dipped rail joints, produce large dynamic increments, and to allow for this, the RSR is magnified several times. For instance, the design RSR for F23 concrete sleepers was 28 tonnes. This was reduced to 24 tonnes for the F27 sleeper, taking account of both the withdrawal of steam locomotives and the decision to increase the density of laying from 24 to 26 per 18m rail length. In more recent revisions the design RSR has been reduced still further, relecting the greater concern over reverse bending moments.

In addition to the vertical loads described above, lateral loading arises from centrifugal and steering effects on curved track, from vehicle hunting, and from the effects of isolated or cyclic track irregularities.

For steady state stress analysis and design of sleepers, the following cases are amongst those which must be examined to identify the worst stresses occurring at critical sections of the sleeper:

Case I   Equal RSR's acting alone

Case II   Equal RSR's with equal and opposite gauge spreading forces (ie, leading wheels on a curve at equilibrium speed)

Case III   Unequal RSR's with equal and opposite gauge spreading forces (ie, leading wheels on a curve - general case)

Case IV   Unequal RSR's with lateral force applied to the rail having the higher RSR (ie, trailing wheels, curve at high cant deficiency)

Case V   Unequal RSR's with high lateral force applied to the rail having the lower RSR (ie, derailment conditions)

Case VI   Overturning - all the forces applied on one rail.

7.3.2   Distribution of pressure on the ballast beneath the sleeper

Since the ballast under the sleeper is springy, a monobloc sleeper, like the rail, is a beam on an elastic foundation. If the ballast were to be packed evenly under the whole length of the sleeper, so that it behaved as a homogeneous, elastic, uniformly compressible solid, (ie, the delection of the top surface of the ballast at all points is proportional to the applied pressure), then impact loads of the magnitude described in the previous section would break the sleeper at its mid-section half way between the rails. To eliminate this problem, monobloc sleepers are always packed over an area limited to 375mm on either side of the centre of the rail and ideally there should be no pressure between the sleeper soffit and the ballast in the centre section. In fact, in designing concrete sleepers, it is assumed that the objective of packing 375mm each side of the rail is only partly achieved. The assumption made, is shown in Figure 7.4. The bending moments induced in the sleeper, and hence the performance of the sleeper in service, crucially depend on achieving this pressure distribution, and therefore in turn, on the faithfulness with which the staff responsible for the laying and maintenance of the track carry out their duties.

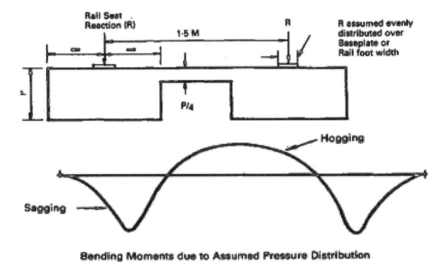

Bending Moments due to Assumed Pressure Distribution

Figure 7.4 Pressure distribution and bending moment diagrams for monobloc sleeper

This problem does not arise with twin block sleepers, since no significant upward load from the ballast can be imposed on the tiebar. If it is also assumed that the blocks themselves are completely rigid and that the underlying ballast is a homogeneous elastic solid, then the pressure gradient on the block/ballast interface is proportional to the angle through which the block is rotated by the algebraic sum of the moments acting on it. Since the tiebar is built into the block, the angle through which the block rotates is the same as the slope of the tiebar as it emerges from the block. This slope is proportional to the bending moment in the tiebar, the ratio between the two being a function of the length and lexural stiffness of the latter. These relationships allow the pressure distribution on the block to be calculated directly for any particular loading case, and "rule-of-thumb" pressure distributions are unnecessary.

For both types of sleeper, lateral wheel/rail forces which are not internally reacted against one another across the gauge (see Cases IV and VI above), are resisted by forces partly on the sleeper soffit and partly on the vertical faces of the sleeper or blocks, and the couple produced by these forces is counter¬balanced by a transfer of ballast pressure from one end of the sleeper to the other, as shown for a monobloc sleeper in Figure 7.5. Account must be taken of this in the bending moment analysis.

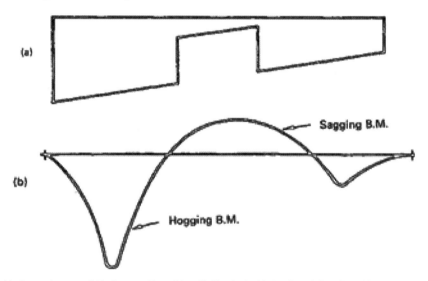

(a)   Assumed pressure distribution as per Figure 7.4 modified by wheelweight transfer and effect of lateral force

(b)   Resulting Bending Moment diagram

Figure 7.5 Pressure distribution and bending moments at high cant deficiency

The above considerations enable the steady state bending moment diagrams for any sleeper for any loading case to be computed, and also enable the forces on the rail fastening systems, baseplates etc, to be estimated. Typical bending moment diagrams for Cases I and IV for a monobloc sleeper are shown in Figures 7.4 and 7.5.

7.3.3   Sleeper length

The sleeper must be long enough to ensure that the resultants of vertical and centrifugal forces on the track, lie well within the middle third of the track form width. At the same time, the longer the overhang of the sleeper, the greater is the likely proportion of the wheel load to be taken outboard of the rail. The consequence of this is a larger rail seat bending moment. In the UK, with its restricted inter-track spacing, the use of shorter sleepers makes single track relaying easier, and this was a factor in the decision to shorten the most recent standard design of concrete sleeper (the F40) to 2420mm. Interestingly, on some European railways where there is not the same problem of restricted inter-track spacing, the tendency is in the other direction.

7.4   MONOBLOC CONCRETE SLEEPERS

a)   Types of Monobloc Sleeper

The difference between monobloc pretensioned and post-tensioned sleepers is related essentially to their method of manufacture. In pretensioned sleepers the bond between the tendon and the concrete is continuous throughout the sleeper, and provides the mechanism for the transfer of force from tendon to concrete. By contrast, in post-tensioned sleepers there is no bond between the tendon and the concrete at the time the prestressing force is established, and the transfer of force is effected through anchor blocks at the ends of the sleeper.

In the pretensioning method, a long line of moulds is used, usually more than 25 moulds per line, and sometimes up to 75. The wires are positioned and tensioned before the concrete is cast. After the concrete has hardened, the wires are cut between the sleepers, thus transferring part of the tension in the wires into the concrete as a compressive load.

The internal arrangements are shown diagrammatically in Figure 7.1.

Figure 7.6 Newly installed G44 concrete sleepers with Pandrol Fastclip® fastenings (Courtesy of Cemex Rail Solutions)

In the post-tensioning method, the sleeper is cast in a mould and is provided with a duct in which the post-tensioning tendon is inserted after demoulding. When the concrete has reached sufficient strength, the tendon is tensioned and anchored to the end of the sleeper by some form of locking system, onto bearing plates. Finally, the steel is protected from corrosion in the duct and at the anchorages by pumping in cement grout. The internal arrangements of a post- tensioned sleeper are shown diagrammatically in Figure 7.3.

As already explained, post-tensioned sleepers are extensively used in Germany and elsewhere, and they are manufactured and used by CIE. The method of manufacture is less capital-intensive than is the case for pretensioned sleepers, and lends itself to relatively small-scale production.

b)   Design Theory of Monobloc Sleepers

Steady state sleeper moments are calculated for vertical and lateral
loadings, using the bearing pressure diagram shown in Figure 7.4. The
design rail seat reactions (RSR) vary from 22 – 28 tonnes.

It has been found that monobloc concrete sleepers are subject to
bending in the opposite sense to that which the steady state analysis
predicts, and it has come to be realised that this is due to the sleeper
"ringing" like a tuning fork under vertically-applied impacts. Figure 7.7
shows the harmonics produced in a sleeper in this way.

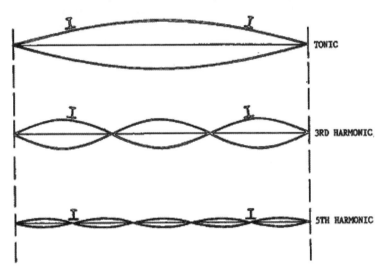

Figure 7.7 Observed vibration modes of sleeper under impact load

The third harmonic results in appreciable hogging bending moments at
the rail seat (always subject to sagging bending of course, in the steady
state). These harmonic effects are more pronounced at high speech As
a result, the largest hogging rail seat moments under normal traffics are
likely to be experienced under fast-moving passenger or lightly- loaded
freight vehicles, even though the largest sagging moments are produced
by slow, heavily-laden freight. This has been observed from strain-
gauged sleepers in track. A further problem is that very large impacts are
observed to be produced by out-of-round wheels. Due allowance

must be made for this effect when determining the design moments of resistance of the sleeper.

Having determined the required moments of resistance of the critical sections of the sleeper, it is possible to determine the cross-sectional dimensions of the concrete, the number and size of the prestressing tendons, and their location within the section. Information about the strengths of concrete and steel used will be obtained from the appropriate specification or code.

Codes for the design of prestressed concrete have been in existence from the early 1950s and have been used for many designs in building and bridge construction. The codes use limited concrete compressive and tensile stress levels at the working load which do not give an unmanageable loss of prestress through creep and shrinkage. There is also a requirement that the ultimate strength of a unit has a sufficient factor of safety, usually in the order of 2.0, over the working load.

Concrete design strengths are rarely more than 60 N/mm2 and until relatively recently, strengths greater than 40 N/mm2 were not codified. Concrete working stresses are in the order of 0.5 x strength at transfer and 0.3 x design strength in service.

The transfer strength is a minimum concrete strength, usually in the 30-40 N/ mm2 range, which results in acceptable loss of prestress for creep, shrinkage and elastic movement of the maturing concrete. The concrete continues to mature and gain strength after transfer, usually resulting in final strengths of 60-75 N/mm. Typical concrete stresses for pretensioned sleepers at transfer and in service are shown in Table 7.2.

Some form of accelerated curing is used in most climates so that transfer of stress to the concrete may be achieved on the morning following casting, ready for the production to be repeated each day.

Originally, all pretensioned sleepers were made with wire tendons but prestressing strands are now increasingly used because fewer strands need to be run out and stretched for a given force (6 No. 9.3mm diameter strands are equivalent to 18 No. 5.0mm diameter wires).

TABLE 7.2 Concrete Stresses in Pretensioned Sleepers

	Concrete Stress (N/mm2)			
	At Transfer	In Service		
Sleeper Type	Compressive	Tensile	Compressive	Tensile
Standard	12	Nil	22	4.5
Shallow Depth	17.6	Nil	24	4.5

Concrete sleepers are tested during the period of their initial development, and subsequently on a routine basis during mass production as part of the Quality Assurance process. The following tests are standard:

i.   Flexural strength of section under rail, in the positive and negative direction.

ii.   Flexural strength of mid point.

iii.   Bond test.

iv.   Torsion test.

v.   Side load (gauge spreading) test.

vi.   Fatigue test.

The routine quality control test usually involves the testing of a sleeper to check that it remains uncracked under a defined sagging moment. Finally, all sleeper production must be tested for soundness, finish, dimensional accuracy etc. Some reference has already been made to the history of the development of concrete monobloc sleepers. The earliest concrete sleeper still to be found in service in the UK is the Type E, made for use with BH rail and cast iron chairs.

All subsequent FB sleepers were designed to be used without baseplates, necessitating the use of Type 2 or Type 3 resilient fastenings and resilient rail pads.

Other developments in the 1960's, included the design of sleepers for use in unusual situations, including broad, shallow sleepers for use where available depth was limited, or a low soffit pressure was considered desirable, sleepers for third rail electrified track, and sleepers for use in check-railed track and in S&C work. Generally the length of sleeper was similar to the standard timber sleeper of 8ft 6 inches (2590mm). The width is in the range 264 – 290mm and the depth in the range 165 – 203mm. The latest designs are somewhat shorter to allow for gantry type relating and normally 2420mm.

On third rail electrified track, which makes up about 15% of the UK the conductor rail is supported at every fourth sleeper by an insulator. To support the insulator, the end of the sleeper must have a lat surface, and four 25mm diameter holes are cored out during manufacture to take the securing bolts.

7.5   PRETENSIONED SWITCH AND CROSSING BEARERS

The increase in use of concrete sleepers led to the development of a concrete bearer to support switch and crossing layouts.

In 1980, the forerunner of the present concrete bearers emerged with the use of a shallow-depth bearer supporting the baseplated switches, and a number of standard layouts have since been installed to this design. The UK now adopts the policy of using concrete bearers in preference to timber where standard designs can be accommodated and where site conditions are suitable.

Developments in concrete bearer designs were based on a variety of thicknesses from 160 to 205mm. Split beaters with splice joints are a common feature in modern installations.

Concrete bearers must be absolutely lat to ensure that the rails can be laid to the required geometry, and to ensure satisfactory driving and detection of switches, switch diamonds and swing noses. For this reason, the bearers are cast on their sides to assist correct location of the tendons in relation to the top and bottom surfaces of the bearers, and to enable full control over the thickness of the bearer by having these surfaces moulded.

Some notes on the installation of prestressed concrete S&C bearers will be found in Chapter 14.

7.6   TWIN BLOCK CONCRETE SLEEPERS (DUOBLOCK)

The twin block sleeper consists of two reinforced concrete blocks joined together by a steel tiebar. The tiebar is rolled from normal quality rail steel to provide resistance to both corrosion and fatigue. The concrete blocks are cast around each end of the tiebar which is surrounded by a steel spiral. Reinforcement grids are positioned above and below the tiebar, providing each block with the necessary bending resistance moment. The general arrangement is illustrated in Figure 7.2. A variety of types are available according to the design traffic loading. These sleepers have ranges of 2240 – 2410mm in overall lengths with block sizes from 680 – 840mm in length, 170 – 220mm in depth and interestingly a standard 290mm in width.

Twin block sleepers are produced in steel moulds in specially designed casting machines. The moulds have recesses which house the rail fastening insert. The reinforcement grids, steel spirals, and tiebars are then positioned and the moulds filled with a semi-dry concrete mix which is thoroughly vibrated. The filled moulds are removed from the casting machine, after which the newly-cast sleepers are instantly demoulded and the moulds returned to the production line for cleaning, oiling and re-use. After demoulding, the sleepers are cured for 20 hours indoors, prior to a 28-day maturing period outdoors and subsequent despatch to site.

Procedures for design, for development and production testing are similar to those used for pretensioned sleepers, and it may be noted that the great lexibility of the tiebar appears to obviate the problems with harmonics described above.

7.7   CONCRETE TRACK FORMS

Track can be supported or laid on a solid foundation which normally comprises of a concrete track structure. This can be constructed in a number of ways but there are generally only two types; the encasement of traditional concrete sleepered track or the construction of a pavement to which the rail is either attached or embedded. The UK generally adopts pavements for its main lines and tram systems.

a)   Monobloc Sleepers in non-ballasted track

Monobloc concrete sleepers are used,-either with prestressing or with plain reinforcement, and in various shapes, in non-ballasted track, by embedding the sleepers in a matrix of concrete which takes the place of the ballast. In some applications, the sleeper is manufactured with dowels in the form of short lengths of reinforcement projecting from the sides of the sleeper (these are often known as "hedgehog" sleepers). When the sleeper is embedded in concrete, the dowels ensure that the sleeper behaves monolithically with its surround. Another way of achieving the same end is to form holes in the sleeper, through which pass reinforcing bars laid parallel with the rails. If it is desired to ensure that the sleepers are free to delect vertically in response to wheel loads, it is possible to lay the sleepers in a trough of rubber or neoprene to prevent the sleeper and surround from acting monolithically.

b)   Embedded sleepers in non-ballasted track

Twin block sleepers can be adapted for use in non-ballasted track in the same way as mentioned above for monobloc sleepers, by embedding the blocks in concrete so that they act monolithically with the surround, or by encasing each block separately in a rubber or neoprene trough, usually termed a boot in this application. Typically, a thick pad of foamed rubber is placed in the boot under the soffit of the block to provide the necessary vertical resilience, and the walls of the boot are grooved to give a degree of lateral flexibility.

Monolithic embedment is used for such applications as LRT street running, in association with the use of grooved rails, whilst the booted form of construction is used widely where, as in tunnels, a positively fixed, and relatively maintenance free, track form is required. In the latter application it is possible to do away with the tiebar, as in the trackform used in the Channel Tunnel.

Figure 7.8 HLS Zuid, The Netherlands - finished track with the RHEDA 2000® system.
(Courtesy of Rail One)

c)   Paved Concrete Track

Paved Concrete Track (PACT) is a continuously reinforced, profiled concrete pavement laid by a purpose-designed and built slip-form paver. This runs on guide rails, which not only provide physical support for the plant, but also control the location of the completed concrete pavement alignment and elevation, in both absolute and relative terms. For this reason, the guide rails must be accurately positioned before concreting starts. The guide rails are usually the long welded rails which will subsequently be repositioned and used as the permanent running rails.

In their final position on the slab, the rails may be continuously-supported on 10mm thick strips of resilient elastomeric pad material, and secured in position using PANDROL brand shoulders, insulators, and rail clips. Alternatively baseplates may be used. These may either be conventional in design or may incorporate some form of resilience. The baseplates will be secured by bolts or screws driven into inserts of some kind which in turn, are placed into holes drilled into the slab after the latter has hardened.

A typical cross-section of a PACT slab is shown in Figure 7.9.

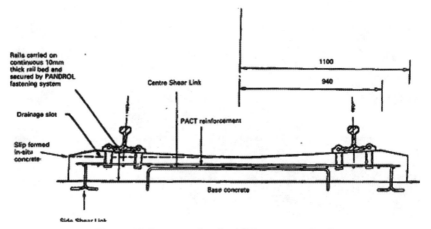

Figure 7.9 Cross section for 1432mm gauge track

The system is of particular use in tunnels, especially existing tunnels where the shallow overall construction depth may permit achievement of the increased clearances required for 25kV electrification, or for the passage of large containers, without significant alterations to tunnel profiles.

d)   Floating slab track

In locations where it is particularly important to minimise the intrusion of noise and vibration from railway operation into adjacent properties, resort may be had to what are described as loating slabs.

A second slab is added to provide channels in which to locate the rails.

In the case of Manchester Tram Track the Ri 59 (59 kg/m) grooved rails were manufactured in Luxembourg were supplied in 18 metre lengths and were pre-curved on site to suit the different radii, the smallest horizontal radius being 25 metres.

Once lined and levelled the rails are embedded in a pourable polymer which on setting maintains the track geometry and provides a resilient support for the rails resulting in much reduced vibration and noise levels. Permanent steel shutters are then fixed to the top of the RC slab and a second pour of polymer bulked with sand is made. The polymer has extremely good mechanical properties combined with a high electrical resistance and is designed to confine the traction return current to the rails. The polymer can be removed with a high pressure water lance to facilitate future maintenance. The grooved rails were laid vertical due to the difficulty in holding down the rails at an inclination around sharp curves. The required 1 in 20 inclination was achieved by subsequent grinding of the rail heads.

The 42 mm wide groove in the Ri 59 rail section together with a special wheel profile eliminated the need for gauge and langeway widening on tight curves which would have required several different rail sections with varying groove widths.

A further feature of interest from a track point of view is that in order to facilitate shared running between mainline and trams in outer reaches of the system, both the wheel tyres on the tram vehicles, and the S&C in the localities where shared running occurs, incorporate specialised design features. These consist of special raised check rails on point work on ballasted track, and the adoption of a modified version of the Standard P8 wheel profile in which the lange back is machined to reduce its width. This modification allows the wheels to negotiate 29 mm langeways in switch and crossing work on the street running section and as previously mentioned eliminates the need for langeway widening due to the reduced width of the lange itself.

In this form of construction, the trackform, which may be either ballasted (as in the LUL installations under the Barbican and elsewhere), or non-ballasted (in which case any of the non-ballasted trackforms mentioned above, may be employed), is placed inside a large and heavy precast concrete box. This box in turn is carried by soft resilient bearings on the tunnel invert.

The design of this type of slab track is outside the scope of this book.

e)   Tram track construction types

Tram systems are usually standard gauge 1432 or 1435mm and powered by 750V DC overhead catenary systems.

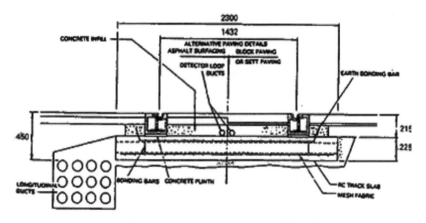

Figure 7.10 Typical construction details of Tram Track

A typical tram construction in a city centre usually involved removing part of the road surface which is excavated to a depth of approximately 0.5 m. After levelling and compaction a blinding layer of concrete is placed in preparation for a 225 mm deep structural concrete slab. Two layers of mesh reinforcement not only provide structural support but also act as a conductor drawing off stray current associated with the traction current return. This will reduce the quantity of stray current that would otherwise find its way into underground services resulting in corrosion or electrical interference. The structural slab is designed to span a 2 metre opening in the event of a collapsed sewer or fractured water main.

7.8   TIMBER SLEEPERS

Despite advances in the design and manufacture of concrete and steel sleepers, experience has shown that timber as a raw material for sleepers and bearers still possesses all the essential features to perform the function required, and indeed for some purposes, it is still preferred. Reasons for choosing timber rather than one of the other materials include, cost, resilience, corrosion resistance, workability, ease of handling, re-use potential, and insulation.

The sleepers used in the UK include:

- imported hardwood such as Jarrah
- imported softwood such as Douglas Fir
- homegrown softwood such as Scots Pine.

The standard dimensions of a timber sleeper are 254mm wide by 127mm thick in cross-section by 2600mm long, for plain line, and 305mm wide by 127mm thick in cross-section for Jarrah S&C bearers (Softwood timbers are 152).

a)   Life of timber sleepers

Jarrah is normally used in its seasoned, but otherwise untreated form, and in this condition is an extremely hard and durable timber with a life in track of up to 35 years or even more in a favourable environment. It usually reaches the end of its life due to weathering rather than decay.

Softwoods must not only be seasoned, but must also be impregnated with preservative before installation if they are to enjoy a reasonably long life in the extremely arduous environment presented by the railway.

Impregnation involves:

1)   incising the timber to assist penetration of the preservative into the fibres of the wood;

2)   placing the sleepers in a steel cylinder, from which air is first exhausted, and then pumping preservative (typically hot creosote) into the cylinder, under pressure.

Since even with incising, many types of softwood are only penetrated to a depth of 10mm or so, it is important that if sleepers are machined in any way (eg adzed or bored) after receipt, the exposed surfaces must be treated with creosote, otherwise decay will soon start from the machined surfaces.

Given that precautions such as this are taken, a life of from 15 to 25 years may be expected from softwood sleepers before the sleeper is so weakened by decay as to be unserviceable, and this life can be extended by the timely use of the "Timbershield" process (qv below).

The lives mentioned above are those that should be achieved provided that the sleepers are laid in a good environment and well cared for. Wet and/or fouled ballast, contamination from spillage of oil or chemicals, and neglect will all shorten the life of sleepers.

b)   Strength of timber sleepers

The mechanical properties of timber vary considerably with species, but generally speaking it is a relatively weak material, as the following statistics for Scots Pine (Pinus sylvestris) in the green (ie, unseasoned) condition show:

Modulus of rupture	40N/mm2
Mean modulus of elasticity	7500N/mm2
Compression parallel to the grain	20N/mm2
Hardness on side grain	2000 N Energy

consumed in bending to maximum load 0.07N.mm/mm3

The best Douglas Fir, and Jarrah, are of course likely to be considerably stronger than this.

From these figures it can be deduced that timber sleepers are something like six times as lexible as concrete monobloc sleepers. For a similar quality ballast bed, this implies that the reaction on the soffit of a timber sleeper is confined to a smaller distance either side of the rail axis than is the case with a concrete sleeper. At the same time, the rail seat reaction is distributed over the width of the baseplate rather than over the width of the rail foot. These effects combine to reduce the maximum sagging bending moment under the rail seat, compared with concrete, so that even allowing for the lower intrinsic strength of the timber, a timber sleeper in good condition on firm ballast is not likely to break under the heaviest load likely to be imposed by trains.

However, as the sleeper ages, it is likely to lose effective cross-sectional area at the rail seat due to wear of both upper and lower surfaces, and impact forces may increase due to loosening of fastenings. The intrinsic strength of the sleeper will decline due to weathering and decay, and the quality of the ballast support may deteriorate. Under these circumstances it is quite possible for the sleeper to become overstressed to the point of breakage, particularly under the high impact forces produced by high speed trains and/or heavy axle load vehicles.

c)   Uses of Timber Sleepers

Bearing in mind the above considerations, it is understandable that softwood sleepers are restricted in use to the slower and less heavily-trafficked portions of the network.

It is understood that there is no technical objection to the use of Jarrah sleepers in some higher categories of track, subject to approval in special circumstances, and Jarrah timbers are authorised to be used in S&C work in even the highest categories of track. The quality of sleepers and bearers laid down is specific in requirements for acceptable species, limitations on acceptable defects such as knots, wane, shakes, sloping grain, and gum pockets, on the proportions of sapwood to heartwood, tolerances on size, and on distortions.

7.9   STEEL SLEEPERS

Figure 7.11 Finished steel sleeper section

Figure 7.12 Improved rolled steel sleeper section

Developments have been made with plate and rolled tough designs and these are shown in Figure 7.11 and 7.12. The latter became the standard due to the lat rail seat area and the heavy version (W400) has the following dimensions:

Plate weight	22.82 kg/m
Rail seat thickness	8.75mm
Rail seat width	168mm
Leg thickness	6.75mm
Section width	260mm
Overall height	100mm

Moment of Inertia	409.47 cm^4
Height of Neutral Axis	65.67mm
Section Modulus (top)	119.27 cm^3
Section Modulus (bottom)	62.35 cm^3
Sleeper length for 1432 gauge	2450mm
Finished weight for 1432 gauge	68.24kg

Spade ends, slots etc., are formed in these sleepers by cold pressing.

a)   Properties of Sleeper Steel

Steel sleepers and bearers are normally produced from mild steel to either BS 500 or UIC 865-1 Specifications. Its typical chemical composition is:

Element Percentage by weight

Carbon	0.15-0.19
Silicon	0.20-0.30
Manganese	0.55-0.75
Sulphur	0.035 Max
Phosphorus	0.035 Max

Steel of this grade has a specified minimum tensile strength of 400 N/mm^2. On a basis of test results, the average strength of the steel as rolled, is 455 N/mm^2. For many environments, the coating of mill scale and rust which the sleepers acquire, provides its own protection against corrosion. The sleepers can however be given any protection specified, such as dipping in bitumen or oil, during the production process.

b)   Rail Fastenings and Insulation

Provision can be made for every major resilient rail fastening type to be adapted for use with steel sleepers. Baseplates or housings of rolled or cast steel can be welded onto the sleeper, or lugs can be pressed up, or holes punched in the steel through which hooks or bolts can be inserted.

Insulation is only necessary if track circuiting is to be used. If insulation is required, not only must the clip be insulated, but also a pad is needed under the rail. It may be noted that pads for steel sleepers have only the one function to fulfil, ie, to provide electrical insulation. They are not needed, as is the case with concrete sleepers, to reduce dynamic interaction between the rail and the sleeper.

CHAPTER 8

PLAIN LINE

8.1 INTRODUCTION

In designing the trace of a railway, the engineer has to satisfy the basic requirement of the users of a railway - that vehicles will not derail or overturn, and that journeys will be accomplished in comfort and/or without damage to merchandise. These aims are achieved at the time of construction by appropriate design of the longitudinal and vertical alignment of the track and of the relative levels of the two rails, and by the specification of a speed limit corresponding to the design values of these parameters. This chapter is concerned with these matters, and it is convenient to deal with longitudinal and vertical alignment in separate sections. Further sections deal with the process of maintaining and updating the plan alignment of existing railways.

8.2 LONGITUDINAL ALIGNMENT

8.2.1   Definitions

Most railways tend to follow a more or less sinuous course imposed by the need to avoid obstacles or to follow the general line of a river valley. In terms of rational design, this sinuous course is obtained by a succession of:

- Straight Lines,
- Circular Curves, and
- Transition Curves, in which the radius is constantly changing according to rules which will be defined later.

Curves are said to be "handed" according to whether the direction of travel deviates towards the left (a left hand curve) or the right (a right hand curve).

These elements are combined to produce:

- simple curves, in which two straights are joined by one circular curve of constant radius,

- transitioned simple curves, in which transition curves are added at one or both ends of the circular curve,

- compound curves, in which two straights are joined by two or more circular curves of differing radii (with or without transitions), but with the same hand of curvature,

- reverse curves, in which two or more curves of differing hand follow one another without any intervening straight track.

8.2.2   Centrifugal force and speed limits on curves

Since in general the cost of construction will be minimised by using curves with a small radius, it is necessary to define relationships between radius and speed which the designer can use, bearing in mind the need to ensure comfort and safety.

As a train goes round a curve, two force systems come into play. One of these concerns the forces necessary to rotate the vehicle in plan and to accommodate the differences in distance travelled by the inner and outer wheels, and results in lateral forces at the wheels of which the most important occurs between the outer leading wheel (sometimes called the guiding wheel) and the outer rail of the curve. These forces (which it is convenient to call "GUIDING FORCES") are dependent upon the mutual geometry of vehicle and track rather than speed, and the influence which they have upon the choice of cant in a particular location will be discussed later in this chapter.

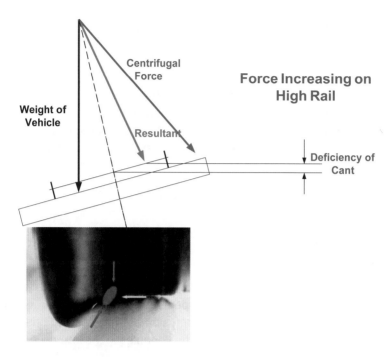

Figure 8.1 Forces associated with a Railway Vehicle Traversing Curved Track

The second force system is the centrifugal effect (see Figure 8.1). The cant required to neutralise the centrifugal force is given by the expression:

$$E_q = \frac{11.82\ V_e^2}{R} \quad\text{......................}\quad 8\text{-}1$$

where:

$V = Train\ Speed\ (km/h)$

$R = Radius\ of\ curvature\ of\ track\ (m)$

$E_q = Equilibrium\ Cant\ (mm)$

If the applied cant E_a differs from E_q, then the net centrifugal force is proportional to the difference between E_a and E_q. If the difference between E_a and E_q be defined as the "DEFICIENCY OF CANT", which is referred to as D, we have $E_a + D = E_q$ and we can write equation as:

$$[E_a + D] = \frac{11.82\ V_e^2}{R} \ldots\ldots\ldots\ldots\ 8\text{-}2$$

Since for reasons of safety and comfort it is necessary to place limits on E_a and D, equation 20-(2) implies limitations on V, so if we define the maximum allowable value of D as D_{max}, then the maximum allowable speed can be found by substituting in equation 20-(2) as follows:

$$[E_a + D_{max}] = \frac{11.82\ V_{max}^2}{R} \ldots\ldots\ldots\ldots\ 8\text{-}3$$

Where V_{max} represents the maximum train speed which can be permitted under the given conditions of radius and actual cant. If now we further define E_{max} instead of E_a the maximum amount of cant which can be applied at a particular location, we can insert this value in equation 8-3 to find the maximum speed potential of a given radius of curvature. It is important to avoid confusion between the various interpretations which are put upon the variables in the above equations so they are listed in detail below:

E_a = *Cant actually applied to the track (mm)*

E_{max} = *Maximum cant permitted to be applied to the track in a given situation (mm)*

D = *Actual deficiency of cant resulting from a given combination of train speed, radius, and cant (mm)*

D_{max} = *Maximum deficiency of cant permitted to be applied to the track in a given situation (mm)*

V_{max} = *Maximum train speed allowable for a given combination of E_a or E_{max}, with D_{max} (km/h)*

8.2.3   Evaluation of maximum allowable values of cant and cant deficiency

A train travelling at excessive speed round a curve may overturn. Quite obviously, whatever may be the value of cant deficiency required to overturn a train, this for practical operation is of little value, since long before that stage is reached, the limit of reasonable comfort either for driver or passengers, will have been overstepped, and any unsecured cargo in a freight train will be shifting laterally with consequent risk of damage.

Since passengers are the most sensitive part of the payload of a train, the measure of the maximum speed usually adopted is the degree of discomfort which it is considered that passengers will, for one reason or another, be prepared to accept.

Experimental investigations into these comfort limits have led the UK to adopt a maximum allowable cant deficiency of 150mm on plain line circular curves, where the track is CWR and there are no features contributing to lateral misalignments. In other circumstances values between 75 and 110mm are specified.

Another possibility is that the vehicle may derail, and we need to ask whether such an eventuality is likely in practice. The factor which determines whether a wheel will climb a rail is the ratio between the lateral force Y and the wheel weight Q. The combination of the guiding and centrifugal forces at the guiding wheel never reaches a value which is sufficiently large in relation to the wheel weight for it to be likely that derailment might occur. In fact derailment is more likely at speeds below equilibrium. Therefore for the purposes of determining maximum permissible curving speeds, the cant deficiency criterion is adequate.

For the reasoning already outlined, the applied cant E_a is controlled by the need to ensure passenger comfort and to avoid risk of derailment at speeds lower than equilibrium and indeed down to zero. The maximum allowable cant on curved plain line is normally specified as 150mm.

8.2.4 Choice of cant to apply

The formulas above in combination with the limiting values of track parameters enable the track engineer to determine the minimum radius he can afford to use in any given circumstances if he wishes to avoid placing any restraint on the speed of trains. Alternatively, if the topography is such that speed restrictions are inevitable, or an existing layout is required to be optimised, they enable the options to be studied.

For instance, if a curve of, say, 600m radius occurs for historical reasons on a route with a general line speed of 200km/h, the designer may have no option but to go for maxima of all parameters, and even then the curve would be restricted to no more than 130km/h. On the other hand if a curve of that same radius were situated in a suburban area where top speeds were only 80km/h, then the choice would be quite wide, between using a cant of 125mm, which is the equilibrium cant for 80km/h, and a cant of say 20mm which would give a cant deficiency of 106mm. In this case it would usually be wise to specify a cant such that cant and cant deficiency were about the same, say E_a equal to 60mm.

There are a number of reasons for this. Firstly, as will be seen when discussing transition curve design equalising E_a and D enables the shortest transition length to be designed. Secondly, having a high cant does not necessarily minimise side wear of the high rail. Indeed the contrary can be the case.

Similarly, if trains often travel at speeds substantially below equilibrium speed on curves below about 600m radius, damage to the low rail from crushing and metal flow can occur, and this would have to be taken into account if it were known that trains frequently travel much slower than the required line speed (eg because of conflicting movements at junctions, freight train movements, etc).

As the radius of the curve increases, the need for cant deficiency to enable the curve to be traversed at line speed will become less urgent, and at the same time, the significance of the possibility of high rail side wear (see Figure 8.2) and low rail metal flow decreases, so it becomes less important to keep the speed above equilibrium.

Figure 8.2 Example of Sidewear on Plain Line Track

8.2.5   Transition Curves

If the track on a simple curve is canted, and the adjoining straight track is transversely level, a cant gradient must be introduced across the join (ie the tangent point), and at that point the passenger will experience a sudden lurch. A considerable improvement in ride comfort, and also in safety, will be obtained if the circular curve is joined to the straight by a length of track having both a steadily changing radius and a matching cant gradient. Such a curve is called a *transition curve.*

In mathematical terms a transition curve is an example of a spiral. The key characteristic of the kind of spiral (the CLOTHOID) from which a transition curve is made is that when moving from a straight to a curve, the curvature (ie the reciprocal of the instantaneous radius) is proportional to the distance along the curve from its tangent point with the straight.

If such a curve is taken to its logical conclusion, it looks like Figure 8.3. The methods of calculation involved in designing and setting out a clothoid were somewhat cumbersome in the days when surveying and setting-out calculations were done by hand using seven-figure logarithms and for this reason the traditional method of designing a transition curve relies on the close approximation between the portion OA of the clothoid shown in Figure 8.3 and a nonspiral curve known as a CUBIC PARABOLA, as in Figure 8.4.

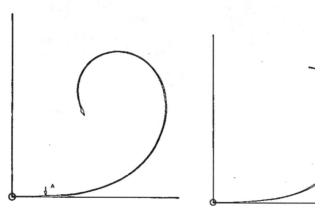

Figure 8.3 Clothoid Spiral

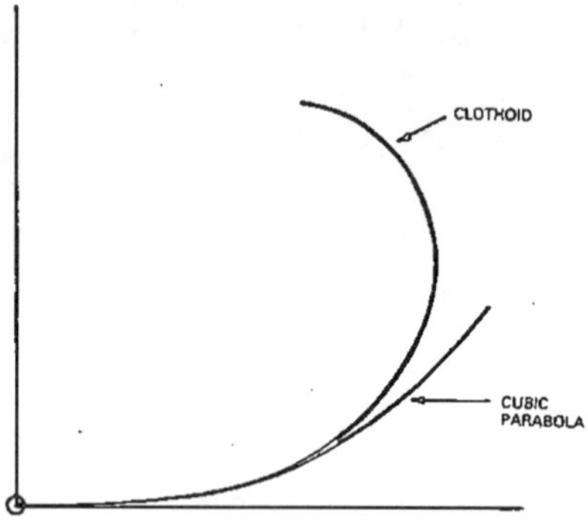

Figure 8.4 Comparison between Clothoid and Cubic Parabola

The vital difference between a clothoid and a cubic parabola is that whereas a clothoid goes round and round, a cubic parabola can never turn through more than a right angle. The inaccuracies involved in using the cubic parabola are small but perceptible, and are summarised as follows:

(a)   there is a difference between the length as measured along the curve and the length along the tangent. The clothoid calculation distinguishes between them, whereas the cubic parabola calculation assumes they are the same;

(b)   the calculated offset from the tangent to the cubic parabola at the curved end of the transition will be perceptibly less than the correct (clothoid) value if the transition is long and the curve of small radius, the difference being greater as the radius diminishes and as the cant gradient is made flatter;

(c)   the true radius of curvature at the curved end of the cubic parabola is appreciably less than the radius as calculated by the approximate formulae upon which the cubic parabola theory is based.

This last feature is probably the most significant one, and it has become more so with the tendency to allow higher values of cant and cant deficiency. It can be shown that the difference between the true radius at the curved end of the transition, and that calculated by the cubic parabola method, depends on the chosen values of cant, cant deficiency, and rate of change of cant, and not on the radius of the circular part of the curve. The effect is to build in at the join between the transition and circular curve, an instantaneous change of radius which can result in a perceptible lateral jerk.

This effect becomes increasingly serious as the radius diminishes. For this reason it is recommended that in designing and setting out new work, particularly if the track is sharply curved, and/or nonballasted and/or in tunnel, the clothoid calculation should be used.

Oddly enough, because the chords are set out along the curve, and not along the tangent, the "Hallade" realignment method actually sets out a clothoid spiral. Hence any curve which has been realigned by the Hallade method will actually be a clothoid and not a cubic parabola. Thus all the rules developed for dealing with transition curves by the chord and versine method can be used without amendment on layouts designed using clothoids.

Finally, the geometry and practical calculations for clothoid and cubic curves are similar.

It is relatively uncommon for the practising track to have to design and set out new work, so for this reason the calculation methods involved in designing clothoids and cubic parabolas from first principles are normally given. The two methods are there set out side by side. On the other hand, in designing realignments using the Hallade method or any computerised derivative therefrom, it is essential to know the length of the transition, and the following section is concerned with the determination of this parameter.

8.2.6   Determination of Transition Length

In determining the length of a transition curve, both the rate of change of cant, and rate of change of cant deficiency have to be taken into account. Whatever the curve radius and traffic needs, the designer should always try to keep these to a minimum compatible with the available space. Maximum and exceptional values should only be used when absolutely necessary. The effect of rates of change of cant and cant deficiency is one of the main aspects of transition curve design, as will be clear from the following clauses.

Distance-related cant gradient

To avoid the risk of derailment due to wheel unloading of freight vehicles with unsophisticated suspensions, it is usual to specify a limit on the cant gradient expressed as gain or loss of cant per unit distance along the track. Depending upon the rolling stock and speed requirements, this value is typically between 1:400 and 1:1000. For a cant gradient of 1:N:

$$L = E_a . N .... 8\text{-}4$$

On UK mainlines the limiting value is commonly 1:400.

Time-related Cant/Cant Deficiency Gradient

For passenger comfort purposes the rate of gain or loss of cant and/or cant deficiency has to be restricted. If:

- V is the train speed (km/h)

- λλE is the change of cant over the transition curve (mm)

- λλD is the change in cant deficiency over the transition curve (mm)

- L is the length of the transition curve (m)

then the time Δt, required for the train to traverse the length of the transition curve is given by:

$$\Delta t = 3.6\, L/V \; seconds \dots\dots\dots\dots \; 8\text{-}5$$

The rate of change of cant E' in mm/s is obtained by dividing $\lambda\lambda E$ by $\lambda\lambda t$ so that:

$$E' = \frac{\Delta E\, V}{3.6L} \dots\dots\dots\dots\dots\dots\dots 8\text{-}6$$

Transposing:

$$L = \frac{\Delta E\, V}{3.6E'} \dots\dots\dots\dots\dots\dots\dots 8\text{-}7$$

Similarly if the rate of change of cant deficiency is D', again in mm/s:

$$L = \frac{\Delta D\, V}{3.6D'} \dots\dots\dots\dots\dots\dots\dots 8\text{-}8$$

The values of V used in the calculation of L equations will usually be V_{max}, the maximum speed for the circular part of the curve computed from equation 8-3. The rates of change of cant and cant deficiency must not exceed the permissible maxima as given in Section 8.5. If the maximum values allowed in confined situations without special dispensation (i.e. 55mm/s) are substituted for E' and D' respectively, we obtain the relationships:

$$L = \frac{AE\, V_{max}}{198} \dots\dots\dots\dots\dots\dots\dots 8\text{-}9$$

or

$$L = \frac{AD\, V_{max}}{198} \dots\dots\dots\dots\dots\dots\dots 8\text{-}10$$

Whichever is the longer. Similar expressions can be obtained if D' and E' are 35mm/s (the 'normal' value) or 85mm/s (the 'exceptional' value), the coefficients then being 126 and 306 respectively.

The values of ΔE and ΔD are determined by the situation in which the transition curve is used, and the possible cases are described below.

Case 1 - Simple curve

Consider a transition between a straight with no cant, and a circular curve on which the cant is E_a and the cant deficiency is D. V_{max} can be calculated from this data using equation 8-3. In this case, ΔE is equal to E_a, and ΔD is equal to D. Equations 8-9 and 8-10 become:

$$L = \frac{Ea\ V_{max}}{198} \quad \text{...} 8\text{-}11$$

or

$$L = \frac{D\ V_{max}}{198} \quad \text{...} 8\text{-}12$$

whichever is the longer. In the event that:

$$E_a = D = 0.5E_q$$

equations 8-9 and 8-10 become, by a further substitution from equation 8-2:

$$L = \frac{V^3_{max}}{33.5R} \quad \text{...} 8\text{-}13$$

This expression gives the shortest possible transition curve length for a given combination of V_{max} and R, and it is valid provided that the EQUILIBRIUM CANT is not more than twice the PERMITTED MAXIMUM value of either the cant or the cant deficiency, whichever is the less. A further check is necessary to ensure that the distance related cant gradient does not exceed the maximum permissible (see equation 8-6).

Figure 8.5 Curved Track on High Speed Line (HS1) in Kent, UK (courtesy LCR Ltd.)

Case 2 - Compound and Reverse curves

Consider two adjacent curves of radii R_1 and R_2 with applied cants E_{a1} and E_{a2} and cant deficiencies D_1 and D_2. If the curves are of the same hand, these quantities can all be considered to be algebraically positive, and there will be one spiral connecting the two circular curves. If they are of opposite hand, then there will be two spirals, one to change the radius of the first curve from R_1 to infinity, and a second to change it from infinity to R_2. They will have a common tangent point. One set of values (say R_{i5} E_{al}) will be positive and the other negative. In view of these differences the calculations for compound and reverse curves will be set out separately.

Case 2a Compound curves

Let R_x be less than R_2, and E_{a1} be greater than or equal to E_{a2}. Then the difference in cant AE in going from one curve to the other will be:

$$\Delta E = E_{a1} - E_{a2} \dots\dots\dots\dots\dots\dots\dots\dots 8\text{-}14$$

Similarly, assuming that D_1 is equal to or greater than D_2:

$$AD = D_1 - D_2 \dots\dots\dots\dots\dots\dots\dots 8\text{-}15$$

From equation 8-2, we have for curve (1):

$$E_{q1} = [E_{a1} + D_1] = \frac{11.82\,V_{max}^2}{R_1} \dots\dots\dots\dots 8\text{-}1$$

and for curve (2):

$$E_{q2} = [E_{a2} + D_2] = \frac{11.82\,V_{max}^2}{R_2} \dots\dots\dots\dots 8\text{-}$$

By subtraction and substitution:

$$E_{q1} - E_{q2} = \Delta E + \Delta D = \frac{11.82\,V_{max}^2\,(R_2 - R_1)}{R_1 R_2} \dots\dots 8\text{-}18$$

Applying equation 8-7:

$$E' + D' = \frac{11.82\,V_{max}^3}{3.6\,L} \cdot \frac{(R_2 - R_1)}{R_1 R_2} \dots\dots\dots\dots\dots\dots 8\text{-}19$$

Hence the transition length L is given by:

$$L = \frac{3.283\,V_{max}^3\,(R_2 - R_1)}{(E' + D')R_1 R_2} \dots\dots\dots\dots\dots\dots 8\text{-}20$$

The shortest possible transition length is obtained if the rate of change of both cant and cant deficiency both have their maximum permissible values. If these are taken to be 55mm/s, equation 8-20 becomes:

$$L = \frac{V_{max}^3\,(R_2 - R_1)}{33.5\,R_1 R_2} \dots\dots\dots\dots\dots\dots 8\text{-}21$$

A further check is necessary to ensure that the distance-related cant gradient does not exceed the maximum permissible (see equation 8-4). The procedure for the design of the shortest possible transition curve in a compound curve would be along the following lines:

Step 1

Decide upon E_{a1} and D_1 for the smaller radius curve, and compute the design speed V_{max}.

Step 2

Compute E_{q2} for the larger radius curve (see equation 8-17). The difference between E_{q1} and E_{q2} gives $(\Delta E + \Delta D)$ (see equation 8-18). For the shortest transition curve, ΔE and ΔD must each be equal to $0.5(\Delta E + \Delta D)$. This enables E_{a2} and D_2 to be worked out.

Step 3

Compute L from equation 8-21. Check that N>400 from equation 8-4. If N<400, recalculate L from equation 8-4 to complete the design.

If for any reason it is essential that the rates of change of cant and cant deficiency are different, ΔE and ΔD are estimated according to the desired requirement, and E' and D' are worked out by proportion (eg if D' is to be 3/5 of E', and E' is required to be 55mm/s, then D' will be 33mm/s), and the appropriate values substituted in equation 8-20 in Step 3.

Case 2b Reverse Curves

In general, the two spirals required to connect reverse curves must be treated separately as detailed in Case 1 above. However it isdesirable for passenger comfort that the two spirals should be geometrically similar, in which case the rates of change of cant and cant deficiency on the two spirals will be the same (ie $E'_1 = E'_2 = E'$ and $D'_1 = D'_2 = D'$).This assumption enables the two curves to be treated as one spiral of length L, and we can write, as in Case 2a;

$$\Delta E = E_{a1} + E_{a2} \ and \ \Delta D = D_{a1} + D_{a2}$$

Then, using equations 8-16 and 8-17, equation 8-19:

$$E' + D' = \frac{11.82\ V_{max}{}^3}{3.6\ L} \cdot \frac{(R_2 + R_1)}{R_1\ R_2} \ldots\ldots\ldots 8\text{-}22$$

If as previously we assume that D' and E' are both 55mm/s, equation 8-22 becomes:

$$L = \frac{V_{max}{}^3\ (R_2 + R_1)}{33.5\ R_1\ R_2} \ldots\ldots\ldots\ldots\ldots 8\text{-}23$$

This gives the shortest possible aggregate transition length between two reverse curves, and is as previously subject to the check on distance related cant gradient.

The computational steps are similar to those outlined in Case 2a, with the addition that because there are two spirals, L must be shared between curves (1) and (2). The lengths of the individual spirals are obtained from equations 8-11 and 8-12. This also locates the point of contraflexure. It is perhaps worth emphasising that although the two spirals must have the same rate of change of cant and the same rate of change of cant deficiency, and these must be equal to one another for minimum transition length, this does not imply that on each curve E_a and D are equal to one another. For minimum transition length, AE and AD must be equal to one another, but the method enables a transition length to be computed even if they are unequal.

Summary of Transition Length Design

Equations 8-3 to 8-23, in combination with a knowledge of the limitations on E and D enable a maximum speed to be defined for any curve or combination of curves if the radius is known, and then enable determination of an appropriate length of transition curve for that curve. These formulae form the basis of the graphs which appeared in previous editions of this book. The universal availability of pocket calculators, however, renders the graphs obsolete and relatively cumbersome to use. They have therefore been omitted.

Similar principles can be applied if the radius, cant and transition length are defined in advance, to determine the allowable speed on the curve, or a curve can be designed for a specified speed, and the next section is concerned with this aspect.

8.2.7   Transition Curves of Limited Length

It sometimes happens that the length of a transition curve must be limited to some predetermined value. In this event, the task is to establish the values of E_a and D which will optimise the design speed V_{max} within the specified limitations on rates of change of cant and cant deficiency. There are two criteria which have to be considered. The first is the limitation on rate of change of cant on the transition curve, and the second is the limitation on the actual cant which can be applied to the circular curve.

8.2.8   Virtual Transitions

There are some situations where it is necessary to design a layout without transition curves. This arises most often where cant is zero, such as in switch and crossing layouts. In such cases the computational device known as the VIRTUAL TRANSITION is resorted to. This is based on the concept that inside the passenger compartment of a vehicle which is passing an instantaneous discontinuity of curvature, the change in cant deficiency associated with that discontinuity takes place in the time lapse between the arrival of the leading and trailing axle or bogie at the tangent point. Thus for bogie stock the virtual transition length is equal to the bogie king pin spacing. This is a special application of the general case of the transition with a short fixed length, discussed in the preceding section.

8.3   VERTICAL CURVES

Gradients are determined from the lie of the land and the performance characteristics of the rolling stock. This section concerns itself with the geometry involved where two gradients meet. The question of the effect of irregularities in longitudinal level is covered in Chapter 2.

Where two differing gradients meet, they are joined by a vertical curve, whose purposes are:

- to limit the vertical acceleration experienced by passengers, to a comfortable value,

- to limit the wheel unloading of freight vehicles,

- to prevent excessive compression of the springs of the inner axles of a bogie or locomotive having three or more fixed axles,

- to maintain under-clearance,

- to maintain over-clearance.

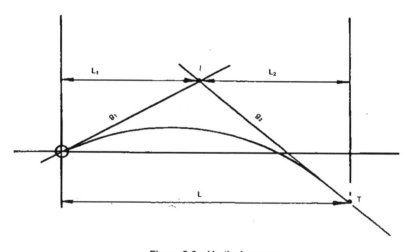

Figure 8.6   Vertical curves

8.3.1   Vertical Curves to limit vertical acceleration or wheel unloading

Figure 8.6 shows a vertical curve connecting two gradients forming a hump or summit, but the formulae derived are equally applicable to a valley, or to the join between two different grades in the same sense.

Let the grades be expressed algebraically as the TANGENT of the angle of slope (eg an upwards slope of 1:100 in the direction of travel would be +0.01, and a downwards slope of 1:80 would be —0.0125). Let the starting grade be **g₁** and the finishing grade be **g₂**, as shown in Figure 8.1, and let the speed of the train along the x-axis be **v**. Let the desired vertical curve joining the two gradients be the line OT. If the downwards acceleration of a particle travelling along OT is assumed constant and is limited to **f,** then the equation of the line OT will be determined from the equations of motion under constant acceleration as:

$$y = x.g_1 - \frac{fx_2}{2v^2} \quad\dots\dots\dots\dots\dots\dots 8\text{-}24$$

Since the slope of curve OT at point T is g_2 we have by differentiation:

$$g_2 = g_1 - \frac{fL}{v^2} \quad\dots\dots\dots\dots\dots 8\text{-}24$$

where L = the overall length of the curve, taken along the x-axis. Hence:

$$L = v^2 \left(g_1 - g_2\right)/f \quad\dots\dots\dots\dots 8\text{-}25$$

It will be seen from equation 8-24 that to a close approximation the departure of the vertical curve from the line of the gradient g₁ is given by:

$$y = \frac{fx_2}{2v^2} \quad\dots\dots\dots\dots\dots\dots 8\text{-}26$$

where x represents the distance along the gradient g₁ produced. Strictly speaking equation 8-26 defines a parabola, but by comparison with:

$$0 = \frac{T^2}{2R}$$

which is the commonly used approximation to a circle R where T is the distance along the tangent and O is the offset, for values of T which are small compared with R it is seen that to a close approximation, the vertical curve between the two gradients may be represented by a circle whose radius R is defined by:

(1)   *Three-axle bogie or vehicle*

Suppose that the rigid wheelbase is 5m long, and the maximum allowable spring travel is 40mm. The minimum allowable radius will be given by:

$$R = \frac{C^2}{8V} \quad\text{..} \quad 8\text{-}29$$

where C is 5.000 and V is 0.040, ie R = 78.1 m

If the positive and negative spring travels are the same, this figure applies to either a hump or a dip.

(2)   *Under-clearance*

This applies on a hump only. If the distance between bogie centres is 15.5m and the allowable loss of under-clearance is 25mm, then applying equation 8-29 the minimum radius R will be 1201m.

(3)   *Over-clearance*

This applies on a dip only. If the distance between bogie centres is 15.5m and the allowable loss of over-clearance is only 25mm, then the minimum radius is 1201m as in example (2).

Whilst the examples above are illustrative only, the limitations suggested are reasonable representations of real situations, and the minima arrived at are at least an order of magnitude less than the minima arrived at from passenger comfort considerations. Hence on a mixed traffic railway, vertical curvature will almost always be controlled by these latter requirements.

8.4 REALIGNMENT

8.4.1 General and historical

There are essentially two ways of setting about the task of realigning an existing railway. In the first method the railway is surveyed by theodolite traverse, with the details picked up from survey lines, or by tacheometry (these days this implies the use of electronic distance measuring equipment). The data collected is used to plot a survey from which the tangents are computed. The new curves are then redesigned on the principles set out above, and the new track set out by reference to the survey lines. Such a radical procedure is only necessary in comparatively rare cases such as those where slews of several metres are involved, placing the track on a new formation at a substantially increased track radius, usually to give a major improvement in speed limits. The techniques involved are those of conventional surveying, the principles of which are well known and do not need to be repeated here.

In the second method, the railway is surveyed by taking offsets from the chord to the running edge of the rail at the centres of successive overlapping chords laid out along the outer or high rail of the track which it is desired to realign. These offsets are usually termed VERSINES. The data is collected in the form of a listing of a more or less irregular series of versines. The realignment process consists of smoothing the versines in a more or less organised way, and determining from the new versines a series of slews, to which the track is then set.

The principle used is that if a chord of length c is stretched between two points A and C (see Figure 8.7) on a curve, then from the offset or versine v between the midpoint of the chord D and the point B on the curve half way between A and C, the radius of the curve can be calculated from equation 8-29. In this chapter versines, will be denoted in formulae by an italic capital V to distinguish them from 'speed', which when expressed in km/h, is denoted by a roman capital 'V'.

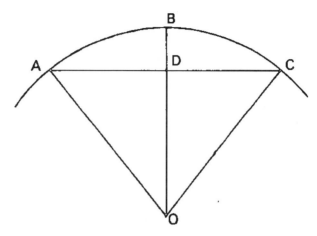

Figure 8.7   The Relationship Between a Versine, a Chord and a Curve

Another way of looking at the same piece of information is that if we regard the subchord AB, whose length is approximately c/2, as a base line, the location of point C, relative to the base line AB can be worked out by drawing a little circle of radius *V* around point B. Chord AC may then be drawn in as a line of length c, starting from A and touching the little circle at D. C will of course be at the end of that line. If we then regard subchord BC as another base line, the same process using the versine taken at point C will enable a further point to be located.

Obviously the attempt to carry out these operations at a scale suitable to a drawing board would be ludicrous, but the simplicity of the principles involved is attractive, and this led an engineer named Shortt in the early years of the 20th Century to devise two methods by which this difficulty could be overcome. The first of these was simply to plot the curve on drawing paper using the principle outlined above, but using one scale for the chords, and another, much larger one, for the versines. It was found that this method still had disadvantages, and he developed a further method, still known as "Shortt's No 2 Method", in which the slews are obtained directly from the differences between existing and amended versines, and the results are expressed directly on graph paper.

Later on, this method was adapted for use under circumstances in which drawing office facilities were not available, and became known as the "Hallade" method, because it was originally used to correct misalignments revealed by a patent track geometry recording machine of that name. Under that name the method was described in detail in a booklet entitled "Hallade Handbook - Theory and Design", published by the LMSR in 1946. The method was used manually and later on by computer developed algorithms through until the 1990's.

The whole process becomes more complex when considering compound curves as shown in Figure 8.8. In these cases, each curve, having a different radius will by definition have a different versine which must be designed in to give smooth transitions between the curves.

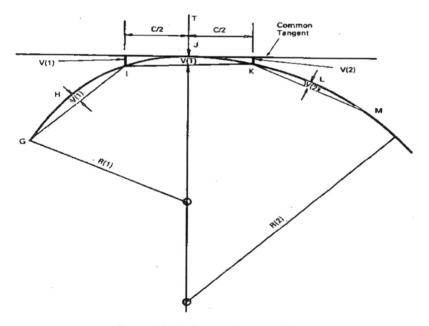

Figure 8.8 Versines on Compound Curves

8.4.2   Computers in Realignment

It will be evident that whilst quite sophisticated approaches to Hallade realignment can be developed from the basic mathematics of transition curves, their application in practice is likely to be dogged by the cumbersome arithmetic involved, which is both time-consuming and error-prone. As a result, designers are unlikely to have the time and/or energy to recycle trial realignments to explore, e.g., the effect of small relocations of tangent points.

This problem can be much relieved if the process can be computerised, and software is now widely available in railway offices to do this. It seems the more important, this being so, that designers should be aware of the effects of small changes in versine, and of the general implications of the process.

8.5 RECOMMENDED MAIN LINE CURVE DESIGN PARAMETERS

8.5.1   Basis for selection of limiting values

Limiting values for all parameters are determined from considerations of both safety and passenger comfort. Results from tests demonstrate that with standard coaching stock, acceptable standards of passenger comfort will be obtained provided the limits laid down are not exceeded. It has also been shown that provided passenger comfort limits are not overstepped, an adequate factor of safety against overturning, flange climbing etc will be ensured. The values indicated in the section relate to policies adopted on main line railways in the UK and are a combination of theory and extensive practical experience.

8.5.2   Use of maximum values

It will be noted that for plain line and for the through tracks of S&C, three limiting values are specified for each of the following parameters:

- Rate of change of cant

- Rate of change of cant deficiency

These limiting values are described as:

- NORMAL

- MAXIMUM

- EXCEPTIONAL

It will also be noted that for the absolute values of cant, negative cant and cant deficiency, exceptional values are quoted, which may be applied under certain specific conditions.

Wherever site conditions allow, curve design must be based on the NORMAL values for all parameters. Higher values up to the MAXIMUM may be used if the speed requirements cannot be met using the normal values. It should be recognised that only an authorised engineer or equivalent has the authority to permit the use of design parameters between the maximum and exceptional values, or to authorise the use of values beyond the limits shown at experimental sites. Such values will only be sanctioned when all the conditions at the individual site have been considered.

8.5.3   Record keeping

The values of all parameters and the design speeds obtained therefrom must be recorded on all realignment scheme plans.

8.5.4   Cant

Cant on curved plain line track shall not exceed 150mm except:

- UNDER EXCEPTIONAL CIRCUMSTANCES and on an experimental basis only, cant on plain line CWR may be permitted up to 200mm.

The cant on switch and crossing work shall not exceed 150mm except:

- On fixed obtuse crossings, cant must be RESTRICTED to 110mm.

Cant on curved plain line track or through any switch and crossing work in station platforms must not exceed 110mm except:

- UNDER EXCEPTIONAL CIRCUMSTANCES and provided that the platforms are to correct gauge and level and ALL Regulatory Requirements are met, cant ON PLAIN LINE ONLY in platforms may be permitted up to 130mm.

8.5.5   Negative cant

Negative cant can only exist in association with S&C in contra-flexure, where the main line is canted. In such cases, the negative cant on the track in reverse curvature shall not exceed 80mm. This applies throughout the S&C, and applies also to swing nose crossings, and to any adjacent plain line having the same conditions as the turnout, except that on fixed obtuse crossings, negative cant must be RESTRICTED to 65mm.

8.5.6   Cant deficiency

Cant deficiency on Plain Line and on the through tracks of S&C

On plain line CWR:

- Shall not exceed 110mm.

- EXCEPTIONALLY, rolling stock may run with a cant deficiency up to 150mm PROVIDED no S&C, catch points, adjustment switches, level crossings, longitudinal timbers, or other feature likely to contribute to a lateral misalignment is situated on the entry transition where 110mm cant deficiency is exceeded, or on the curve itself.

- ENHANCED PERMISSABLE SPEEDS, where tilting trains are operating on curves greater than 700m radius cant deficiencies between 225mm and 300mm may be allowed subject to special standards on such routes.

On plain line jointed track:

- Cant deficiency shall not exceed 90mm.

- EXCEPTIONALLY, passenger rolling stock may be permitted to travel at a speed which would result in a cant deficiency of 110mm.

Cant deficiency on the diverging tracks of turnouts and through curved diamonds

At switch toes:

- Cant deficiency on a curve of radius equivalent to that obtained by calculation from the versine on a chord 12.2 metres long placed symmetrically across the switch toes shall not exceed 125mm.

On the body of the turnout curve, including swing nose crossings, fixed common crossings and switch diamonds:

- Cant deficiency shall not exceed 90mm.

On fixed obtuse crossings:

- Cant deficiency shall not exceed 75mm.

Figure 8.7   Pendolino Tilting Train traversing reverse curves on West Coast Main Line (courtesy RailPictures.Net)

8.5.7 Cant gradient

It is considered neither necessary nor desirable to design cant gradients FLATTER than 1:1500, and this limit applies on plain line and switch and crossing work.

- On Plain Line and on the through track of S&C cant gradient shall not exceed 1:400.

- On the diverging tracks of turnouts and through curved diamonds Ccant gradient shall not exceed 1:400.

- Swing nose crossings MAY NOT be placed on a cant gradient.

- At Switch Toes cant gradient shall not exceed 1:600.

Note that if a cant gradient of 1:600 is applied on switch toes, the effect, combined with the dip of the wheel at the point of transfer, will be to produce a cant gradient of 1:400. This limitation therefore extends for 3 metres each side of the mid-point of the switch planing.

- At fixed common and obtuse crossings and switch diamonds cant gradient shall not exceed 1:1200.

Note that if a cant gradient of 1:1200 is applied on a crossing, the effect, combined with the dip of the wheel at the point of transfer, will be to produce a cant gradient of 1:400. This limitation therefore extends for 3 metres each side of the crossing nose.

8.5.8   Rates of change of cant and cant deficiency

Rates of Change of Cant and Cant Deficiency on curved plain line and on the through tracks of curved S&C:

- NORMAL conditions - Rate of change of cant or cant deficiency should not exceed 35mm/ second

- MAXIMUM conditions - Rate of change of cant or cant deficiency in confined situations must not exceed 55mm/second unless special dispensation is obtained.

- EXCEPTIONAL conditions - Rate of change of cant of 85mm/sec may be allowed and Rate of change of cant deficiency of 70mm/sec may be allowed.

- ENHANCED PERMISSABLE SPEEDS - where tilting trains are operating on curves the maximum design value for rate of change is 75 mm/second and exceptional value is 85mm/second. The rate of rotation of the car body can vary from 140 – 200 mm/second.

Rates of Change of Cant on the diverging tracks of turnouts and through Diamonds

The "Normal" and "Maximum" rates of change of cant quoted for plain line apply.

Rates of Change of Cant Deficiency on the diverging tracks off turnouts and through diamonds

At switch toes:

- Since they incorporate an angular discontinuity the rate of change of cant deficiency at switch tips is theoretically infinite, and this is accepted.

Through the body of turnouts and diamond crossings, including fixed common and obtuse crossings, switch diamonds and swing nose crossings:

- On vertical S&C - Rate of change of cant deficiency shall not exceed 80mm/sec.

- On inclined S&C - Rate of change of cant deficiency shall not exceed 55mm/sec.

8.5.9   Line speed limits and permanent speed restrictions

Line Speed Limits and Permanent Speed Restrictions (PSR) will be calculated in km/h and then converted to mph (1 mile = 1.609344km). If the calculation has been done on a basis of "NORMAL" rates of change of cant and cant deficiency, then the speed in mph should then be rounded to the nearest 5 above or below the calculated figure. After rounding, the rates of change of cant and cant deficiency must be recalculated, and must not then exceed the "MAXIMUM" limits quoted above. If the calculation has been done on a basis of "MAXIMUM" or "EXCEPTIONAL" values then the speed in mph must be rounded DOWN to the nearest 5mph below the calculated value.

CHAPTER 9

SWITCHES AND CROSSINGS

9.1 INTRODUCTION

The fundamental concept of any railway is the configuration of its trackwork. Plain line is a term given to stretches of open railway with simply one direction of travel in a longitudinal sense. There will always be a need to connect railways to each other, divert onto or cross other routes and this is done by the construction of switches and crossings (S&C). Each unit has a set of points which are usually electrically operated to switch from one track to another.

This chapter will cover the basic theory of S&C and explain the various components involved; there will also be a description of the various manufacturing methods used both in the past and in the modern railways in the UK.

Figure 9.1 Switch and Crossing layout on a High Speed Line (courtesy LCR Ltd.)

9.2   Crossing Design and Manufacture

9.2.1 General Description

A Switch and Crossing layout (S&C), however complicated it may appear, is constructed to meet two requirements, either singly or in combination. These are:

i.   The requirement for one track to cross another, and

ii.   The requirement for one track to diverge from or to merge with another.

Requirement (i) is met by the DIAMOND, whilst requirement (ii) is met by the TURNOUT. These are illustrated in Figure 9.2. Turnouts are sometimes called "leads", "single leads", or even "half-leads". Diamonds are sometimes called "diamond crossings" or just "crossings", but so far as possible the terms "turnout" and "diamond" will be used exclusively in this book.

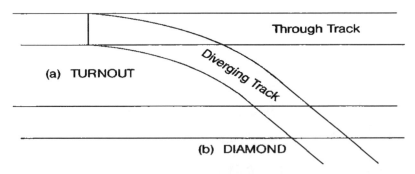

Figure 9.2 Turnouts and Diamonds

Any assembly of turnouts and/or diamonds is called a layout. S&C layouts are configured as "vertical" or "inclined" which indicates the orientation of the running rails. The symmetrical profiles of a flat bottom rail indicates that when its base is on a horizontal level the orientation is "vertical". This assists in geometry, movement of switches and the planning of the components. Inclined S&C is the track in its normal position, inclined at 1 in 20 as all plain line is configured. There have been many trials around the world of both types, but a key feature of pass through the rails of the track to be crossed. If the two tracks are now twisted round so that they cross at an angle other than a right angle, to become diamond shaped as in Figure 9.4(b), it will be seen that the L-shaped pieces change shape. One opposite pair join at an acute angle, and the other opposite pair join at an obtuse angle.

The two crossings having the L-shaped pieces joining at an acute angle are usually referred to as COMMON CROSSINGS. Alternative names are ordinary, acute, or vee. The two crossings whose L-shaped pieces join at an obtuse angle are called OBTUSE CROSSINGS. Alternative names for obtuse crossings are diamond, or K crossings. In practice, the short lengths of rail between the L-shaped pieces are extended and supplemented by other rails, the purposes of which will be discussed later in this chapter.

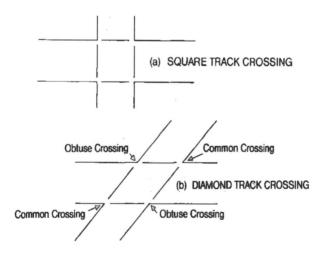

Figure 9.4 Common and Obtuse Crossings

It is convenient however to include here as Figure 9.5 a diagram of a typical diamond with the full complement of rails shown, together with their usual names. The student should study this and memorise the names of the constituent parts.

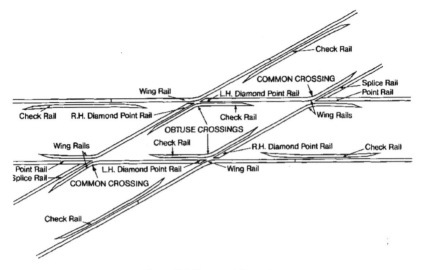

Figure 9.5 Diamond Crossing

An important point to note here is that for various practical reasons, the point rail of a common crossing does not terminate at a sharp point at the theoretical intersection point of the two gauge lines but is cut short to terminate in what is referred to as its BLUNT NOSE as shown in Figure 9.6. Similarly at an obtuse built up crossing, the wing rail does not bend at a sharp angle where the two gauge lines meet, but is radiused. The practicalities involved lead to differences between theoretically calculated and actual lengths of the various parts of the crossing, which have to be taken into account in designing and setting out layouts.

9.2.2 Crossing angles

The most important geometrical parameter of a crossing, by which it is usually described, is the angle at which the two rails cross one another. The method used to describe the magnitude of an angle in S&C work is called the CENTRE LINE METHOD (CLM).

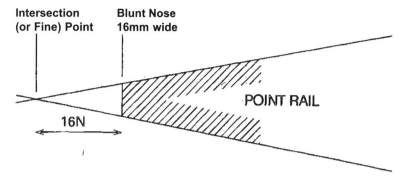

Figure 9.6 Relationship between Intersection Point and Blunt Nose

Curved Crossings

In the simpler designs of turnout, the common crossing is designed with both its legs and wings straight. However this places severe limitations on the speed potential of the turnout, and in modern UK practice the diverging rail is in certain circumstances curved up to and perhaps beyond the place where it has to cross the opposite rail of the through track. There is a need to design curved crossings, both common and obtuse when a curved track crosses any other track.

Geometry and Curved Crossings

In order to define a crossing angle for such an arrangement it is necessary to introduce the geometrical concept of the angle between a curved line and a straight line which crosses it. Consider first Figure 9.7(a). In this figure a straight line AB just touches a circle at point X. This line is called the tangent to the circle at point X. Note two important features here:

The direction in which the tangent AXB is pointing coincides with the direction in which a particle travelling round the circumference of the circle is pointing at the instant it passes X. The line connecting X with the centre of the circle O, is at right angles to the line AB. This radial line is called the NORMAL at point X.

Now let another line CD cross the circle at X, as in Figure 9.7(b). The angle between the circle and line CD at X is now seen to be equal to the angle between the tangent AB and the line CD, since the circumference is pointing in direction AB at X. Hence the angle we are looking for is angle AXC.

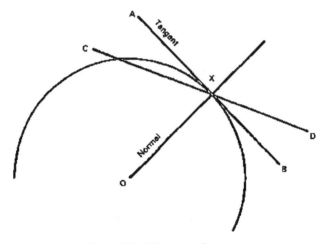

Figure 9.7(a) Concept of tangent

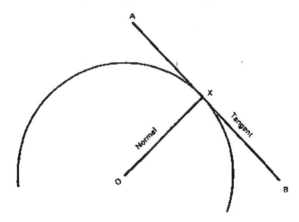

Figure 9.7(b) Concept of the angle normal between a straight line and a circle

However, since angle AXO is 90 degrees, we say Angle AXC = 90 - Angle OXC

Now, if we know:

- The location of the centre of the circle, O,
- its radius, and
- the location of point X,

we can calculate the direction in which the Normal OX is pointing. By comparing this with the direction in which CD is pointing, we can calculate angle OXC, and hence the crossing angle AXC.

In this construction the method of determining the angle between a straight line and a circle has been demonstrated. If the two rails which cross are both curved, the same principle applies, but now it is necessary to know the locations of the centres of both circles, and their radii, and the calculation involves finding the angle between the two normals at the point of intersection. Since the angle between the two normals is the same as the angle between the two tangents, the problem is then solved.

9.2.3   Manufacture of Built-up Crossings

Built-up common crossings are made from four pieces of rail as shown in Figure 9.5:

- the POINT Rail, which as its name implies, is the piece from which the actual tip of the crossing nose is formed,
- the SPLICE RAIL, which is joined to the point rail some little distance back from the nose,
- and two wing rails.

The hand of the crossing is determined from the relative positions of the point and splice rails. Viewing from the throat of the crossing, if the splice rail is to the left of the point rail, the crossing is said to be left-handed, and vice versa.

The four rails are first cut to length and the holes for the bolts are drilled at predetermined positions. The various rails are then bent, machined and fitted together. The process for the point and splice rails is quite complex as is shown in Figures 9.8, 9.9 and 9.10. Finally all the parts are fixed together with high tensile steel bolts or multi-groove locking (MGL) pins, with spacing (wing) blocks to form the flangeways.

Obtuse crossings are made from two wing rails and two check rails as shown in Figure 9.5.

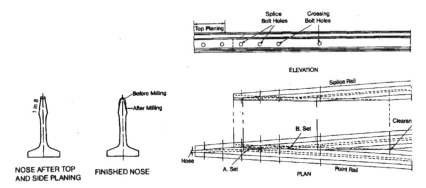

Figure 9.8 Section through Crossing Nose   **Figure 9.9 Plan and Elevation of Common Crossing**

9.2.4   Standard Built-up Crossings

Standard "vertical" BS113A crossings are made with all the running faces straight. If required to be installed in a curved layout, the curvature is achieved by curving the rails before assembly and selecting suitable baseplates as described later. The angle of a crossing is determined in principle by a combination of site circumstances and mathematical calculation.

For common built-up crossings of angles in the range of N between 4 and 8, standard parts are available. As the angle gets larger (i.e. N gets smaller),problems arise which get progressively worse as the N value decreases. When the N value drops below 6, it becomes necessary to angle the rail drilling in order to accept the through bolts.

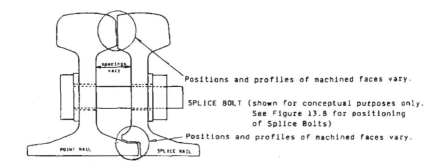

Figure 9.10 Diagrammatical cross section through a common crossing vee

If the N value goes below 4, standard parts are no longer available, so the cost of the installation increases greatly. Furthermore, if transverse supporting timbers are used, the angle between the rails and the timbers departs more and more from a right angle, and the baseplate detailing becomes correspondingly more and more awkward, particularly where the supports revert to conventional sleepers at right angles to the rails. In fact for N values below about three, it becomes normal practice to lay the diamond on longitudinal timbers. When this measure is resorted to, the timbers under one of the two tracks involved must be very short, and this raises other problems. Thirdly, as the N value decreases, hammer blow and vibration increase because the wheel tread is no longer supported as it crosses the flangeway gap from leg to wing rail and vice versa, and because the impacts produced by contact between the back edge of a worn wheel tyre and the rail head (the so called false flange effect) increase.

Hence the use of tracks crossing at angles giving N values less than 4 is to be discouraged, and when it is absolutely necessary, specialist advice needs to be sought, and cast crossings would probably be used.

Switch Diamonds and Swing Nose Crossings
When the N value of an obtuse crossing exceeds 7.5, other problems arise. Figure 9.4 shows that in a diamond not only do the two obtuse crossings occur almost exactly opposite one another, but also the wing rail of one crossing is the same shape as the check rail of the crossing

on the opposite side and parallel to it. For this reason it would be theoretically possible for the wheels of a wheelset which is free to yaw, to twist themselves around in the gaps between the left hand and right hand point rails of the opposite rails, so that the flanges could take the wrong side of the running-on point rail, thus causing derailment. For N values less than 7.5, and wheels of conventional diameters, this cannot happen in practice because the leading edge of the flange of the wheel reaches the running-on point rail before its trailing edge is released by the running-off point rail. Above this value wheels of relatively small diameter can be momentarily free to yaw, and could possibly take the wrong direction. For this reason, fixed obtuse crossings with N values greater than 7.5 on straight track are not allowed. If the track is sharply curved these problems start to arise at N values less than 7.5.

If the tracks must cross at angles flatter than those specified above, or if negative cant occurs through one track of the diamond, the fixed obtuse crossing must be replaced by an assembly known as a SWITCH DIAMOND, in which the point rails are driven like switches and close the flangeway not in use, thus providing continuous guidance to the wheel as it traverses the obtuse crossing.

Standard switch diamonds can be made at angles of 1 in 7.5, to 1 in 28. For N values larger than about 35, a fixed common crossing is no longer adequate because the wheel transfer area becomes too long and the nose too thin, and a SWING NOSE CROSSING must be used. In a swing nose crossing the vee of the crossing is designed to be driven like a switch, to lie against either wing rail, closing the unused flangeway gap, and presenting a continuous support to the wheel through the crossing.

Because there is no gap between the nose and the wing rail, the noise and vibration associated with fixed common crossings is reduced in swing nose crossings. Hence when S&C has to be installed in particularly noise sensitive areas, swing nose crossings with angles considerably sharper than the limit mentioned above may usefully be employed.

They may also be installed in very high speed lines either to reduce noise or to reduce maintenance cost, although it is observed that impacts on fixed crossings do not rise as speeds increase beyond 175km/h. Swing nose crossings of relatively small N values may also be found to be cost effective on routes having large flows of very heavy axle load freight traffic at relatively high speed.

Some indication of the requirements for the manufacture of switch diamonds and swing nose crossings are given below.

9.2.5   Other problems with Built-Up Crossings

It will be evident from the chapter that the manufacture of built-up crossings is a highly skilled task, involving a lot of machining and care in assembly. Moreover, when completed, the machining involved will have appreciably weakened the crossing rails, and the number of working faces, fasteners and other points of weakness leads to a short life and a considerable maintenance liability. Built-up crossings are not therefore cost effective in tracks taking fast and/or heavy traffic. Furthermore, it is difficult to make them strong enough longitudinally to resist the thermal loadings associated with continuous welded track, and they are therefore not usually welded to the rails on either side, even though the rail steels may be compatible from a welding point of view. The following sections describe alternative forms of construction which obviate some or all of these problems.

9.2.6   Part-Welded Crossings

A part-welded crossing consists essentially of the same four rails as a built- up crossing, and is usually made from standard rail. The assembly is however designed to be strong enough to take thermal loads and consequently it can be welded into CWR leaving only the flangeway gap as a source of wheel/rail impact. Compared with ordinary built-up crossings, and most AMS crossings (see below) this is a distinct advantage.

The vee of a part-welded crossing is prepared by machining two pieces of rail into a symmetrical straight splice with a weld preparation milled into the head and foot. The electroslag welding process is used under carefully controlled conditions to produce a continuous homogeneous weld without slag intrusions or porosity. The weld deposit is laid down by means of an automatic welding machine in which the top and bottom welds are done simultaneously so that distortion is kept to an absolute minimum. The weld material is carefully matched to the parent metallurgy in order not only to provide good penetration and freedom from cracks but also to give a hardness in the weld metal which slightly exceeds that of the parent rail. The extra hardness slightly improves the wear resistance at the nose.

Cast iron blocks are used to maintain the appropriate flangeway clearances through the crossing and to hold the vee and the wing rails together rigidly enough to resist traffic and thermal forces while at the same time offering a degree of resilience.

The crossing is held together by high tensile steel bolts, nuts and fitted cast washers or pre-tensioned MGL pins which give a higher bolt tension than is possible with conventional torque-controlled bolts.

The welded crossing is versatile in terms of both rail section and running edge geometry. Special angles can be produced without difficulty and leg lengths can be supplied either to length or over-size for cutting-in. The legs can be straight, or curved to suit main line or junction conditions. It is however difficult to curve the vee through the welded zone, which is longer than the length of a built-up splice.

9.2.7   Crossings in Cast Austenitic Manganese Steel

Manufacture of AMS Crossings

The metallurgical properties of austenitic steel are described in Chapter 3. A cast crossing in this material can look in plan very much like a built-up crossing of the same geometry, in that there are the same rail heads, and the same grooves through which the wheels can pass. However, there the resemblance ends. Firstly and most obviously, the whole is in one continuous piece of metal, hence the often used term "monobloc crossing". Secondly, it is hollow on the underside, as is shown in Figure 9.11.

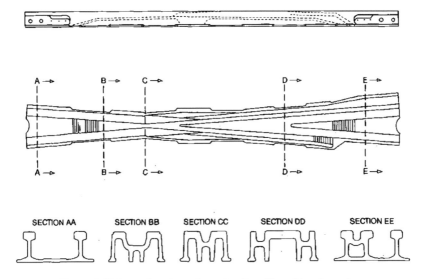

Figure 9.11 Cross Sections through a Cast Monobloc Crossing

It is made by pouring molten steel into a mould. The mould (which is itself contained in a heat resisting box), has the image of the external shape of the finished casting, only lasts for one cast, and is made from a permanent pattern. To form the hollow interior of the casting, a core is placed in the mould, and held in position by being attached to the lid of the box.

The steel is produced in an electric arc furnace to the specification laid down pearlitic steels which remain at their initial hardness throughout their lives.

The running table profile can be designed to support the wheel as it traverses the gap between the throat and the nose of the crossing, and this can reduce the impact and wear problems resulting from worn tyre profiles.

AMS crossings can, if required, be designed to incorporate direct fastenings to the substructure without baseplates. This is not favoured for high speed tracks because the rigidity of the casting results in a hard spot where unavoidable wheel impacts cause voiding.

They require fewer blocks or bolts to assemble and eventually maintain.

Several crossings can be combined (e.g. the obtuse and common crossings in a scissors crossover can be made from a single casting). The advantage of this in confined spaces is obvious.

On many railways, AMS crossings are work hardened before installation, notably by explosive hardening. This involves detonating a layer of plastic explosive applied to the traffic bearing surfaces of the component. More than one treatment can be applied. The process produces some very localised deformation of the metal, but a layer of hardened material results which can extend up to 25mm into the metal.

Limitations of AMS Crossings

In spite of its advantages, AMS is not a completely ideal material from which to produce crossings. Amongst its disadvantages, is the fact that the nature of the manufacturing process itself entails limitations on what can be produced. Since the casting must be water quenched, the longest crossing which can be produced is limited to the length of the available quenching bath. This problem has led to the development of so-called "cast centre" crossings in which the main casting is confined to the nose/throat area of the crossing. The non-running ends of the legs and wings are then bolted or pinned in position so that splayed, left-hand or right-hand or double parallel-winged crossings can be made from the same pattern. Because the casting is so short, it is possible to achieve slight variations in geometry without having to make special patterns.

In spite of rigorous quality control, it is not possible to eliminate manufacturing defects altogether from the large and complicated castings required for crossings with high N values, and this can result in premature removal, or the need to repair the crossing in track. Nevertheless castings are guaranteed against premature failure for three years. This is a feature not generally found in other components.

Cast crossings are of course very heavy, and it is sometimes very difficult to handle them in confined spaces or busy layouts.

Because of the limitations on heating already mentioned, AMS castings cannot be welded to any other rail, and indeed they can only be repair welded under strictly controlled procedures designed to limit heat input to a minimum. Consequently, at some point a fishplated joint is required to connect the AMS rail to the adjacent (pearlitic) BS 11 rail. Where AMS crossings are put into CWR layouts, this joint is fitted without an expansion gap.

Developments in cast crossing technology

Over recent years, a variation to the standard composition has been developed (reducing C to around 0.8% and increasing Mn to 14-17%) which retains its austenitic structure through normal cooling to room temperature, without quenching. With this steel, the restrictions on welding are not so stringent, and in particular, it can be welded to itself. This can be of value, particularly in that it makes it possible to produce cast centre crossings which are effectively "monobloc".

Flange Running Crossings

As mentioned above, crossings which intersect at angles approaching a right angle (at which N is equal to 0.5) present special problems because the wheel flange is not continuously supported as it passes from the vee to the wing rail or vice versa. Clearly if the crossing is a true right angle there will be no support at all as the wheel crosses the opposing flangeway, and the effect is the same as if there were a rail joint with a gap of 42mm. This exposes both the wheel and the crossing to very large impact forces, which materially shorten the lives of both. This effect can be minimised if cast crossings are used by making the flangeway groove much shallower than normal, so that the tip of the wheel flange actually rides on the floor of the groove as it traverses the flangeway gap. The floor of the groove is gently ramped up and down on either side of the intersection so that as little vertical impulse as possible is imparted to the wheel as it passes over the crossing. This device is much commoner on Light Rapid Transit (LRT) layouts than it is on heavy railways as in street running, the tracks have to follow street alignments which more often than not intersect at angles of nearly 90°, but it is equally valid as a device for both.

9.2.8   Check Rails and Flangeways in S&C

Check rails are provided opposite all fixed common crossings, and form part of the construction of all obtuse crossings. Their function is to control the alignment of the wheelset so that it is not possible for the wheel moving across the gap in the throat of the crossing, to strike the nose of the point rail. To do this the flangeway side face of the check rail must come into contact with the back of the wheel flange, before the swept contact area of the flange of the opposite wheel starts to encroach on the rail head profile.

Flangeway Width

The distance between the wheelbacks is nominally 1362mm, so that the wheelback with the wheel in its neutral position is 35mm from the gauge face of the rail. Since dimension B with the wheel in its neutral position on straight track is 8mm, the wheelset could theoretically be allowed to move by this distance before adverse contact was made, leading to a flangeway width of 43mm. However this would leave no allowance for tolerances, wear, or the effects of curvature, to allow for which factors, the standard flange way in vertical S&C is normally set at 41mm. It is further reduced in high speed checks to 38mm.

The critical dimension is not so much the flangeway width, as the distance between the running edge on the crossing side, and the working face of the opposite checkrail. This dimension must be maintained as nearly as possible 1392mm (1394mm for high speed checks). This is the reason why check rail supports on many railways are often connected across to the crossing rather than being fixed to the opposite rail, as is done in the UK. Flangeways of similar widths are provided between the vee rails and wing rails of common crossings, and between the wing and point rails of obtuse crossings.

Check rails have traditionally been made in the UK from machined sections of running rail, and fixed to the running rail by distance blocks and bolts, but some designs are now available which use UIC 33 section rail. Flange ways are flared towards the ends as described above for plain line check rails, the angle of flare and method of achieving it, depending on the details and circumstances.

Check Rail Height

Control over the alignment of the wheelset can be assisted by raising the top of the check rail significantly above the plane of the running rails, thus enlarging the segment of flangeback which is in contact with the check rail. Raised check rails are the norm in many parts of the world, and some administrations use specially rolled tall angle sections of rail for this purpose. UIC 33 rail is quite commonly placed in a raised position on a specially designed bracket. Raised checks are to be found in many places, although they are no longer being installed in new layouts.

Shared Running with Light Rapid Transit Vehicles

Very tall raised checks (up to 35 or 40mm above the running table) are used in some places to enable LRT vehicles with the small flanges required to fit into Ri section rails to run over S&C with 41mm flangeway gaps. The tyre profile used by such LRT vehicles incorporates two flangebacks, one to fit the grooved rails at low level, and one at 1362mm at a higher level, with a tapered transition section between them. No special modification of "heavy rail" vehicles is then needed, although the tall check rail may inhibit use of the shared track.

9.2.9   Timber and Concrete Bearers

The support for switch and crossing units is usually known as bearers which are longer and deeper than standard sleepers. They are made from timber (usually hardwoods) with cast iron baseplates or reinforced concrete with cast in housings.

The normal timbers used are Jarrah or Karri, of 307mm x 127mm section, or possibly softwood 307mm x 154mm, in lengths varying in 150mm steps from 2450mm to 6350mm, and then in 300mm steps up to 10250mm. Lengths of 2450mm are not used on third rail electrified areas as they do not allow sufficient space for the insulators. Timbers longer than 6050mm are unusual and generally require a longer lead time (up to one year) for ordering. Also available are timbers of different cross section to the normal, although these also have a long lead time.

On crossovers, through timbers are used throughout the crossing portion (5900mm long with a standard six-foot of 1970mm, or 6050mm in third rail electrified areas). These may however be split into two separate lengths broken in the crossover road four-foot every alternate timber, (leaving a maximum of 1500mm without a gauge tie) as long as the timbers supporting the crossing noses are not split. This method of timbering is not normally encouraged these days, and many designers would reject this arrangement for main line work. Separate timbers are used where tracks opening out exceed 900mm between running edges. This figure has to be increased to 1200mm when concrete sleepers support the separate tracks.

Timbered layouts where third rail electrification is present require a minimum of 545mm from the running edge to the end of the timber to allow sufficient room to position the insulators.

Concrete bearers are specially made pre-tensioned, pre-stressed beams. They are usually supplied in two standard depths, and in lengths with predetermined fixing positions to order. Standard bearers as illustrated in Figure 9.12 are available for many standard turnout configurations. Only the movable parts of switches need baseplates, the rest of the layout being fixed directly to the concrete in the same way as on concrete sleepers.

The design of concrete bearer layouts requires much more discipline than if timber is used because each bearer has a special mould plate which cannot be varied. Consequently the designer must ensure that changes in radius occur at the vee end of the crossing, and not at the intersection point, as is the traditional method.

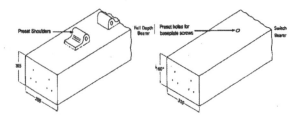

*This depth applies to bearers for full depth switches. For shallow depth switches, bearers 168mm deep must be used.

Figure 9.12 Concrete Bearers

It is usual to limit the length of concrete bearer to 4500mm and use splicing connectors (see Figure 9.13 below). Where greater lengths are required its delivery to site is usually in modular format using special tilting wagons. The lengths of bearers are often limited to 4600mm.

Concrete bearer crossover designs are available for many track intervals. If crossovers are required for intervals greater than 3500mm they can be made from two independent turnouts. Layouts with non-standard geometrical features are to be avoided with concrete bearers as the large number of specials required create great difficulty both in initial supply, and as regards spares. Complicated geometrical arrangements sometimes used to be best catered for in timber but many standard concrete bearer designs are available.

Figure 9.13: Split concrete bearer showing a splice connection
(Courtesy Steven Pearson)

Specialist baseplates are needed for S&C to cater for multiple rails.

Figure 9.14 shows vertical baseplates. VC and VD allow for wing rails. On inclined S&C, these are a far more complicated set of baseplates to enable the rails to be inclined at 1 in 20.

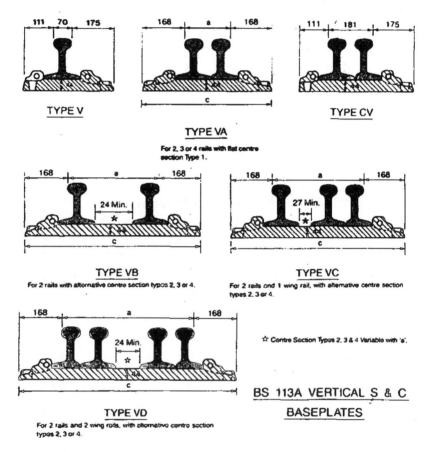

TYPE V

TYPE VA

For 2, 3 or 4 rails with flat centre section Type 1.

TYPE CV

TYPE VB

For 2 rails with alternative centre section typos 2, 3 or 4.

TYPE VC

For 2 rails and 1 wing rail, with alternative centre section types 2, 3 or 4.

TYPE VD

For 2 rails and 2 wing rails, with alternative centre section types 2, 3 or 4.

☆ Centre Section Types 2, 3 & 4 Variable with 'a'.

BS 113A VERTICAL S & C BASEPLATES

Figure 9.14 BS 113A Vertical S&C Baseplates

9.3 TURNOUTS - GENERAL DESCRIPTION

9.3.1 Definitions

A turnout enables a vehicle to be diverted from one track to another, and it consists of a pair of switches and a crossing, connected by closure rails.

Referring to Figure 9.2(a), the reader should satisfy himself by sketching the wheel flange trajectories that the type of crossing required for a turnout is of the COMMON variety. Common crossings indeed get their name from the fact that since they occur in every turnout there are many more acute than obtuse crossings.

To 'hand' a switch stand at 'A' and fixed and one movable rail (e.g. Parts number 6 and 7 in Figure 9.15). The fixed rail is called the STOCK RAIL (Parts number 6 and 9) and is of more or less the same section as the plain rails to which it is connected. The movable rail is called the SWITCH RAIL (Parts number 7 and 8). The switch rail is machined to a sharp point at one end. The tapered portion is referred to as the SWITCH TONGUE and is designed to fit closely against the stock rail, and when in this position a wheel will run smoothly from the stock rail onto the switch rail or vice versa. The tapered end of the switch tongue is usually called the TOE (sometimes the tip) OF THE SWITCH. The toe of the switch is connected to some kind of driving and locking mechanism which can move it away from the stock rail, and when so held, the wheel passes smoothly along the stock rail. Clearly two switch rails and two stock rails are required in any turnout, and the two together are termed a SET of switches. One of the switch tongues must for safe operation always be held firmly at its toe against its stock rail while the other must be held away from the opposite stock rail. The minimum separations required between the stockrail and an open switch are shown as dimensions "4" and "5" in Figure 9.15. The two switch tongues are held in their correct relative positions by at least two STRETCHER BARS (parts number 2 and 3 in Figure 9.15).

If the turnout is so arranged that in the predominating traffic direction, the tracks diverge, the turnout is described as a FACING TURNOUT. If the main traffic direction is such that the two lines merge, it is a TRAILING TURNOUT.

When viewing the turnout looking towards the divergence (i.e. from point A in Figure 9.15), the rail joint connecting the stock rail to the plain line is called the STOCK RAIL JOINT, and the portion of the stock rail between the stock rail joint and the switch toes is called the SWITCH FRONT or the STOCK RAIL FRONT. The joints at the opposite ends of the switch and stock rail are the SWITCH HEEL JOINT and the STOCK RAIL HEEL JOINT respectively.

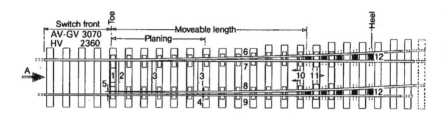

Figure 9.15 Component parts of a typical set of switches

1. Soleplate

2. 1st stretcher bar

3. 2nd and subsequent stretcher bars

4. Minimum switch opening 50mm

5. Toe opening 108mm (105mm min, 108mm max, for Clamp Lock operation)

6. Left hand stock rail

7. Left hand stock rail

8. Right hand stock rail

9. Right hand stock rail

10. Slide baseplates

11. Heel baseplates

12. Stress transfer blocks

* To 'hand' switch stand a 'A' and look as shown

The point where the S&C rails terminate beyond the crossing is the HEEL OF THE TURNOUT.

The detailed geometry of a turnout is affected by the side of the "through" track on which the diverging track lies and for ordering and general descriptive purposes there are a series of conventions about the handing turnouts and switches. Viewed from point A in Figure 9.15, these are respectively:

- Right hand turnout - The turnout diverges to the right of the through line.
- Left hand turnout - The turnout diverges to the left of the through line (as drawn in Figure 9.15).
- Right hand switch - The switch rail lies to the left of the stock rail, and both rails are to the observer's right.
- Left hand switch - The switch rail lies to the right of the stock rail, and both rails are to the observer's left.

In an ideal world, the diverging track of the turnout would be precisely tangential to the through track. However it is generally considered that a switch tongue machined to the slenderness required for a truly tangential turnout would be impractical to install and maintain, and the toe of the switch tongue is therefore always wedge shaped. The resulting angle between the line of the stock rail and the running edge of the switch tongue is called the SWITCH ENTRY ANGLE.

If the through track of the turnout is straight, then one of the stock rails will also be straight. The portion of the switch tongue which rests against it is machined so that the face presented to the wheel flange when the switch is in the closed position may be curved or straight. In modern designs of turnout this face will be curved and the radius of this curve is called the PLANING RADIUS. The thick end of the planing is called the HEEL OF THE PLANING, and occurs at a point where the distance between the switch curve and the stock rail, measured radially to the switch rail, is equal to the width of the rail head.

Between the toe of the switch and the point where flexing ceases, the switch tongue is supported on SLIDE BASEPLATES (parts numbered 10 in Figure 9.15) which support both rails when vehicles run over the switches and enable the tongue to be moved sideways when required with the minimum of friction. Between the point where flexing ceases and the heel of the switch are placed anchorages of some kind. In modern designs these consist of at least four substantial tapered distance blocks (parts numbered 12 in Figure 9.15), through each of which pass two bolts. These anchorages have two functions. Firstly they restrain the switch blade against any tendency to creep relative to the stock rail under the action of traffic, and secondly if the turnout is welded into a long length of CWR they transmit the thermal loads from the closure rails of the turnout, to the stock rails.

9.3.2   Design for Standard Switches

From the description given in the last paragraph, it will be evident that a switch/stock rail assembly is quite complicated to design and manufacture, and for this reason, in exactly the same way as there exists a limited range of standard crossings (see above), there is also a range of standard switches. A standard half-set of switches consists of everything between the stock rail joint and the switch heel joints, complete with all distance blocks, anchor blocks, and bolts, ready to be joined to the closure rails and placed onto the bearers. Further information can be found in the PWI book (2009) Switch and Crossing Design (BRT 7th Ed Vol 1 Part 2).

9.3.3   Designs for Standard Turnouts

As already noted there are standard switches, and standard common crossing designs suitable for incorporation in turnouts. These are combined to give either NATURAL ANGLE or COMPOSITE turnouts. A natural angle turnout is one where the switch radius and turnout radius are the same. In a composite turnout however, the turnout radius is either larger or smaller than the switch radius. Such a range of designs is unusual, and is provided to give as great a freedom as possible to the designer in the location of turnouts in relation to the optimum position of the diverging tracks, and to enable a choice of switch entry angle to be available for a given set of circumstances (see below).

9.4 SWITCH DESIGN AND MANUFACTURE

9.4.1 Switch Entry

In principle, the sudden change of direction introduced by the finite angle formed by the toe of the switch must produce impact loadings which cause localised wear on the contact faces of the switch or stock rail, depending on the direction of travel. These effects will be related to the speed and to the size of the entry angle, and it would seem logical that these factors should be taken into account when designing a new range of switches, or in applying an existing range of turnouts to an unconventional track situation.

In practice, however, present day UK switch designs have evolved from earlier designs in such a way that the new turnouts would replace the old with as little need for layout alteration as possible. The geometry was in each case therefore worked out on a basis of:

- determining the radius, crossing angle and theoretical lead length to match the pre-existing design.
- locating the heel of the planing.
- deciding on a convenient planing length to give as nearly as possible the desired overall lead length.

The "permitted speed" of the switch was then deduced using the rather artificial "effective switch tip radius" method. It recognises that an average wheelset in good condition will run centrally in gauge with a clearance of at least 6mm to the running edges of the stock rails. The point of first contact is therefore likely to be at the position where the switch is 6mm thick, and therefore the geometry of the switch between that point and its toe does not affect the running of the wheelset.

It recognises that when switches are manufactured on conventional machine tools, it is impossible to pre-curve the switch rail to its very end. Beyond that point, its planing radius is equal to its switch radius, thus eliminating a disturbing change of radius at the heel of the planing.

9.4.2   Rail Sections for Switches

Plain Line Section

The traditional rail section used for producing switch rails is the same section as that used for the plain line. Although switch rails machined from the plain rail section are more than adequate for most applications, for higher speeds, where long switches are required, or where high density traffic occurs, many railways are now using specially rolled rail sections for switches in these situations. These rails have thicker webs than plain line rails and may be shallower in depth.

The standard rail steel for S&C rails is BS 11 Grade A (UIC 860-0 Grade 900 A). S&C rails can also be made from AMS, or fully mill-heat-treated steel. The use of AMS, which is highly work-hardenable, causes some manufacturing problems. The fully mill-heat-treated rail has an enhanced hardness and a more straightforward machining and welding capability.

Shallow Depth Section

There are a number of shallow depth rail sections available, and all of them have advantages over the plain line rail sections, but they all require that the ends of the rails are forged to the same profile as that of the plain line rail section in order that standard rail jointing methods may be used. Alternatively a forged or cast transition piece may be welded onto the switch rail to change from the shallow cross-section to that of the plain line rail.

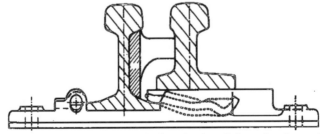

Figure 9.16 Cross-section through shallow depth switches, showing method of securing stock rail.

9.5 SWITCH AND STOCK MACHINING AND ASSEMBLY

9.5.1 Switch Rail Profiles

Machining of the switch rails and stock rails may be carried out to a number of different profiles, and these are described below.

Head Machining

There are three profiles for machining the rail head:

i. Undercut:

This profile, shown in Figure 9.17(a), is achieved by machining the side of the head of the switch rail to enable it to fit snugly against head of the stock rail, which is left untouched. The switch toe is consequently a very thin piece of metal which is easily bent over and often breaks off, and because of this, this profile is now obsolete.

ii. Straight Cut:

This profile, shown in Figure 9.17(b), provides additional thickness for the switch rail and is achieved by machining the side of the head of the switch rail to enable it to fit against the side of the head of the stock rail. However, since the switch rail protrudes into the gauge, the stock rail must either be machined as well (inset switches) or be set (joggled switches) in such a way that the switch rail running edge lies in its correct position. This profile also produces a discontinuity of the running edge for wheels travelling along the stock rail on the open switch side. Straight cut switches should not be specified if the turnout is to be used for more than the occasional trailing traffic movement.

iii. Chamfered:

This profile, shown in Figure 9.17(c), also gives additional thickness to the switch rail but requires the machining of both the switch rail and the stock rail. It does however provide continuity for wheels travelling from the stock rail to the switch rail. It is achieved by machining the side of the head of the stock rail at an angle of 1 in 4 (14°) to the vertical.

Foot Machining

Two profiles are currently used for machining the rail foot for full depth switches:

i.   Undercut:

This profile, shown in Figure 9.17(d), is produced by machining the foot of the switch rail to fit over that of the stock rail. This results in an eccentric loading being applied to the remainder of the rail foot which now has only a small area bearing on the baseplate. Because of this, this profile has been obsolete since about 1970.

ii.   Straight cut:

This profile, shown in in Figure 9.17(e), requires the adjacent flanges of both the stock rail and the switch rail to be machined vertically. Each rail is therefore well supported under each web.

Where shallow depth rail sections are used for the switch rail, the stock rail foot does not require machining. Only the switch rail foot is machined, as shown in Figure 9.17(f).

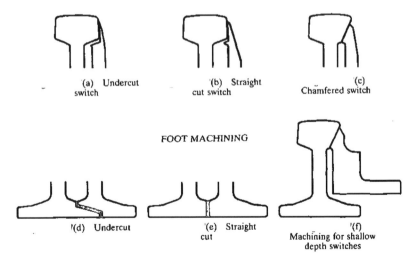

(a) Undercut switch

(b) Straight cut switch

(c) Chamfered switch

FOOT MACHINING

(d) Undercut

(e) Straight cut

(f) Machining for shallow depth switches

Figure 9.17 Foot Machining for Various Design of Switches

9.5.2 Facing Points

In the early days of railways, facing turnouts were avoided in passenger lines due to the high risk of accident inherent in the then doubtful track standards and the lack of a reliable means of ensuring safe working over the points. In the UK, we now have thousands of facing turnouts and a variety of apparatus for their safe operation.

Standards specify that facing points should be locked and that the position of each switch and the fact that they are locked should be detected. When points were mechanically operated by rods and levers, a Facing Point Lock (FPL) was positioned in the middle of the soleplate (see item 1 in Figure 9.13)

The FPL was adjacent to a lock stretcher bar connecting the two switch toes together and contained a lever operated plunger which could pass through slots in the FPL stretcher bar and lock the points in either position. Mechanical detection was also used, on the principle that signals could only be released when slotted detector blades connected to the points were in the correct alignment. Nowadays operation of switches is more often achieved by point machines and these can take several forms. The two most common in the UK are the combined electric point machine, so called because it combines the operating machinery, lock and detection in one box, and the rail point clamp lock.

All point machines perform a similar sequence of operations. In the case of points operating normal to reverse the sequence would be:

a) Points standing locked and detected normal.
b) Points called to reverse.
c) Provided that the presence or approach of a train is not detected by the track circuits, the points unlock. Movement commences and detection is broken.
d) Points reversed.
e) Points lock in reverse, and detection is re-established.
f) Points standing locked and detected in reverse position.

9.5.3 Point Machines

Combined electric point machines comprise of an electric motor, a drive mechanism, a detector mechanism capable of detecting the position of the switches and the lock and the lock device itself. The lock device normally consists of lock dogs passing through slots in a lock blade coupled to the lock rod. Normal and reverse dogs and slots are different to prevent accidental locking in the wrong position. Combined point machines are heavy and bulky and they require extended sleepers.

The rail point clamp lock has a separate lock for each rail. It is designed to retain the closed rail firmly to its associated stock rail. The locking member is pivoted on a long pin to accommodate switch creep. The pivot is mounted as high as possible consistent with wheel flange clearance to minimise the twisting effect on the switch rail. Detection of the lock and switch rail position is contained in the rail mounted unit.

The clamp lock is hydraulically operated via a compact power pack which may be mounted as convenient. Work on strains in signal engineers' track mounted equipment have shown, that the equipment is very susceptible to track vibration.

Further development in switch activation systems is the HPSS (High Performance Switch System) – a heavy motorised enhancement. The Hy-Drive system is an in bearer clamp lock at the switch toe and hydraulic supplementary drives. Rollers to assist in switch movement have been featured, also Teflon based non-grease slide baseplates.

Point machines are more affected by voiding which places heavy loads on the long rods. Clamp locks, whilst affected by voiding are more susceptible to dipped rail joints in advance of the points. Track Engineers should be aware of the effect of poor track maintenance on signal engineers' equipment.

Figure 9.18 Left Hand Turnout and HPSS Point Machine (courtesy Network Rail)

9.5.4   Two-Levelling

Two-levelling is the process whereby the thickness of the base of baseplates is altered to give different cants for the two tracks on the same timber. The rake of the timber is the actual cant over gauge taken by the timber when it has no two level baseplates on it. There are three basic rules for two-levelling:

- The timber rake should not vary under a cast manganese crossing, and only very small variation is permissible under built-up crossings. This is because they cannot be physically twisted.
- The longest practical distance should be used to change the rake and cant, to enable the effect on the train to be minimised.
- When a two-levelling scheme has been completed, a relative rail level diagram should always be drawn, taking the low rail of the whole layout as base. This will show up any unacceptable bumps in rail levels which might not have been apparent.

If there are longitudinal level difficulties at the site, a complete level scheme should be undertaken, and the relative rail level diagram plotted to give true levels. This is particularly important when there are changes in gradient. The longitudinal section must always be produced where overhead electrification exists. Care must be taken to ensure that the physical cant gradient on any track does not exceed the limits laid down in standards.

Two level switches are not available in vertical S&C, and in all cases special baseplates are required. Normal practice today is to make the cant the same on the diverging tracks until they are clear of the connecting long timbers.

9.5.5   Switch Heaters

The majority of switches in running lines have some form of switch heating to keep the switches operational in very cold weather. There are two main types - gas and electric. Most current types are obstruction-less and do not have to be removed for tamping operations.

Electric switch heaters, are of the following types:

- Baseplate pads
- Cartridge heaters in jaw blocks
- Strip heaters clipped to stock and switch rails

The last mentioned type is the current standard and most used design in the UK.

Gas switch heaters are usually the responsibility of the Track Engineer. Modern types are fully automatic, with the operation of the heater controlled by a thermostat which switches on a gas and an electricity supply when the rail temperature falls to 0°C. The electricity is usually supplied from batteries but may be supplied from the mains. The gas may be supplied from storage tanks or from the mains. The gas is directed to burners where it is ignited by spark electrodes. When the rail temperature rises to 3°C the thermostat extinguishes the burners and turns off both the electricity and the gas supplies.

For switches not fitted with any form of switch heating, it may be necessary to apply approved de-icing compounds in order to keep them working in very cold weather. Salt should not be used since it could contaminate and affect signalling equipment.

9.6 THE DESIGN OF SWITCH AND CROSSING LAYOUTS

9.6.1 Introduction

Given that manufacturing details of the components are determined, as described earlier, the design of S&C work has two aspects. One is concerned with the speed potential of the layout under consideration, and the other is the precise determination of its geometrical parameters. The two aspects are related, if for no other reason than that the choice of switch (which means essentially the choice of switch entry angle) will dictate the speed potential of a turnout. It will therefore be convenient to describe first how the geometry of an S&C layout is calculated, and then go on to deal with speed considerations. The information contained in this Chapter is based on the reference material and can be found in the PWI book (2009) Switch and Crossing Design (BRT 7th Ed Vol 1 Part 2).

9.6.2 Geometrical Design of S&C

In order that a switch and crossing layout can be manufactured, assembled, and laid into the track, every part of it must be very accurately specified in dimensional terms, usually on a drawing. Scales of 1:100 are used for this purpose, and 1:50 or even larger scales may be used on occasions. It is undesirable to attempt to scale off even from the largest scale drawings dimensions to the accuracy necessary to obtain parts which properly fit together. Hence every line on the drawing must be defined by dimensions, and these must be confirmed by calculation. Only then can it be reasonably certain that the manufacturer will produce components which correspond with the engineer's intentions. It is to this end that it is necessary to define all layouts in geometrical terms.

Until recently all this work had to be done by individual calculation, and a substantial body of formulae has been developed over the years to make the task easier. Much of this technique has been made obsolete by the coming of Computer Aided Design (CAD), and it is wise to clarify at this point what CAD involves as applied to permanent way layouts.

The computer is loaded with:

- Suite of software to enable it to carry out geometrical and trigonometrical calculations.
- Library of data which consists of the dimensions (stored both in terms of radii, crossing angles, and other leading dimensions, and as coordinates) for every possible S&C component, and for the range of turnouts and standard diamonds. These form the building blocks with which the designer will work.
- Specialist software to enable it to bend the individual building blocks described above, from the standard "straight main line" configuration in which the data appears in the data library, to any radius desired by the designer.

The designer starts with a survey of an existing layout, or of the area where the layout is to be installed, and this is converted into digital form and fed into the computer memory (most probably as x-y co-ordinates). The designer then defines the main outlines of the layout, such as the alignments of tracks which are to remain and cannot be altered. The next step is to consider the brief, and command the computer to take from the library what is considered suitable building blocks to use for the initial parts of the layout at any position on the plan, and any desired curvature.

For example, if it is required to design a crossover between two non-parallel curved main lines, we can locate one of the turnouts on one of the tracks, and instruct the computer to bend the "through" track to the curvature of the track at the desired point. The computer will do this, and will respond with information about the resulting radius of the turnout curve, and the direction in which the rails will be pointing at the common crossing nose. The designer can then do the same on the other track, resulting in two turnout lines which will probably not meet in a convenient way. However, the facility will exist to enable him to move one or other of the turnouts about until a suitable configuration is obtained, and the computer will work out the radius and tangent points of any curve necessary across the six-foot. The operation can be repeated as many times as necessary, and alternative schemes can be placed in the computer's memory, and/or printed out as hard copy for study by other members of the project team as required.

It is however still important that track engineers embarking upon the science and art of S&C design should grasp the essential characteristics of the basic building blocks, and the geometrical and trigonometrical principles upon which all the calculations performed by the computer are based, and this is the comparatively limited objective of sections above. As far as possible, the formulae used in the main this chapter are expressed in CLM terms.

In the design of any S&C layout the restrictions imposed by the surroundings must be taken into account. Clearance requirements are laid down and a more detailed account of clearance problems is given in Chapter 12.

After the drawings have been completed and the material ordered and supplied it must be set out on the ground, first in the assembly depot and then on the site itself. This is only achieved by giving sufficient information on the drawing. This consists of two parts; the details of the alignments, and the position of the S&C on those alignments. The alignment details are normally given in diagrammatic form on the drawing, rather like a page from a survey book, giving baseline measurements and offsets, directly obtained from the calculations for the design.

It is desirable for track designers to be present on site when the job they have designed is laid in. This will not only broaden the designer's knowledge of relaying methods but could also demonstrate where snags are likely to occur, and would therefore help to reduce these in future designs. They will also be able to assist if required in setting out. It is also helpful to inspect the site some two or three weeks after relaying to see whether the design has been fully implemented, and if this has not been achieved, to determine why not.

9.6.3 Turnout Geometry

In general terms there are two kinds of standard turnouts, natural angle turnouts, and composite turnouts. Standard turnouts are designed out of straight main lines, and their geometry will initially be described as such, followed by the method of modification to deal with a curved main line. The turnouts used in transitioned crossovers are a specialised form of composite turnout.

The Theoretical Geometry of a Natural Turnout

In the simplest possible geometry for a turnout, the diverging track consists of a segment of a circle which is tangential to the through line at the start of the divergence. This geometry, which is in essence that adopted for many modern designs of turnout in the UK and other leading world railways, will first be considered. It must be remembered that in practice the switch tips do not coincide with the tangent point and the effect of this in geometrical terms will be explored below.

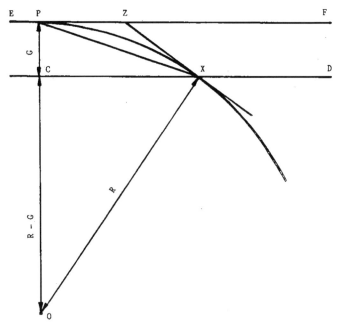

Figure 9.19   Theoretical Geometry of a Natural Turnout

The arrangement is shown in Figure 9.19, in which for clarity only the two straight through rails and the outer of the two diverging rails are shown. The two straight rails are labelled CD and EF, whilst the diverging or turnout rail is labelled PJ. The track gauge is G, and the radius of the outer turnout rail PJ is R, with its centre at O. The turnout rail PJ and the straight through rail CD cross at X and a line is drawn from O to X. The tangent to the circle at point X, and line PX are also drawn in. The tangent at X is line XZ, so that the crossing angle is angle CXZ. We have to work out the value of this angle.

Since we have constructed the curve PJ tangential to through line EF at P we know that radius OP must be at right angles to line EF. Since line CD is parallel to line EF angle XCO must also be a right angle. Hence triangle XCO is a right angled triangle and we can apply the theorem of Pythagoras to it.

Since by construction the length OP is R and the length PC is G the length OC must be (R-G). Also length OX is R. Hence by the theorem of Pythagoras:

This length, CX, is the Theoretical Lead Length;

$$ex = (R^2 - (R-G)^2)^{0.5} = (2RG - G^2)^{0.5} \ \ldots\ldots\ldots\ldots 9\text{-}1$$

From the geometry of the circle, it can be shown that triangle ZPX is isosceles. Hence angle ZXP is equal to half the crossing angle, and it follows that angle CXP is also equal to half the crossing angle. The tan of this angle is PC/CX, so we can write:

$$tan\frac{[CXZ]}{2} \ = \ \frac{G}{CX}$$

If we compare our equation with CLM method they are respectively identical, so that if we regard "N" as the CLM magnitude of angle CXZ, we can write:

$$\frac{1}{2N} \ = \ \frac{G}{CX}$$

Transposing:

$$CX = 2GN .... 9\text{-}2$$

Putting equations 9-1 and 9-2 together, and transposing:

$$N= \frac{[(2R\text{-}G)]^{0.5}}{4G} .... 9\text{-}3$$

Which gives the desired result. Alternatively if N, the crossing angle is known, the turnout radius is obtained from another transposition as:

$$R = \frac{(4N^2 + 1)\,G}{2} .... 9\text{-}4$$

(N = crossing angle (1 in N); R = turnout radius; G = track gauge

Using formulae 9-3 and 9-4 the reader can prove for himself that a turnout with R equal to 184.012m would have a crossing angle N equal to 8, and vice versa. It will be noted that this corresponds to a UK Standard Turnout with a 1 in 8 crossing and a BY switch.

There are a total of nine combinations whose geometry can be calculated in the same way. These are AV – 7; BV – 8; CV – 9.25; DV – 10.75; EV – 15; FV – 18.5; SGV – 21; GV – 24; HV - 32.365

These combinations are all called "NATURAL TURNOUTS" because their radii are the same throughout the length of the turnout, except where modified to accommodate the planing at the switch toe.

The distance along the straight switch rail from the toe of the switch to the nose of the crossing is called the LEAD LENGTH. The theoretical lead length CX, derived in formula 9-1, is slightly longer than the actual lead length. The difference between the actual and theoretical lead length depends on:

- the switch entry angle and switch planing radius
- the thickness of the blunt nose of the crossing.

The way in which the actual lead of a natural turnout to the intersection (or fine) point of the crossing is determined, is described in the PWI book (2009) Switch and Crossing Design (BRT 7th Ed Vol 1 Part 2). For each UK switch from AV to GV, the planing radius is constant throughout, and starts at the point where the back face of the rail head of the switch rail first touches the stock rail.

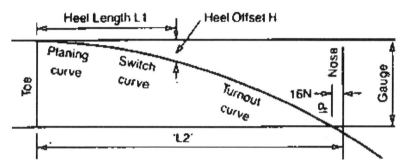

Figure 9.20 Standard Drawing Dimensions for UK turnouts

Actual Lead Length of a Natural Turnout (to Intersection)

The lead length of a natural turnout (LN) is shown in Figure 9.21. It is made up of two parts, the lead length Lp from the toe of the switch to the heel of the planing, and the lead along the stock rail from the heel of the planing, to the fine point of the crossing (Lx).

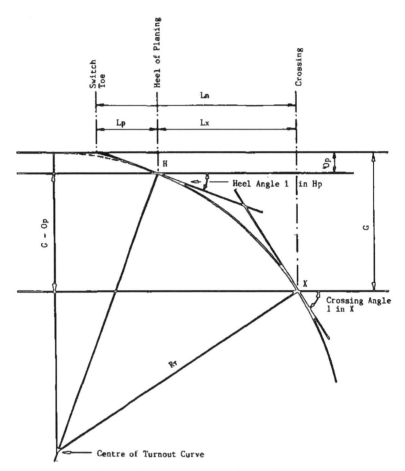

Figure 9.21 Lead Length of a Natural Turnout

It will be recalled that the turnout radius of a natural turnout is the same as the switch radius, and the relationship between the turnout radius and the crossing angle have already been defined. Thus for a given turnout radius $R\rho$, the crossing angle X (CLM) will be known. Since the planing heel offset and the gauge are known, we also know the offset across the track between the heel of the planing and the crossing rail. Thus we now have expressions to enable both Lx and Lp to be calculated from known data, and the lead length for the natural turnout from the switch toe to the theoretical intersection point of the crossing or FINE POINT is given by:

$$Ln = Lp + Lx.....................9\text{-}5$$

This completes the required geometrical information for a natural turnout.

Distance between Blunt Nose and Intersection Point

Since the standard width of the blunt nose of any UK FB crossing is 16mm, this must be added to the lead to the intersection point (eg the length Ln obtained in 9-5 in order to arrive at the practical length of the turnout as it will be built.

The Switch as a Building Block

In a composite turnout, the radius changes at the heel of the switch. For this reason, a standard "half-set of switches" as described earlier forms a building block in the design of a turnout, and it is defined by the following parameters:

- Switch Radius
- Toe to origin of Switch Curve
- Heel Offset
- Toe to Heel Length
- Heel Angle
- Entry Angle

The heel of the switch is much further from the switch toe than the heel of the planing, and the two must not be confused. The geometry of that part of the switch between the heel of the planing and the switch heel, is similar in principle to the geometry of the natural turnout between the heel of the planing and the crossing.

Composite Turnouts

The design of a composite turnout thus comprises a set of switches, whose geometry is known, a crossing of arbitrary angle X (but different from that of the natural turnout associated with the chosen switch), and a connecting curve. The design unknowns at this stage are the lead length and the radius of the connecting curve, which is called the TURNOUT RADIUS. Transitioned turnouts are used instead of natural or composite turnouts when designing running line crossovers. The object of the design is to improve the passenger comfort factor for the crossover.

9.6.4   Diamond Crossing Geometry

Calculations for a Straight over Straight Diamond

Consider Figure 9.22, in which two straight tracks cross one another at an angle 1 in N. The information required to enable the diamond to be either fabricated or set out on the ground is:

- the lengths of all four sides of the diamond
- the total length along diagonal AD.
- the distance AP, where BP is perpendicular to line AC. This distance is referred to as the LEAD of crossing B over crossing A. This distance is of course equal to the lead of crossing D over crossing C.

Since the gauges of the two tracks are the same, the four sides AB, AC, BD and CD are all equal; also BC and AD are at right angles and bisect one another.

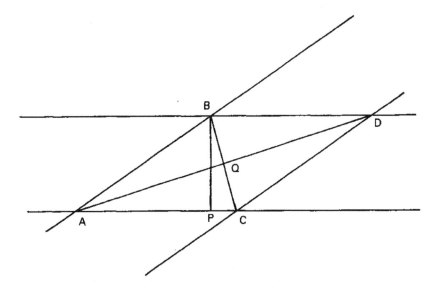

Figure 9.22 Straight over Straight Diamond Track Geometry

Junctions on Curves

The track designer will often face a complex set of circumstances where the S&C is located on curves with often varying radii, intervals and flexure. If a curved track crosses a straight track, or an inter-track space between two parallel tracks, then each crossing will have a different angle, and each leg will have a different length. A common case to be solved is that of the curved diamond crossing in a double junction. Such a diamond is represented in Figure 9.23. The CLM value of either crossing will first be determined to facilitate the use of standard crossing angles, from the geometry of the turnout and the inter-track space. It should be noted that the crossings will not be identical. In order to arrive at a design in which all the crossings can be taken from the standard range, it is necessary to modify the turnout radius as it proceeds around the curve. In current practice this would be done by complex trigonometrical methods.

A range of designs of double junctions incorporating standard switches and crossings for use with a 1970mm inter-track spacing, together with their speed potentials, is available. It may be assumed that these designs ensure a minimum six-foot space of 1970mm throughout, but designers must remember to check for possible clearance problems between kinematic envelopes, allowing also for end and centre throw for the most critical rolling stock for the particular site.

The only way of solving the problem of a diamond with both tracks curved is to go back to first principles of trigonometry and work from the circle centres.

In the case of one curved track crossing two other tracks, both curved, but not concentric, this problem can be solved by trigonometry if the radii, and the positions of the centres, of all three circles are known.

Figure 9.23 Diamond with a curved track of constant radius crossing a curve

(courtesy Network Rail)

9.6.5 The Assessment of Speed Potential of S&C Layouts

The principles from which the rules and formulae for plain line in Chapter 8 were developed apply equally to the curves forming S&C. However, by comparison with plain line, S&C presents three problems when designing for speed. These are:

- transition curves are absent from the diverging track of a turnout where indeed there is the further difficulty of an instantaneous change of direction at the switch toes;
- features which interrupt the smooth passage of the wheel, such as flangeway gaps, and the necessity for the wheel tread to pass transversely over rail heads;
- difficulties in applying cant due to the need (with some exceptions) for the heads of all four rails to be in the same plane.

When considering the speed potential of layouts, due note must not only be taken of the limits on cant and cant deficiency as summarised in Chapter 8, but also of current line speeds and permanent speed restrictions affecting the locality where the work is to be done. Whilst these restrictions can be altered (indeed it may be an objective of the upgrade to enable this to be done), nevertheless any proposed alterations must be agreed with the operational and signalling departments, and due notice given of their implementation.

Speed potential of the diverging track off a turnout with a straight through track

The absence of a transition curve at the toes of the switches of a turnout would, if the principle of virtual transition curves described in Chapter 8 were applied, prevent advantage being taken of the full allowance of cant deficiency available. The situation is made worse by the existence of the switch entry angle, however, it was found that the lateral ride through a turnout, as experienced by the passenger, would remain within acceptable limits provided that the following empirical guideline was observed. Cant deficiency through the body of the turnout must be limited to that applied to jointed track at that time, i.e. 90mm.

In applying this rule the speed potential of each successive radius encountered during passage through the turnout has to be assessed and therefore the minimum radius controls the speed. For the range of natural angle turnouts, the speed is controlled by the common switch/turnout radius, since the planing radius is the larger. For composite turnouts, the controlling radius may be either the switch or turnout radius. In addition, the rate of change of cant deficiency must be checked at each instantaneous change of radius using the virtual transition standards.

The rate of change of cant deficiency rule does not apply at the switch toe. Instead a parameter known as the theoretical cant deficiency at the switch toes must be used. The value of this parameter must not exceed 125mm.

Speed potential of turnouts with curved through tracks

The principles used in the design of turnouts where the through track is curved have been described in above. In assessing the speed potential of such a turnout clearly it is necessary to consider both tracks.

Consider first the through track. By definition, both rails of this track will be on a curve which will be continuous from before the switch to beyond the crossing. There is therefore no geometrical difference between this track and a plain line curve of the same radius. The difference if any lies in the possibility of disturbance to the trajectory of the wheel which has to negotiate the crossing. That wheel also has to be transferred from being carried on the stock rail to being carried on the switch or vice versa as it moves past the switch. This can result in disturbance to its trajectory if the wheel tread is worn hollow. Disturbance may also be caused by changes in effective rolling radius caused by the lateral movement of the contact patch in the same location. Until recently these influences were not taken into account and full plain line cant deficiencies were allowed on the through track of a curved turnout. However on very high speed routes where the new cant deficiency limit of 150mm applies on plain line CWR, the effect of the disturbances mentioned above is taken into account, and the maximum allowable cant deficiency on all S&C welded in to CWR track remains at 110mm.

Contrary flexure turnouts with cant applied to the through track

If cant is applied to the through track, this cant will (in the absence of any provision of two-level baseplates) also be imposed on the turnout track, and will be in the adverse sense. The UK places a limit of 80mm on this adverse, or negative, cant in S&C, and hence the cant on the through track of a contrary flexure turnout is effectively limited to this value. Also it must be noted that whereas on a curve with cant the right way, the speed potential is obtained from the sum of the cant and cant deficiency, the speed potential when the cant is negative must be obtained from the excess of permitted cant deficiency over applied negative cant. The effect is to impose a marked reduction in the permissible speed on the turnout.

Equal split turnouts

The equal split turnout is a special case of the contrary flexure turnout wherein:

- the track leading to the toe of the switch is straight
- the switch entry angle is divided equally between the two arms of the turnout
- no cant is applied to either track
- the radii of the two arms of the turnout are identical.

When this design is used, both arms have similar characteristics, and the limiting cant deficiency of 90mm is then appropriate for both tracks. A considerable improvement in the speed potential of the "turnout" track is obtained, albeit at some cost to the potential of what would otherwise be the "through" track. For example, the speed potential of an EV/15 laid in straight track is 70km/h . If laid in contrary flexure on a through track curve of 1290m radius, the speed potential on the through track is l00km/h, but on the turnout, although the radius is flattened from 645.116m to 1290m, the speed is limited to the speed for the switch, which is only 70km/h. By making the turnout into a symmetrical split, the switch is effectively redesigned with the switch entry angle half that of the parent design, so that its potential on an "effective switch toe radius" basis is raised to 113.6km/h. The speed of both arms then becomes l00km/h.

The classic case of the application of this principle is Colton Junction, at the north end of the Selby Diversion, where a split double junction using HV/ 32.365 turnouts with turnout radii of 6000m, enables trains to take either the Leeds or Doncaster arm of the junction at the full line speed of 200km/h.

Figure 9.24 High speed curved double junction on 3rd rail electrified route with HPSS point machines. (Courtesy of Cemex Rail Solutions)

Similar flexure turnouts

When a given turnout is bent inwards to form a similar flexure turnout, the radii of all the constituents of the turnout are reduced according to the rules and in principle, the designer needs to check the speed potential of each section in succession in order to arrive at the speed of the turnout as a whole.

Speed potential off a crossover with straight through track

In normal designs of crossover the crossover track is straight between the two crossings. Hence its speed potential is limited by the instantaneous change of radius from the turnout curve to straight, as derived from the formulae in the section on virtual transitions applied to reverse curve without transitions but with a short length of straight between the tangent points in Chapter 8.

Speed potential of a crossover between two curved tracks

By application of the principles outlined above it will be clear that a crossover between two curved main lines will equate to a standard crossover between two straight tracks, the whole being bent to the radius of the main lines. This leads typically to a layout in which one turnout is in similar, and the other in contrary, flexure. Such a layout is more likely than most to require some cant on the through tracks, and the method of tackling the design, incorporates this feature. It also illustrates the checks required to ensure that the limiting rates of change of cant deficiency on the reverse curve across the six-foot are not exceeded.

Thus, as for the crossover across straight main lines, the parameter which controls the speed potential is the reverse across the six foot, and the permitted speed for the crossover is 35km/h. We can now reconsider the question of what cant to apply on the through tracks. The factor which limits the cant which can be applied is the requirement for the speed on the turnout track of the contrary flexure turnout (which will be subject to a negative cant), to be at least 35km/h. The cant deficiency must not exceed 90mm, hence the cant, being negative, must not exceed (90 - 53) or 37mm. Hence a satisfactory and convenient cant to apply to the complete layout would be 35mm.

Transitioned Crossovers

The limitation on the speed potential of a circular-curved crossover as described above due to the reverse curvature across the inter-track space can be eliminated by replacing part or all of the turnout curve between the switch heel and the crossing, by a transition curve. By suitable design, the speed potential of the crossover can then be made equal to the speed potential of the switch. The design process is complex and leaves much to the imagination and initiative of the designer. The variables involved are:

- Rates of change of cant deficiency.
- Length of transition curve (these two being interdependant).
- Crossing angle
- Whether the transition curve will be confined between the switch and crossing or whether it will extend into the inter-track space.
- The location of the tangent point between the circular curve and the transition curve.

Double Junctions

The speed potential of a curved track through a double junction is that of the turnout, or of the succession of curves through the diamond, whichever is the less. Each instantaneous change of radius must be checked to ensure that the rate of change of cant deficiency on the virtual transition does not exceed 80mm/s.

CHAPTER 10

BALLAST, SUBGRADE, EARTHWORKS AND DRAINAGE

10.1   INTRODUCTION

Over the past 150 years track engineers have been constructing and maintaining trackbed for the prevailing speeds and axle loadings according to the local climate and geological conditions. They have accumulated a body of experience concerning the requirements of depth and layers of a ballasted track for minimum maintenance and good running to derive features of good practice which are continually under revision. As described in Chapter 8, special machines now simplify many previously labour intensive operations and both the installation, cleaning, compacting and preparation of trackbed for high speeds are achieved quickly. New materials and systems cope with the requirements of trackbed for accurate level and alignment.

"Trackbed" comprises the ballast and sub-ballast layers including any geotextiles, geomembranes, meshes, or other materials for separating, filtering, waterproofing, drainage, reinforcement or other strengthening of those layers. "Subgrade", on the other hand, is the term for the natural soil stratum (or embankment soil) after trimming, on which the trackbed is constructed. The term "natural soil" includes whatever material may be found at the bottom of a cutting through which the railway passes, and may be anything from a soft, compressible material such as peat, to a hard rock. This is sometimes known as the term "formation". A typical section, showing a twin track railway in cutting is shown in Figure 10.1b.

Figure 10.1a Typical view of track support (courtesy Network Rail)

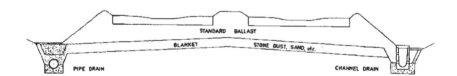

Figure 10.1b Typical section of blanketed formation

The function of trackbed is to support the track, to drain water away from the bottom of the sleepers and to distribute load to the subgrade in such a way that the position of the rail. There are a few features that are also required which include the objective that the loaded condition does not change with time, it returns to its original position after the passage of each train and moves elastically under vehicle loading within tolerable limits for the mechanical design and use of those vehicles.

This chapter will deal the design and maintenance of the ballast and the sub-ballast layers, the subgrade and earthworks. This will include a methodology for diagnosing the causes of various kinds of ballast, subgrade, and earthworks failure.

10.1.1   TRACKBED DESIGN

A design to satisfy the loading conditions has been sought through different theoretical and empirical approaches but the results in practice in many countries are reassuringly similar (allowing for differences of geology and climate) and suggest that the currently recommended ballast depths represent an optimum for the loads imposed by the rolling stock. Such design methods relate to the subgrade and include the California Bearing Ratio (CBR) test, mathematical modelling of stress distribution, "quick" soil tests, static or dynamic plate loading tests (EV_2 and Westergaard), threshold stress for over consolidated soils and a subgrade classification method. The results of all these methods are in turn modified by sleeper spacing and type, width of ballast shoulder, blanket layers, geomembranes, etc., as well as the local geology and hydrology.

It can be shown that the stresses in the rails, the rail seat reaction, and the stresses in the sleeper, are all to some extent related to the springiness of the cushion of ballast under the sleepers. This quality of springiness is given the technical name "TRACK MODULUS". The Track Modulus relates the pressure imposed on the ballast by the sleeper to the deflection of the top table of the rail. Since in structural terms the track is thought of as a beam on a continuous elastic support it is defined as "The uniformly distributed line load required to produce unit deflection of the support". Its units are N/m/m, or N/m^2. Since realistic values of the track modulus are very large it is common to express it in Meganewtons /m/m, ie MN/m/m, or alternatively

N/mm/mm. These units are similar to those in which Young's Modulus relates the deflection of a steel member to an axial load in it, as explained in Chapter 2.

All other things being equal, the more springy the ballast, i.e. the lower the track modulus, the more the track will deflect under load, the less will be the rail seat reaction, and the greater will be the bending moment in the rail itself. Since it is more important in most locations to limit deflection, it is desirable that the track modulus should be fairly high. One possible contra indication to this general statement is that the stiffer the track, the more likely it is to transmit vibration into the surrounding ground.

The trackbed should be free draining to at least 25 mm below the underside of the sleeper at all times. In practice the track bed will probably be free draining to around 200 mm. The trackbed must also be deep enough to ensure that the loading on the top surface of the subgrade is uniform, ie, the "pyramids of support" from sleeper to subgrade should touch or overlap (see Figure 10.2).

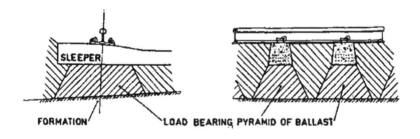

Figure 10.2 Load on formation

10.1.2   BALLAST DEPTH

Ideal ballast depths are specified according to the mix of speed and tonnage shown in Table 10.1.

These are arrived at by theoretical and practical assessments of the various loadings and their frequencies, but by necessity are general in their application. It remains a distinct possibility that poor subgrades may still be highly stressed even with full ballast depths, and although ballast achieves the required load spread it also adds more dead load into the track substructure. In this respect analysis of the whole track and substructure condition is necessary when faced with a problem.

TABLE 10.1

Recommended ballast depths

LINE SPEED (km/h)	LINE TONNAGE (tonnes/yr)	BALLAST (mm)
Above 170	7 million upwards	280
	Below 7 Million	230
130 to 170	15 million upwards	280
	Below 15 Million	230
Below 130	All	150

10.1.3   PROVISION OF LATERAL STABILITY TO THE TRACK BALLAST WIDTH

If ballast specifications are derived primarily to satisfy vertical loading conditions it is no less important to ensure longitudinal and lateral stability using the same material. The former is achieved by filling cribs between sleepers with ballast up to sleeper top level. Good frictional contact at ballast/sleeper interface also assists in this respect.

Lateral stability now receives more attention with the widespread use of continuous welded rail, with the result that ballast shoulders are now specified in the UK given in Table 10.2. As ballast has a bulk density when tipped which is much less than solid in-situ stone, full opportunity was also taken to develop heavier concrete sleepers in this same period to assist in overall stability. Areas between tracks are also now ballasted.

10.1.4 MATERIALS FOR BALLAST

Ballast is by weight and volume the largest component of the track. The cost of buying and distributing it is a significant part of the entire civil engineering budget. The enormous volume of ballast in the UK and the need for a cheap material capable of packing demand a natural (cheap) and granular (packable) material. The strength characteristics necessary to distribute stress from the underside of the sleepers without crushing, breaking etc. lead to rock being selected. The necessary frictional and interlock properties to resist longitudinal and lateral movement require a crushed rock rather than smooth round gravel.

The optimum particle size distribution for ballast is influenced by a number of factors. If the ballast is too large there may be only two or three particles between sleeper and subgrade and insufficient ballast particles to distribute the stress. Large particles will also lead to difficulties in obtaining a level surface and to difficulties in tamping and may lead to rocking of the sleepers. With small particles there will be small voids and as a result the ballast will be susceptible to clogging by fine particles (from ballast breakdown, spillage etc) with consequent poorer drainage. If water is allowed to pond in the ballast then the attrition process (ballast particles rubbing against each other and the sleeper) will accelerate and the ballast will be worn down relatively quickly resulting in Dirty Ballast Pumping Failure. The uniformity of ballast particles is a choice or balance between a very uniform aggregate which will have a higher percentage of voids and larger voids but lower strength properties, and a well- graded aggregate which will be lower in the former but higher in the latter.

TABLE 10.2 Recommended widths of ballast shoulders

TRACK TYPE	WIDTH OF BALLAST SHOULDER (mm)
CWR – Straight	375
CWR - Curved track, radii flatter than 800m	450
CWR - Curved track, radii 800 m and less	525-600
All jointed track	300

Note: In all CWR track shoulders should be heaped 125 mm above sleeper top level

10.1.5 THE BALLAST SPECIFICATION

From these fairly simple considerations and experience gained over the years the grading requirement in the present specification has evolved. This requirement and the other quantitative requirements are given below:

(i) Grading requirement:

The ballast shall consist of a mixture of sizes mainly between 50 mm and 28 mm to conform to the limits of the following table.

Square mesh sieve	Percentage to pass
63mm	100
50mm	97 – 100
28mm	0 – 20
14mm	0 – 2
1.18mm	0 – 0.8

In the wagon these sizes must be evenly distributed.

    (ii)   Shape requirement:

        (a) Flakiness Index - Maximum Permissible Value 50%

        (b) Elongation Index - Maximum Permissible Value 50%

or

    Not more than 2% by weight of particles shall have a dimension exceeding 100mm.

    Not more than 25% by weight of particles shall have a dimension exceeding 75 mm.

    (iii)   Mineral properties requirement:

    The Wet Attrition Value (WAV) shall not exceed 4% for any ballast supplied, except by the previous written agreement of the civil engineer who shall specify if and when inferior ballast, having a WAV not exceeding 6%, may be supplied.

It will be noted that the vast majority of ballast should be between 28 mm and 50 mm with a 2% dispensation given at the 14 mm sieve. The 1.18 mm requirement is to guard against all of the 2% dispensation being supplied as a rock flour or similar which would cause almost immediate drainage problems. The strict 0.8% limit includes any allowance for errors in sampling and testing. The Elongation Index and Flakiness Index figures are stipulated to control the proportion of particles which are tending in shape towards "carrots" and "potato-crisps" respectively. The Wet Attrition Value requirement is discussed below. The specification also calls for the ballast to be natural. Slag was used as ballast but the quality was variable and it is no longer used.

10.1.6 WET ATTRITION VALUE TESTING

The ballast particles themselves can suffer degradation due to the action of traffic and to maintenance operations in one of two ways. Either they become rounded at the corners (rather than sharp edged) or the particles break due to crushing and/or impact loading. All types of stone are affected by these actions to some extent, but some are worse than others. In particular, some stones suffer very badly from attrition in the presence of water. This syndrome is called *wet attrition*, and it is one of the factors which predispose towards the development of wet spots (qv section 10.5.1(a) below). It is obviously undesirable to use as ballast a stone which behaves badly in this way, and to enable proper selection, the old test known as the Wet Attrition Value (WAV) Test was chosen from others as giving the best correlation with real life.

The WAV test involves placing a given weight of ballast along with a given volume of water in a metal cylindrical drum which is then sealed and which rotates 10,000 times round a skew axis. During the test the particles wear against one another and the drum, and are subject to impacts against the top and bottom of the drum as it rotates round the skew axis. The percentage of ballast which has worn down to a size less than 2.4 mm is the WAV.

The WAV test gives a slurry similar to that found in wet spots (except where there is a layer of ash near the sleeper soffit). The test distinguishes clearly between most limestones (which tend to have poor WAV's) and most granites (which tend to have good WAV's), and links have been found in experience, between frequency of wet spots and WAV.

10.1.7 OTHER MEASURES OF BALLAST QUALITY

Whilst it might be expected that tests to measure resistance to impact and crushing should be carried out as well as WAV tests, impact and crushing tests were abandoned some years ago because it was found that a ballast which satisfied the WAV stipulation almost invariably satisfied the crushing and impact criteria. Recent laboratory research work suggests that it may be worth re-introducing the crushing value at a more stringent level than previously. The role of the impact value is awaiting investigation. The aim of this research is to ascertain which ballast properties are important in determining ballast life and to fix them at levels to prolong ballast life as far as possible.

Features such as resistance to chemical or frost attack are not normally regarded as significant in UK, except perhaps that in particularly badly polluted areas, the effect of aggressive chemicals may need to be considered. Resistance to chemical attack is tested for in a "soundness" test devised by ASTM and currently being considered for use in UK. The rock is repeatedly soaked in a solution of magnesium sulphate and dried.

10.1.8 BALLAST SUPPLY

Traditionally ballast has been taken from a large number of quarries spread across Britain except for the South-east where the geology is such that there are no hard rock quarries. In recent years the number of quarries in England and Wales supplying ballast has fallen. This is partly because of a fall in the amount of ballast ordered, partly due to a policy of withdrawing from quarries with unsatisfactory WAVs and partly due to a change in the strategy of using the ballast wagon fleet.

Ballast wagons actually perform two functions. They take ballast from the quarry to a siding or marshalling yard near the site and act as a store for the ballast until it is needed on site. A few years ago it was thought that the wagon fleet might be better utilised if dedicated wagons ran ballast from quarries to a large stockpile (known as a 'virtual quarry') at a central location and other wagons, more suitable for shorter distances, took the ballast to sites in the locality as required.

In recent years, the number of quarries used in Scotland has remained fairly constant. Owing to the locations of branch lines in the north of Scotland and the relatively small amount of ballast used, in certain areas it is more economical to use a small local quarry and pay for road haulage to the nearest railhead than to bring ballast over long distances by rail.

10.1.9 QUALITY ASSURANCE

Whilst the specification for ballast is well proven in terms of the size criteria and is becoming research-led in terms of crushing and impact stipulations, and the third aspect of customer satisfaction is to ensure that the product received from a competent supplier is actually within the specification. In the past, this has sometimes not been the case. Often ballast is loaded into wagons in quarries and stays in those wagons until unloaded on site - usually to replace ballast removed in a routine reballasting operation under a possession. If on unloading, it is obviously outside the specification, the option of returning the ballast to the quarry is almost never possible, as a replacement will not be available at short notice, and opening to traffic on time is considered more important than the quality of the ballast that the track is laid on. Accordingly, inspection prior to the wagons arriving on site is needed. The best place to sample is in the quarry. It removes any doubt over the source of the ballast, allows sampling from a conveyor belt (to obtain a more representative sample than from a stockpile), allows the supplier to sample at the same time and emphasises concern over quality.

Quality control operations in which personnel visit quarries, sample, test, and report on the ballast should be applied to all quarries. However, whatever steps are taken to penalise sub-standard deliveries, such as compensation, or dropping the quarry from the list of approved suppliers, quality control has been an issue.

It is against this background that it is often necessary to introduce a system of Quality Assurance for ballast. The nature of Quality Assurance is to foresee and eliminate all the problems of a product or service before that product is manufactured or service introduced. In essence, the change from the previous quality control arrangements to Quality Assurance will mean that instead of monitoring the quality of the actual ballast produced, one should monitor the whole process by which the quarries produce ballast. This will include management (eg named individuals being responsible at every stage of the production and checking process), production (eg rates of progress of rock through crushing machines being in accordance with manufacturer's specifications), monitoring systems, storage etc.

The quarry will have some degree of choice as to how it implements Quality Assurance. The international standard for Quality Management, ISO:9001 2008, gives three possible models which can be implemented. Two of these seem suitable for ballast. The first is used when the system of Quality Assurance used by the supplier covers production and installation processes and the second when the system covers only final inspection and test. The desired objective in both models is the same, i.e. that the ballast should satisfy the Ballast Specification.

In the first model, the supplier will give detailed plans covering every stage from the quarry face to final storage and despatch, as mentioned above. Experts will inspect the plans, visit the quarry and discuss details with the supplier. After any changes agreed have been implemented, the QA system will be installed. The client will subsequently concentrate on inspecting or auditing rather than taking regular ballast samples. (It is to be expected that part of the monitoring system implemented by the quarry will be regular, perhaps, daily, sampling and checking of the ballast - the results from these tests to be available to scrutiny).

In the second model, the supplier will produce ballast by whatever methods suit the particular quarry without reference to the client. The ballast produced will then be stored in batches and sampled and tested by the supplier. Only if the tests show that the ballast is satisfactory will it then be released. In this case the Quality Assurance plans will involve the systems for storing the batches, sampling and testing, certificating the various batches, marking the status of the batches at various stages, procedures for release. The client will satisfy itself regarding the sizes of the batches, the size and number of samples from a batch, the laboratory procedures for each batch etc.

10.1.10 FUTURE DEVELOPMENTS

Quality Assurance will be one of the major initiatives in ballast research, development and supply in the near future but not the only one. A computer program to model the various ways in which fines find their way into ballast (tamping damage, spillage etc.) has been developed and will be used to predict the lives that can be expected from various ballasts depending on the volume and type of traffic that passes over them. Research is planned to see what benefits might accrue from using recycled ballast (life-expired ballast removed from track and washed and sieved to conform to the grading requirement of the ballast specification) on relatively lightly used lines.

10.2 PRINCIPLES OF DRAINAGE OF TRACKBED IN CONTACT WITH NATURAL GROUND

10.2.1 Classification of soil

If the trackbed is laid on natural ground (ie either on level ground or in a cutting) the control of water in the trackbed is a major factor in designing the construction layers in relation to the type of subgrade material. The types of material which are commonly encountered in subgrades are:

Non-cohesive soils	Gravel or Sand
Cohesive soils	Silt or Clay
Organic soils	Peat Organic Clays and Silts
Cemented sedimentary rocks	Sandstone, Limestone
Metamorphic rocks	Slate
Igneous rocks	Granite or basalt

10.2.2 Drainage of Non-cohesive soil

The essential property of a non-cohesive soil is that the individual particles of material are entirely separate from one another and water passes freely through the medium in the interstices between the particles. Such materials normally make fairly good subgrades and can easily be drained so that the water level remains low in the ballast, unless the subgrade/trackbed interface level happens to be below the general water table.

In this case there can be a slow upward migration of fine to medium sand under track vibration combined with water flow and a geotextile is necessary to hold down this sand.

10.2.3   The drainage of cohesive soils

A cohesive soil is one which contains at least 10% to 20% of clay particles. These particles are extremely small, and have a plate-like structure (around 100 microns across and only a micron thick or less). It is this property which in the presence of water gives a cohesive soil its characteristic of mouldability.

The pores between the particles are extremely small, and form a network of fine capillaries, with a very low permeability. By virtue of the phenomenon known as capillary suction, the faces of a hole dug in a body of cohesive soil will remain apparently stable for some time. If rainfall or some other external water supply then permeates the soil, the capillary suction is eventually dissipated, the particles are no longer held together, and the face of the excavation collapses. The length of time taken for this to happen depends on the pre-existing conditions in the pores of the material.

The geological processes which produce clay involve the application of enormous pressures from overlying sedimentary deposits over many millions of years, to a body of very fine grained sediment which is originally saturated with water. Ultimately much of the water is squeezed out, consolidating the material and imposing a highly specific interlocking pattern upon the particles which is characteristic of what is known as clay. The removal by erosion in the course of time of the overlying deposits leaves the clay in an over consolidated condition. When an over consolidated clay is strained, the particles attempt to re-orient themselves and the pores try to expand. Due to the low permeability, an additional suction, over and above the capillary suction, results. This additional suction is termed a negative pore water pressure (PWP). This negative PWP can be very high indeed, and may take many years to dissipate.

This gives a spurious appearance of stability and indeed a cutting formed in an over consolidated soil can remain stable for up to a century. Eventually the PWP reaches equilibrium with the underlying water table, and the slopes become weaker and much more sensitive to rainfall. This needs to be taken into account when designing earthworks in over consolidated clays, and will be further considered in section 10.8.

Normal weathering reduces the effect of over consolidation at and near the surface, so that the top one to four metres of ground below undisturbed agricultural land on an over consolidated clay will have reached equilibrium. Similarly, cohesive soils such as alluvium, which have no history of consolidation by overlying material, will be at or near equilibrium in the undisturbed state. If a cutting is driven through such material, or the material excavated is used to form an embankment, the strength of the material and the stability of the slopes will not change with time.

So far as subgrade and drainage are concerned, the points to note are firstly that the subgrade in a cutting driven through an over consolidated clay will be subject to the same kind of long term deterioration as the slopes, and this is one of the factors leading to Bearing Capacity Failure.

A water table can be identified in a cohesive soil, in spite of its low permeability. In a cohesive soil the water table is the level at which the PWP is equal to the atmospheric pressure. Above this level the soil pores may still contain water, due to capillary action, but the PWP will be negative. Drainage will only be effective below the water table, where the PWP is positive. Even below the water table, because of the small size of the pores and the fineness of the interconnections between them, pore water cannot immediately escape into an excavation, and it follows that drainage channels or pipes installed in cohesive soils will only drain water from their immediate vicinity, and many closely spaced drains are required to drain effectively a body of cohesive soil. This is not usually practical.

On the other hand, if water arrives on the surface of such a body of material, either as rainfall or as run off from adjacent land, it is probably more effective and economic to intercept and remove it before it has opportunity to alter the PWP and weaken the material, rather than to try to drain it off after it has done the damage. This is the main reason for putting surface drains in cohesive soils.

10.2.4 Drainage of Other Materials

Peat is a special case, and has a variety of textures. It will shrink if drained, or if subject to repeated or static loading, but rarely otherwise. Fibrous peat has a high angle of friction which is mobilised fully only after large strains have occurred. Drainage of peat may have a deleterious effect, as it results in large settlements which may affect the fixed installations near the track. The track itself would also require regular lifting. Stiffening of the trackbed may be preferable in this case.

Cemented sedimentary rocks and metamorphic rocks are often permeable by water and are often aquifers in their own right. However, they are usually much stronger in the un-weathered state than cohesive soils and provided suitable means for drainage are adopted, a blanket between ballast and subgrade will normally suffice even if the material is waterlogged.

Igneous rocks are very hard and impermeable, and are not likely to present any problem involving drainage.

10.2.5 Design of Drainage Systems

Drainage systems which collect water from the ballast should be placed at the minimum depth consistent with avoiding ponding of water in the ballast or above an impermeable layer. Drains designed to intercept and collect water (eg from slopes), can be much deeper. Carrier drains may be channels or pipes set at a depth determined by track gradient and the outfall. Tests indicate that 80% of the water falling on track flows along the interface between ballast and sub-ballast layers to the side of the track. Only 20% penetrates the lower layer and, if this is in good condition with proper cross-falls, it will be almost completely drained within an hour after cessation of rainfall.

10.3   DRAINAGE IN THE CONTEXT OF TRACKBED RENEWAL

10.3.1 Control of water falling onto the track

Many trackbeds are renewed down to a level compatible with the site, equipment and possession time available leaving a polluted sub-layer in position under the new blanket or blankets which are placed to proper cross-falls. This old sub-layer must now be considered as part of the subgrade system. It will not drain freely and will generally be saturated with rain water flowing laterally through the ballast. Unfortunately this water satisfies the negative pore pressures in the subgrade on its way to the drain and, if the soil is sufficiently weathered to respond to some threshold conditions of stress, then cumulative deformation can occur. In such locations the layer next to the subgrade should incorporate an impermeable film. The edge of this film should be designed to continue to the drain invert.

10.3.2 Sand Blankets

A blanket in trackbed terminology is a permeable layer of *fine* granular material (as distinct from a blanket in dam construction which is impermeable). The blanket in new construction is placed directly on to a cohesive subgrade which varies from a weak clay to a mudstone. It is not necessary if the subgrade is non cohesive.

In the repair of an old trackbed with erosion pumping failure, the blanket may be placed at any convenient level permitting adequate thickness of ballast. Blanket thickness can be reduced to 100 mm under a suitable geotextile separator or may be up to 200 mm thickness if it is to contain a waterproofing film of plastics (a geomembrane), usually polyethylene, to prevent water reaching a subgrade liable to Bearing Capacity Failure. The geomembrane is not there to control pumping and is therefore not an alternative to a geotextile.

The particle size distribution, or granulometry, of the blanket should be designed to minimise penetration of slurry into the ballast. It is quite possible to have a distribution such that penetration is hardly perceptible (say 3 mm per century). A recommended granulometry is given in Table 10.3. The granulometry in angular sands derived from rock crushing does not conform to this, and if such materials are used slurry penetration of 50 to 75 mm in only a few years is quite possible. To obtain the most efficient blanket a natural sand should be obtained, although suitably designed mixtures of naturally occurring sands and crushed rock dust can be made to conform to Table 10.3, and may be much cheaper. The efficiency of some crushed rock dusts can be enhanced by using geotextile.

TABLE 10.3

Recommended granulometric analysis of blanketing sand

B.S. Test Sieve Nominal Aperture Size	Cumulative % of weight passing B.S. Sieve
	100
2.36 mm	80 - 100
1.18 mm	70- 90
0.60 mm	48- 76
0.30 mm	24- 60
0.15 mm	5-42
0.075 mm	0 - 1 0

Notes to Table 10.3:

(1)   If there is more than 4% of material passing the 0.075 mm B. S. Sieve it should not contain sufficient clay or fine silt to exhibit plasticity.

(2)   If the top of the blanket is within 230 mm of bottom sleeper level, then in addition, the Uniformity Coefficient, Cu, must have a value of at least 2.0, where $Cu = D_{60} / D_{10}$

D_{60} = diameter of 60%   size of sand

D_{10} = diameter of 10%   size of sand

A blanket must stretch across the whole width of the trackbed and measures must be taken to prevent slurry entering sideways above it from adjacent tracks or cutting slopes. When opening cribs for inspection as part of an investigation for likely remedial measures do not remove or penetrate a sand blanket unless there is sand available to make good and a permanent way manager to ensure that this is done. Perfectly good blankets over a century old can be (and have been) destroyed in a decade by enthusiastic engineers taking out trenches at frequent intervals.

10.3.3 Geotextiles in Trackbed

The term "geotextile" means a permeable, woven or non-woven fabric-like material made from polymerised resins used in civil engineering construction in sheet or strip form. Geotextiles must not be confused with geomembranes which are intended to be impermeable, nor with geogrids and such like which have an open, net-like texture and quite different uses. The term covers a wide range of products of varying strength, thickness, permeability (either across or in the plane of the fabric), and pore size. Optimistic claims have been made about the value of geotextiles to replace or augment blanket layers. Such claims have not always been fulfilled and in the light of experience suppliers are becoming more realistic in their approach to their use. Experience gathered via the UIC in addition to experience has enabled the following conclusions related to the use of geotextile in trackbed to be drawn:

(1)   Geotextiles are of no use in cases of dirty ballast pumping where attrition products exist or will occur due to poor ballast *above* the fabric.

(2)   There is great economy to be derived if a geotextile can be placed over a dirty or slurried ballast layer (eg during ballast cleaning or reballasting) without having to introduce an extra granular filter layer. This can only work if the fabric does not disintegrate in the ballast. So far the only geotextiles known to last 10 years or more and which provide substantial dynamic filter protection are non-woven fully heat bonded fabrics having a mass per unit area of 240 g/cm^2 or more. The majority of non-woven fabrics are mechanically bonded by a needle punch technique and it is generally accepted that they soon disintegrate in ballast under traffic loading.

Certain heavy grade mechanically bonded non-woven fabrics having a mass per unit area of over 1000 g/cm^2 with additional bonding by impregnation with resin have shown promise in Canada. These have been found to give adequate performance placed in the ballast without excavating to the subgrade even with the very heavy axle loadings ruling in that country.

If a fabric is laid as part of ballast cleaning or relaying then stray ballast particles standing proud would stretch any fabric placed over them. To withstand this distortion a high degree of elongation before failure is required for geotextiles for trackbed use. This dictates that a non-woven fabric should be used as wovens do not permit such elongation before failure. Some composite fabrics in which a needle punched material is bonded to a woven carrier have been developed but these are more expensive and should be used only on carefully levelled and compacted sites.

(3)   Geotextiles do not filter clays and silts under dynamic loading conditions. The various fabric properties including opening (pore) sizes derived from static filter tests, and permeability both across and in the plane can be established by British Standard tests and included in the manufacturers' specifications: they cannot, however, be interpreted for behaviour in track. Similarly there ais little test data on durability and long-term abrasion resistance in track. Dynamic filter tests in the laboratory using simulated track loading and a silty clay show that clay slurry will pass through all commercially marketed fabrics. In the real conditions of track, rainfall cycles, cant/crossfall and ballast shoulder impoundment it seems that some slurry passes slowly up through the fabric to about 60 mm above it and then practically stops leaving free draining ballast above it to bottom of sleeper level. That is to say, the conditions for pumping track do not occur and the rail holds its position. This probably depends on the rainfall and drainage conditions prevailing, as already mentioned.

The reason why geotextiles with a given apparent pore size pass slurry under traffic loading, whilst sand blankets with a nominally equivalent void size will control it for over a century, is probably the thickness of the sand blankets. In a sand blanket complex processes of pore pressure dissipation and partial clogging with fine particles may occur in the first few millimetres from the clay-sand interface. This is implied by discoloration of this layer of sand by the clay. The thickness of this layer exceeds that of the geotextile once the latter is compressed under use.

(4)   When a geotextile is placed over a sand blanket layer it acts as a separator to prevent ballast particles punching into the blanket. This means that a thinner blanket can be used of 100 mm thickness. To specify thinner than this, 50 or 75mm would be possible, but would not be reasonable, as random stone particles left on the interface *below* the blanket would give locations of no sand where slurry could

pass. The geotextile protects the blanket by being placed *on* it. There is no point in placing the fabric *under* it as it serves no filter function, and does not reduce the required sand thickness.

(5)   A geotextile is not an alternative to a geomembrane in trackbed. Each has a different function, the former to help filter slurry by itself or as part of a sand filter, and the latter to waterproof in order to divert water from a subgrade liable to bearing capacity failure. It is possible to serve both functions, by having the geomembrane *in* the sand blanket, and the geotextile over the blanket.

(6)   The tensile resistance of a geotextile is mobilised at a high degree of strain and it cannot contribute to the rigidity of the trackbed. Even geogrids, with their superior strain resistance, would have to be placed so near to the bottom of the sleeper to have a measurable effect, that they would prevent tamping or similar maintenance techniques.

(7)   Any geotextile, geogrid, or membrane placed in a trackbed, must be positioned sufficiently deeply below base of sleeper to avoid ravelling when the track is subsequently ballast cleaned.

(8)   Techniques are being developed with various on track machines (eg the Track Gopher) to place fabrics quickly in trackbed which offer great economy especially under S&C work.

10.3.4 Geotextiles in Lineside Drainage

Any attempt to filter water entering a drainage system results in accretion of detritus at the sand or fabric face, with reduction of rate of flow where the latter is least desired. High rates of flow to keep the zone immediately beneath base of sleeper free from slurry depend primarily upon a clean ballast shoulder (see below). Any geotextile in the trackbed should be detailed to reach to the end of the ballast and pass to the base of any pipe or channel.

10.3.5 Side Drains in Cutting

Once water has left the trackbed and is past the ballast shoulder, it should be collected in a suitable collector drain, as shown in Figure

10.1b.

Since one object in designing the system of track drainage is that it should minimise maintenance, the use of channel drains is to be recommended. Channel drains are readily accessible for cleaning and deal with rainwater. They can be laid at very shallow gradients, and can discharge at catchpits to deeper piped carriers if necessary. Piped collector drains if used should have open joints, or be perforated to allow water to enter the pipe from the surrounding material. Channel drains are commonly made from precast concrete, while pipes can be of glazed earthenware, galvanised corrugated steel, and, now coming more into use because of ease of handling, plain or perforated pvc or polypropylene.

Generally speaking piped side collector drains in a cutting should follow the track gradient, and their diameter may have to be designed from a knowledge of the hydrology of the area.

10.4   DRAINAGE OF EMBANKMENTS

Since embankments are frequently made from cohesive soil, the interface between the ballast and the subgrade needs to be protected in much the same way as that in a cutting, ie an impermeable membrane may be required.

However, water moving laterally across the fall of such a membrane does not then require to be led into a collector drain. It may be allowed to find its own outfall, as the experience of the last century, in UK at any rate, shows very few cases of rilling or surface erosion of embankment faces, at least once a suitable vegetative cover has been established. One beneficial effect from the depredations of burrowing animals is that their activity reduces the tendency to surface erosion.

In order to avoid feeding water into the core of the embankment, which may lead to slipping, the ballast should be laterally free draining. Since this quality tends to become impaired towards the end of its life. One relatively economical way of extending the effective life of the ballast is to clean the shoulder, as distinct from the whole cross section.

Machines are available to carry out shoulder repair work and reinstate lineside drainage. They are usually of vacuum or mechanical type technology.

10.4.1 SEPARATION OF ADJACENT TRACKS

A useful technique is the placing of a vertical layer of geotextile in the middle of the six foot when one track is blanketed or ballast cleaned and the adjacent track contains some slurry (even if not yet pumping). The fabric will filter slurry to prevent the clean track acting as a drain for the dirty track.

10.5 CAUSES OF DETERIORATION OF CROSS AND LONGITUDINAL LEVEL

Leaving aside special subgrade conditions such as peat, underground fires, springs etc, the modes of failure resulting from ballast and/or subgrade deterioration are of two main types; pumping failures and bearing capacity failures.

10.5.1 Pumping Failure

This is a track movement associated with the presence of a liquid slurry at or above sleeper soffit level. When the slurry is below this level the sleeper is stable, that is to say movement is not due to lubrication of the ballast by slurry. There are two forms of pumping failure:

(i) *Dirty Ballast Failure*

Dirty Ballast Failure (DBF) is formed from attrition products from mediocre ballast and its frequent tamping, from windblown deposits, brake dust, concrete sleeper erosion, old ash, and dirt dropping from vehicles, collecting in the ballast voids. DBF is remedied by cleaning the ballast.

(ii) *Erosion Pumping Failure*

In the case of Erosion Pumping Failure (EPF), the slurry derives from the cohesive subgrade. The contraction and dilation of the ballast voids causes the viscous slurry to be pumped up from the cohesive soil or mudstone to the underside of the sleeper. The permanent remedy is a sand filter layer, forming a blanket of correct granulometry across the whole width of the track. It is not necessary to reach the subgrade in order to install the blanket. EPF is not related to the strength of the subgrade soil.

Each form of pumping failure causes a sudden loss of rail level and of some alignment shortly after the onset of rain. If the level of slurry can be lowered below sleeper soffit level by forming channels in the four foot ballast or by removing ballast at sleeper ends or by installing flexible perforated tubes in the ballast then the problem is alleviated, e.g. at wet spots.

The following are examples of the effects of pumping failures:

(a) *Wet Spots*

These are DBF locations where the water is impounded by a polluted ballast shoulder, sufficiently to allow slurry to form at or above Bottom of Sleeper Level (BSL). Release of this slurry by opening out cribs to the cess gives immediate but temporary relief. A useful technique is the installation of a perforated flexible plastics tube in the crib below BSL to allow local drainage. The tube should pass through the shoulder to the cess from where it may be rodded clean. It will be damaged during subsequent tamping operations but will continue to function through several of these before requiring replacement.

(b) *Slurry*

Slurry may be produced from any cohesive soil however hard and contains proportions of silt and sand which can affect the efficiency of filter layers placed over it. Some old trackbed layers may be of ash placed over half a century ago and this breaks down to form a black slurry mainly of silt size. This is easily identified when dry by its grey matt appearance with many glistening particles. Old polluted trackbed layers which produce slurries have voids containing many silt and sand particles; such layers act as a partial barrier to cohesive slurries if there are such beneath them. To remove such a trackbed layer by cleaning because it is in fact a DBF would be followed by EPF as the cohesive soil arose from the subgrade. A full check on the conditions is necessary to establish the need or otherwise of a blanket.

(c)   *Side Slurry*

Maintenance methods in the past have included digging out slurried ballast by hand and throwing the material on to the cutting slope or into the ten foot way. Rainfall can carry this slurry back to the top ballast, sometimes over the top of a good blanket subsequently installed.

(d)   *Springs*

Tie upward flow of water from springs in the subgrade may be of such a velocity that the finest particles of blanket would be dislodged if not held by the other coarse particles in the layer. The effect of vibration under traffic is sufficient momentarily to lift small particles up to 150μm size over a short vertical distance. The cumulative effect is to alter the granulometry of the blanket and to impregnate the ballast around the sleeper with fine silty sand. A similar effect occurs over soft sandstone without a blanket. This hydraulic condition can be found over weak fissured rocks and some stiff clays, in cuttings and in tunnels without an invert. One approach is to dig and break open fissures to the side drain to relocate water flow. If there is a blanket a geotextile should be placed over it to prevent the upward rise of fine sand.

10.5.2 Bearing Capacity Failure

Bearing Capacity Failure (BCF), sometimes called strength failure, can be defined as a continuous loss of rail level which starts some days after the onset of wet weather, and continues cumulatively until after at least a week of dry weather. In this respect, the weather response is similar to track over an embankment slip, to which the problem is closely allied.

There is a pattern of rail movement, with rail lengths of 3 to 10 sleepers being affected, usually on the low rail or the cess rail. Investigation has shown that the maximum depression of the subgrade generally occurs at a point 150 to 250mm away from the running edge towards the sleeper end.

A common feature is a heave of soil in the cess adjacent to the rail which has lost level. This 'heave' in the cess acts as a counterweight to the soil mass slipping from beneath the track; removal of the heave material causes swifter loss of rail level. Conversely a high cess and a wide ballast shoulder reduces this tendency.

10.6   REMEDIAL MEASURES FOR BEARING CAPACITY FAILURE

Conventional methods for dealing with bearing capacity failure have depended upon excavation which has the effect of:

(a)   increasing subgrade depth to lower pressure intensity on the soil surface;

(b)   changing the Track Modulus by virtue of the new ballast and sub-ballast layers put in, which, in turn, redistributes pressure;

(c)   reducing the access of water to the subgrade when low permeability layers or waterproof membranes are inserted;

(d)   improving the drainage of water from the subgrade surface by trimming the subgrade to fall to the cess or six foot.

As the types of soil subject to this type of failure are among those which can lead to EPF, the opportunity is invariably taken during excavation, to place a blanket layer to prevent the rise of clay slurry. Blanketing as such is not a cure for BCF but for EPF.

Instead of relying on the inefficient method of general drainage of an area to avoid BCF, it is better to prevent access of water at the outset. Water percolating through the trackbed lowers the bearing capacity of the subgrade on its way to the drain: when it gets there it is too late.

The method of placing a polyethylene film in the middle of the sand blanket to prevent local access of water, has been used for many years with success. The film must be extended to the edge of the trackbed, and must reach any drains which may be installed.

There may be an old blanket in the trackbed, functioning well to control erosion, but which nevertheless allows enough water to reach the subgrade to cause BCF. One technique to remedy this situation is to use a ballast cleaner to remove 200 to 300 mm of ballast from beneath the sleeper, and to waterproof the cut exposed ballast surface with a spray of cationic bituminous emulsion. The ballast cleaner can be adapted simply for the purpose, by attaching a spray bar to the cutter guide, and connecting the spray bar to suitable pumps, with a bitumen tank wagon towed by the cleaner. It is not necessary to remove track nor to excavate to the blanket. An immediate improvement in bearing capacity follows treatment.

A good flow of water through the top ballast is necessary to remove attrition products and dust deposits which would otherwise foul the ballast and cause DBF. This is the most common sub-sleeper defect and is made worse if the wide ballast shoulders necessary for stable track are not kept clean and water is impounded. The use of a shoulder ballast cleaner is recommended in this case, as it does not disturb the existing sleeper support and permits high speed running immediately after ballast replacement.

When planning trackbed repairs adjacent to mass retaining walls or similar structures having shallow foundations, it must be remembered that the stability of the wall depends partly on the passive earth pressure from the material in front of it. Removal of the track and ballast etc. for renewal removes at least part of this passive pressure, and it is not fully restored upon track replacement. Soil strains will continue in front of the wall for a year or so and may lead to general failure of the wall. Geotechnical advice should be obtained before such work is undertaken.

10.7   THE DIAGNOSIS OF THE CAUSES OF ALLAST/SUBGRADE FAILURE

It is not always a straightforward or easy task to identify the true cause when faced with a section of track which is subject to the development of lateral and/ or longitudinal defects in level and/or alignment. To assist in this process a diagnostic process can follow an investigation initiated by the observation of geometrical faults on the track at a site.

1. Faults are observed above the bottom of the sleepers, due to the state of the sleepers, rails or fastenings, including wear of the base of the sleeper.

2. Faults in track geometry reappear after correction of defects in track material.

3. The embankment on which the track is laid appears to be unsuitable e.g. the embankment is unstable due to slips; the underlying ground is weak or subsides; the top of the embankment is too narrow to accommodate the ballast bench properly.

4. Faults reappear after the stabilisation/broadening work has been completed.

5. Clear running water in the ballast; the water could be turbid after the passage of a train; the behaviour of the water varies according to whether it springs from fissured rock or filters through permeable gravel and indicates the presence of an underground stream or spring.

6. The underlying ground contains very compressible material such as peat; the embankment material is combustible and there is evidence that underground combustion is taking place

7. Ballast round the sleepers is clean but inadequate as regards quantity and/or quality, e.g.

   Ballast surface well below sleeper tops; Stones worn, spherical or rounded; Stones too small

8. The ballast contains wet or dry fine particles extending below the sleepers and liable to form mud which could rise above the bottom of the sleepers after rain, or there is visible splashing of the sleepers, ballast, and rails. The slurry may include sand and silt particles as a result of wear of the underside of concrete sleepers. The general state of the

ballast should be noted and not simply joints and welds where very localised phenomena may be apparent.

9. Possibility of Bearing Capacity Failure (BCF). Indications are heave of ground in the side path or between tracks; or weakness appears at least two days after rainfall and persists for at least seven days dry weather after rain. Heave of ground in the fourfoot between the sleepers is symptomatic of overloading of the subgrade due to combination of heavy axle loading, shallow ballast, and weak subgrade material.

10.8   GENERAL PRINCIPLES OF EARTHWORKS STABILITY

10.8.1 Preliminary Investigations

Brief reference is made below to the reasons why quite substantial earthworks are sometimes necessary when constructing railways. The characteristics of the naturally occurring materials through which cuttings have to be driven, and of which embankments are made, are mentioned above.

In constructing new earthworks for roads or railways today, detailed geotechnical investigations are undertaken to determine, firstly the safe angle at which cutting slopes may be excavated, and secondly, the correct treatment to be applied when the excavated material is formed into an embankment, so that both cut and fill will remain permanently stable.

At the outset a preliminary study may give adequate information to specify the route corridor from geological maps and memoirs, topographical maps and aerial photographs. Earth satellite imagery with interpretation of selective wavebands by specialists in remote sensing can indicate important features. Water table conditions may vary throughout the year from those obtaining at the time of exploration.

For more localised investigation, the type of equipment (augers, percussion and rotary tools, penetration heads, loading plates, pumps), instruments (piezometers, inclinometer tubes, seismometers, resistivity meters, gravimeters, etc.) must be chosen according to conditions. Relevant disturbed or undisturbed samples should be procured for testing. According to the type of ground, the construction and the design philosophy applied it may be necessary to carry out full scale site testing with long-term monitoring of instruments.

These investigations will reveal inter alia the classifications and properties of the soils through which cuttings must pass, and of the material available for forming embankments. The most important property required to be known in order to design an earthwork is the long term shear strength of the soil.

10.8.2 Shear strength of different types off soils

The shear strength of a soil may be considered as:

- internal friction, which is the resistance due to interlocking of the particles;
- cohesion, which is the resistance due to the forces appearing to hold the particles together in a solid mass.

The shear strength due to internal friction depends upon the pressure across the sheared surface, whereas shear strength due to cohesion is a property which varies with the rate and extent of strain and with drainage during strain, and is independent of intergranular pressure.

Reference was made above to the classification of soils into cohesive and non-cohesive types. Non-cohesive soils such as sand and gravel derive their strength from internal friction, whereas most clay soils, although they may contain material which has both cohesion and internal friction, fail according to their geological loading history. Material NOT previously subjected to over-consolidating pressures behaves for calculation purposes as though it possesses cohesion only (ie the quick undrained strength can be used to assess stability). This assumption is the basis of the analysis of the next section.

10.8.3 Designing Earthworks

The details of the earthwork design process are outside the scope of this book, but it will assist in the understanding of the underlying reasons behind earthworks instability, to have at least a qualitative appreciation of the mechanical principles involved.

Short Term Stability of Normally Consolidated Soils

If the material of which an earthwork is composed is assumed to be purely cohesive and completely uniform in its structure and properties, then for the earthwork to be stable, a state of equilibrium must exist across any and every cylindrical surface within the material, between the mass of material contained by the surface tending to slide outwards and downwards, and the shear resistance across that surface. This situation is shown diagrammatically in Figure 10.3 which shows a part section across an embankment or cutting, and a circular arc ST of radius R drawn at random about a centre O. The centroid of the area bounded by the arc ST and the top and side of the earthwork SVT is at G.

Then the moment of the block SVT per unit length of earthwork about O is Md. The tendency of block SVT to rotate anticlockwise is resisted by the shear resistance along the circular face ST, and this force, totalling Aq per unit length of earthwork has a moment about O ARq. Then for the block not to rotate, the shear resistance q across the face must not be less than:

$$q = \frac{Md}{AR}$$

If the shear strength of the clay is q_{ult}, then the Factor of Safety F is given by:

$$F = \frac{q_{ult}AR}{Md}$$

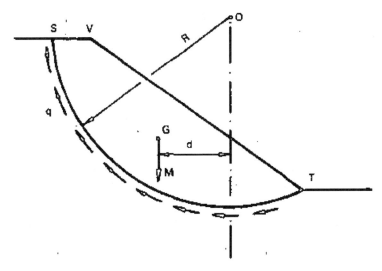

Figure 10.3 Analysis of earthwork stability

M = mass of unit length of the earthwork

d = horizontal distance between the line of action of M and the vertical through O

A = length of arc ST

q = shear resistance along arc ST

For normally consolidated soils, if the slope is stable after the cutting is formed, then with good drainage it should remain so indefinitely.

Internal friction can be taken into account by adding a component equal to the weight of the slice multiplied by the coefficient of internal friction across each element of the arc of shearing. The direction of this force must be calculated tangential to the slip surface if it is circular.

The art of the designer lies in determining for any given slope and depth of proposed cutting, the location of the circle which gives the minimum required value of F. This is known as the critical circle. The operation is potentially tedious, but as a result of experience, design tables are obtainable to give guidance in selecting the critical circle. Such tables may presume a hard layer at depth, to limit the extent of possible failure surfaces: this is a "Total Stress" table. For long term stability with pore water pressure involved (effective stress) there are more complex tables which give a first assessment of F.

Long term Stability of Over-consolidated Soils

For slopes in cuttings and embankments in over-consolidated soils (including mudstones and shales), the position of the water table is used in the calculation of forces in each slice of the moving mass and is represented by a dimensionless quantity, the hydraulic gradient r, in the formula. This will be an *effective stress* calculation, a good example of which is Professor Bishop's method (see Code of Practice on earthworks BS 6031). This can be applied in a simplified or rigorous form. Software is available to solve this problem, and several layers of soil with different parameters can be accommodated in the model. The printout will give various values of F for different centres of rotation, as specified by the user. It may be necessary to run the program more than once in order to find the centre of rotation which gives the lowest F.

For some soils, such as London Clay, and Upper Lias Clay, the strength parameters have been derived from analysis of many old failed sites, and these can be used to give a 50-year slope stability value. These latter should be used with confidence rather than parameters derived from tests on samples taken on site, as test values are usually optimistic and require the assumption of a high value for the Factor of Safety F.

10.8.4 Constructional Detailing

In designing and constructing cuttings the control of water is very important.

Drainage measures should take the form of preventing water reaching the slope and of removing it from the slope. Unlined ditches behind the crest of the slope increase the hazard. Drainage trenches, whether behind or below the crest, should be designed to intercept water and have impermeable membranes below them and on the downhill face to prevent water once collected from re-entering the soil. Modem counterfort or buttress drains differ greatly from the original open forms. Like all drains, they should be lined with a geotextile layer to prevent erosion behind and fouling within them. The top one or two metres should be composed of impermeable material, either compacted clay fill or a system of stone and plastics membrane to prevent surface water reaching deep into the ground, where it could increase the pore water pressure at likely slip surfaces.

Once the cutting has reached its designed depth it is essential that construction traffic is kept off unprotected surfaces that will ultimately form the surface on which ballast is laid, and the drainage of the subgrade should be completed as soon as possible to minimise softening of this all important layer.

The nature of the ground underlying an embankment must be investigated, as, if it is very weak, failure may occur during or shortly after construction. Various means can be employed to prevent this, eg, subsidiary embankments can be constructed as counterweights, a geotextile or geogrid layer can be laid over where the embankment is to be tipped, the underlying ground can be precompacted, or the embankment can be constructed higher than its final level to encourage rapid settlement, and the surcharge removed after the desired settlement has been achieved.

The treatment of fill will include ensuring maximum compaction at an optimum water content or at minimum air content using the Moisture Condition Value test in modern practice, as well as determining the correct slope. Operations often have to be suspended in the winter or during prolonged periods of bad weather.

If the site is constrained so that desirable slope angles are unobtainable, then measures such as retaining walls or reinforced soil are adopted.

10.9   NEED FOR REMEDIAL WORKS TO EXISTING EARTHWORKS

Most railway earthworks were constructed at a time when knowledge of soil mechanics was at best rudimentary and much effort in maintenance and remedial works on railways today result from ignorance and bad practice during construction, which may of course have been up to 150 years ago. The remainder of this chapter is concerned mainly with these problems rather than with those involved in designing new ones. Nevertheless, the need does occasionally arise to widen existing earthworks and to meet this need brief reference will be made in Section 10.16 to the useful techniques of soil nailing and reinforcement.

Maintenance and repair of existing railway earthworks is made harder because of access difficulties, especially in urban areas, and because of the confined space between track and boundary within which the work has to be carried out. Methods not involving excavation are usually to be preferred, such as drainage, grouting, soil nailing, anchoring, and sheet piling.

10.10   TYPES OF INSTABILITY

Non-cohesive soils such as sands or gravels have well defined angles of repose and provided that the slope is made flatter than this, a cutting or embankment made from such material on a sound foundation should remain trouble free indefinitely, if well drained. Protection against surface erosion may be required.

If cohesive soils are involved, as is most commonly the case, the problem is more complex. It is well known that in a stiff clay, an excavation with vertical faces will stand unsupported for some time, whereas as a result of weathering action and the dissipation of negative pore water pressure with time, the slope at which the same material will be stable over the long term may be as flat as 1:6 or even flatter. The effect of pore pressures in promoting this degradation has already been noted. The causes of instability include:

- shear strains such as slips and bearing capacity failures; existing ancient slip surfaces;
- volume change, ie shrinkage or swelling of certain soils or densification of soil material;
- erosion both external and internal (internal erosion is often identified with "piping"

Some of the more significant of these causes are discussed below.

10.11   SLIPS

Any movement of the surface layers of material of a slope as a body towards the toe of the slope, may be classified as a slip. When this occurs the profile of the slope is modified, becoming steeper towards the top, and a tongue of material is forced forwards at the toe of the slope, as shown in Figure 10.4.

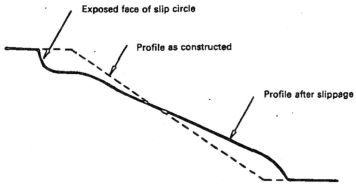

Exposed face of slip circle

Profile as constructed

Profile after slippage

Figure 10.4 Profile before and after a typical slip

Referring to Figure 10.3, it is clear that if there exists a circle where the required value of q exceeds the shear strength quit of the material of which the earthwork is made, then the earthwork will fail by rotation of the material in block SVT along the surface ST. Clearly as the material rotates the disturbing moment is reduced because G moves closer to the vertical through O, and equilibrium is restored when this becomes less than the residual restoring force across the slip circle. When such a slip occurs in an earthwork which may have been stable for years, this is an indication that along some critical surface within the body of the earthwork, the shear strength of the soil has weakened to such an extent that it will no longer support the mass of material outside and above it. Such a surface is known as a slip surface (sometimes a slip plane, but since the surface is as is seen from the analysis in Section 10.8.3, not necessarily, or even usually, a plane in the geometrical sense, the term slip surface is better). Shallow, planar slips are more common in steep, modern, well-compacted embankments made from over-consolidated clay.

10.11.1 Causes of slips

Slip surfaces can develop for a number of reasons, and it is important that the cause is identified before deciding on the remedial measures to be adopted.

In some cases, the local geology can have an influence. If for example a cutting passes through strata which dip across the railway, and the strata include one or more bands of material of significantly different permeability and/or lower shear strength than the main body, there may be a tendency for a failure surface to form, on a plane sloping towards the railway, over which the super-incumbent block of material may slide. Alternatively a change of properties may serve to locate a slip circle. Invariably a rise in the water table triggers movement at the critical surface.

Another very common cause of slipping is surface weathering. In some types of clay the internal structure of the surface layers can be broken down by repeated wetting and drying. This leads to so-called surface slides affecting a layer only a metre or so thick on the surface of the slope.

If a cutting is driven through over consolidated clay, then as explained in 10.2.3, the shear strength of the clay may undergo a slow deterioration over a long period as the pore pressures adjust themselves to the loss of overburden. This phenomenon can lead to very deep seated and troublesome slips.

Deep seated circular slips in cuttings can be encouraged by:

- Deep cracks which form in some clays during prolonged dry weather. These not only shorten the potential length of the slip circle along which restraining shear can develop but also admit water to the clay at a deeper level, thus weakening it. The effect of this can be envisaged from Figure 10.3 If a deep crack forms at S clearly the length of arc A will be reduced.

- Drainage ditches near the top of the slope can perform the same function as the cracks referred to in the previous paragraph.

- Surcharging near the lip of the cutting will add to the disturbing moment and may be sufficient to trigger off movement.

- Excavations at or near to the toe of the slope may alter significantly the location of the critical circle.

On embankments slips can be encouraged by:

- Surcharging.

- Excavations at or near the toe of the slope, often outside the boundary fence.

- Poor detailing in the construction process (eg if top soil and vegetation is not removed from the underlying ground, this can form a plane of cleavage between the tipped material and the ground along which a slip can develop).

- Ballast cleaning allowing ingress of water to a ballast pocket.

10.12   VOLUME CHANGE AND SURFACE EROSION

Alluvial soils with a high water content can subside early in the life of a bank due to the weight of the bank squeezing water out of the pores. The situation usually stabilises itself in the course of time and is not often a problem with older embankments.

Whilst increase in moisture content due e.g. to the felling of trees can lead to clay swelling with serious consequences to sensitive structures, the movements involved are not large enough to cause problems on the track.

Surface erosion is essentially a drainage problem. Once a good vegetation cover is achieved serious erosion of the surface of an earthwork rarely occurs in UK. New surfaces can be protected by a sheet of open-textured geotextile of plastics or natural fibre into which seed is dropped. The seed is held in position until a good root system is established.

10.13   REMEDIAL WORKS

This section mainly concerns the stabilisation of slips. Methods of treating slips may be seen as falling in principle into four categories: reducing disturbing forces by reprofiling increasing soil strength controlling water in the system mechanical support to resist deformation.

Few remedial measures can be classified exclusively as either temporary or permanent although the objective of any investigation into a slip should be to design an immediate and permanent treatment. Sometimes the time element or some local feature may prevent an ideal solution being obtained, and occasionally the scale of movement may be so great that there is no economic justification for the technical measures that would be required. It may then be necessary to consider how the problem may safely be tolerated, or alternatively whether it is necessary to relocate the railway elsewhere.

10.13.1 Detection of movement

The design of measures to prevent or stabilise a slip should ideally be preceded by a site investigation along the lines outlined earlier. However, it cannot be too strongly emphasised that by far the best method of dealing with earthwork instability is to identify and cure the problem before it becomes an emergency which blocks the line or delays trains. With this in mind it needs to be mentioned that there are a number of indicators to earth movements which will often warn track engineers of the possibility of trouble before it develops into an emergency. These include:

- Distorted fence lines
- Fence posts tipping over where there is no sign of animals having pushed them over
- Tree trunks tipping significantly away from the vertical
- Cess heave (Note: this may also be an indication of trackbed problems)
- Loss of cess
- Cracks in the ground surface
- Loss of top and/or line which recurs after correction.

If any such phenomenon is observed the local civil engineer should immediately be alerted, but another device available without recourse to skilled assistance should not be forgotten. This is, to drive a line of stakes across the suspect bank slope. The line should start and finish in sound ground, and should be driven in as straight a line as possible. Movement will be obvious immediately it occurs, and it may even be possible to keep a history of movement, by recording at intervals, the offset of the stakes from a string line connecting the ends of the line.

In all this it must be remembered that the track maintenance staff and section managers are the eyes and ears of the railway. There is no one else to notice if things start to go wrong. The same applies to activities outside the boundary fence that might adversely affect the stability of railway earthworks. It will be realised from what has been written that the digging of drains or ditches, or certain other activities such as the creation of tips, storage stacks, even perhaps building tall buildings, near to the edge of a cutting, may spark off instability, and local staff should not assume that "the management" knows all about developments "in the pipeline". So any piece of intelligence affecting the neighbours' property should be reported.

A more sophisticated process which can be applied to identify a surface of movement if one is present, is to arrange for a pattern of 12mm bore plastic tubes to be driven into the ground over the suspect area. These tubes are inserted into holes formed by forcing into the ground a 30mm diameter driving head. This often consists of sections of pipe with screwed unions and a solid head. When a depth considered sufficient to have penetrated the slip plane is reached the pipes are withdrawn and the tube inserted.

The tube is then plumbed with a short steel mandril attached to a length of cord, to verify that it is clear of obstruction, and its depth and the "reduced level" at its upper end are recorded together with reference data to identify it on subsequent visits. When movement occurs, the shearing action along the slip plane will distort the plastic tube, and the mandril will not now reach its bottom. It will however indicate the level at which distortion is taking place. If a number of tubes are inserted, the three dimensional shape and extent of the slip plane can be determined. The tubes can also be used to determine the level of the water table. This system works for slip movements of 10mm or more. If it is desired to monitor very small movements, rates of movement, and directions of movement as they change with time, then an electronic inclinometer can be used. A borehole is especially lined with plastics guide channels and the inclinometer is lowered and readings taken every 0.5m. Results can be processed manually or by computer. The cost is very much greater than the simple mandril method.

10.13.2 Embankment Slips

As shown in Figures 10.5 and 10.6, if an embankment slip is accompanied by loss of material from the cess, or ballast, then the top width of the bank must be made good by filling before normal traffic working can be resumed, and this will be a major priority for the track staff. This fill will itself tend to restart the instability, unless proper stabilisation measures are taken lower down the bank, and even then it is obviously desirable to minimise the weight of any tipped material. This is why locomotive ash, apart from being cheap, was in the days of its ready availability, such a popular material on slips. These days the only advice which can be offered about material used for this purpose is that it should be of as low a density as possible bearing in mind that it must be cheap, and very quickly available in the quantities required, readily consolidated, that it should not be such as would contaminate the ballast. Free draining material will allow water to percolate and fill ballast pockets over cohesive soils. Measures are therefore required to pipe it away or to displace it by grouting.

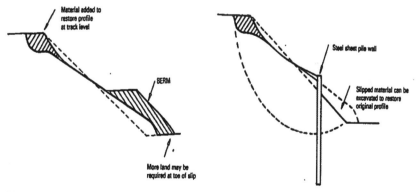

Figure 10.5 Treatment of bank slip
by berming

Figure 10.6 Treatment of bank or
cutting/slip by piling

Embankment slips are stabilised by one or other of the following methods:

- Grouting

- Sheet piling

- Berming and/or gabions

- Reprofiling

Berms and Gabions

In emergency the only one of the above methods which is readily accessible to non-specialist staff is probably berming. This operation consists in depositing a small embankment of material near the foot of the slipped bank, to act as a counterweight, as shown in Figure 10.5. It will be evident from this illustration that for the berm to do any good it must be placed within the slip circle, and as near to its lower extremity as possible. In this case the material used should be as dense as possible, and again should be free draining. Care must be taken in siting the berm to ensure that it does not block any existing drains or overlie any fixed equipment. The berm must also not be placed where it can be seen that water is issuing from the body of the slip. It is a good idea to strip vegetation from the area to be covered by the berm since if left, it would rot and form a slippery surface between the berm and the supporting material, thereby setting up conditions that might be conducive to another slip.

Gabions may be regarded as a kind of berm in which the deposited material (consisting, in this case, of uniform, broken stone) is confined in a wire mesh cage. The value of this is that the material, because of its confined state, is much more rigidly interlocked in position, it can be placed more precisely, and it will stand at much steeper angles. Because rock can be used, and packed carefully inside the cage, it is also possible to achieve much higher densities of deposited material. Gabions are most useful where a substantial weight of stabilising material is required to be placed where space is limited, or where there is a risk of loss of loosely deposited material. In such a case, rows of gabions can be placed vertically above one another if necessary to provide what will virtually be a retaining wall which must be designed as such.

The sort of situation where gabions are of particular value is where a slip has resulted from riverine erosion at the foot of an embankment and a means must be found that will resist the eroding forces. A geotextile should in such cases be placed behind the gabion wall to prevent internal erosion of the soil behind it. Whilst the wire mesh of the gabion cages cannot be seen as permanent, it will last until such time as the contents of the cages are stabilised by root growth and soil penetration. Strengthened or reinforced soil can be used instead of gabions and will stand at very steep (sometimes even vertical) angles.

Grouting

This operation requires specialist equipment, and as a rule is out of the sphere of action of the track staff. In this process large numbers of grouting points are driven successively according to a prearranged pattern across the area of instability. Each point is driven to a depth that will ensure penetration of the slip plane (hence again the importance of knowing where this is). A proportion of the points at the top are driven obliquely so that the section under the track can be treated. Typically a grout formed from three parts of sand or pulverised fuel ash to one of cement is injected.

Quantities injected vary from 1½ to 5 cubic metres per point. The process works partly by expelling water from the embankment material, particularly in the upper parts, partly by forming a seam along the slip plane to increase friction in the re-oriented soil, and to a lesser extent by forming a network of grout veins. The process can be very effective, if properly designed, and many embankments have enjoyed a trouble free life of 30 years or more following treatment.

Piling

The driving of piles (most usually steel sheet piling) into either a cutting or an embankment slip is again a specialist operation demanding heavy civil engineering equipment, and must be designed and executed by experts. In principle, the sheet piles act like a wall cantilevered upwards from the underlying sound clay, to hold back the material above the slip circle, as shown in Figure 10.6. It is only rarely possible to anchor the sheet pile wall into sound material behind the slip.

Clearly the requirements are that:

- the piles must be strong enough in bending to resist the bending moments imposed by the tendency of the clay above the slip circle to continue to rotate;

- the piles must penetrate far enough below the slip circle to ensure that the pressures induced between the pile and the clay in resisting the overturning moment produced by the overburden do not overload the clay (sometimes this is prevented if a very hard layer is encountered at a shallow depth below the slip circle). As a general indication the depth of penetration below the slip circle needed for fixity is about twice the length of pile required above the slip circle. In practice the system is only favoured where the depth to the slip circle is less than 5m.

Bored piles are sometimes used to stabilise slow moving slips. Holes are bored one at a time in an alternate sequence. If the ground has non-cohesive layers and has a high water table either the bore hole must be lined or a bentonite suspension used.

10.13.3 Cutting Slips

If a slip takes place in a cutting the first problem the track staff have to solve is the invasion of the trackbed by spoil pushed outwards at the toe of the slope. Again, the immediate priority will be the removal of this material to enable the line to be reopened, and as it is removed, so may there be a tendency for more material to rotate round the slip circle. If this occurs some immediate way of stabilising the slip must be found, and this must clearly take the form of adding material to act as a counterweight. Gabions, see above, are sometimes used for this purpose. They can be placed clear of the cess and may provide enough stability to allow the obstructing material to be removed from in front of them, as shown in Figure 10.7. Steel sheet piles can be an effective and possibly permanent cure, but the time taken to design a scheme, find a contractor, procure the materials and equipment and mobilise the operation may be considerable. At the same time ingress of water to the slip zone must be minimised by sealing tension cracks and diverting water away from the top of the slope.

The question must always be asked: why did the slip occur? As indicated earlier in this chapter, the track engineer faced with a slip must look for all possible trigger factors, including disturbance above the cutting, drainage difficulties, and weathering, and these must be attended to if a dangerous situation is not to recur.

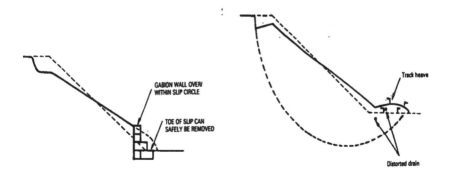

Figure 10.7 Use of Gabions to stabilise slip in either cut or fill

Figure 10.8 Circular slip extending beyond toe of cutting slope

A further feature of slips in cuttings which must be borne in mind is that, depending on the position of the lower end of the slip surface, the lineside drainage may be blocked or severely distorted, as shown in Figure 10.8, or in bad cases there might even be some heave of the track itself. If the track heaves, this will usually be evident, and its reinstatement to its original line and level will be part of the business of restoring normal traffic working. However it must not be forgotten in such cases, that the drains must also be reinstated before the job can be regarded as complete, and that the reinstated drains must themselves be strong enough to resist the horizontal forces associated with the slip. Furthermore, in the urgency of getting traffic running again, the heaved track may have been left with an inadequate depth of ballast under the sleepers. If this is not put right then serious damage to the trackbed will occur in course of time.

10.14   ROCK CUTTINGS

The stratification mentioned under geological factors above occurs not only in cohesive soils, but also in many types of rock, and can be coupled also with cleavage planes perpendicular to the stratification, so that there is a natural tendency for the rock to divide itself into blocks. This tendency is worked upon by weathering actions, notably the freeze-thaw cycle, so that the face of a rock cutting can become unstable, and large blocks of stone may fall foul of the line. If this occurs or is feared, expert advice needs to be sought, and remedies such as rock bolting, or the enclosure of the face in wire mesh, may be sought. In some cases it may be so difficult to ensure that rock falls will not occur, that the only course may be to install trip wires which will automatically set the signals at danger if a fall occurs. A further possibility is the construction of a protective gallery over the railway. Such a gallery these days will usually be a substantial concrete structure.

10.15   MINING SUBSIDENCE

When a support is removed from the roof of a relatively small subterranean cavity, then assuming that the overburden is not intrinsically strong enough to span the opening, it will collapse, and the collapse will extend upwards to the ground surface. The affected ground will be in the form of an inverted cone centred on the cavity, the steepness of the sides of the cone being determined by the shear strength of the ground, and a matching saucer shaped depression will form at the ground surface, the rim of the saucer coinciding with the intercept between the ground surface and the sides of the cone. In this affected zone, each element of disturbed ground is drawn inwards and downwards towards the cavity, and this movement is referred to as "draw".

A railway line running across the affected area will be distorted in elevation so that as well as subsiding, it will also be stretched since the distance along the depressed surface is greater than the distance along the original undistorted plane surface. It may also suffer some alignment distortion, depending upon the shape of the subterranean cavity, the contours of the ground surface, and the geology.

If instead of a small, discrete cavity, a large area is involved, and if in addition, the collapse takes place in a progressive manner along a subterranean front, then there will be an associated frontal disturbance at the ground surface. At places well within the area subject to draw, the vertical, lateral, and longitudinal distortions will increase to a peak, as the surface front passes the railway, and the vertical and longitudinal alignments will tend to revert to their original form after the front has passed, ideally leaving the railway as it was before, but at a generally lower level. At the edges of the subsiding area, the distortion in plan and elevation will remain even after draw ceases due to the differential movement between affected and unaffected territory, and so where the line passes into and out of the area affected by subsidence, permanent distortion may be expected.

These movements are to some extent predictable, and there will be an essential need for the track engineer to liaise with the local mining engineer to obtain information on which estimates of severity and timing of distortion can be based. The extracting authority is obliged by law in the UK to give notice of its intentions, and it is possible under certain circumstances to obtain what is termed a "right of support". What happens in this case is that effectively the railway authority purchases all the coal within an area, under and adjacent to the railway, known as the "pillar of support".

The coal there cannot then be mined. If it is decided not to purchase support, then the mining engineer reports the progress of extraction to the track engineer, and advises on the timing of the onset of subsidence. When subsidence commences, the line has to be kept under observation and speed restrictions are imposed if misalignments and/or differential settlements exceed the tolerances for the class of route involved. In due course appropriate realignment or relevelling operations are carried out as necessary.

If the track concerned is CWR, the effect of the stretching and possibly subsequent contraction as the wave passes may be to alter substantially the stress-free temperature of the rail strings, and buckling may occur. For this reason it has been the practice to replace the long welded rails with jointed rails during the period when draw is active. However, the invention of a device known as the Rail Force Transducer (RAFT) has enabled this course of action to be avoided. A RAFT consists essentially of a pair of stretched wires fixed at right angles to each other, into a hole drilled in the web of the rail. The tensile force in each wire can be measured by plucking it electronically, and recording the frequency of its vibrations. As the longitudinal force in the rail varies, so the tension in the wires varies, according to a known relationship. Thus by monitoring the RAFTs at intervals the behaviour of the rails can be monitored.

This monitoring can be done either by a technician making periodical visits, or the RAFTs can be linked by land line to a multi-channel recorder in a chosen location. The RAFTs then give an indication of the changes in stress in the rails as distortion starts, and enable decisions to be taken about when to carry out restressing operations.

10.16 REPAIR TECHNIQUES

10.16.1 Reinforced Soil

Soil has no tensile strength but this can be provided by including various strengthening fabrics, grids, meshes, strips and rods made of metal or plastics. The plastics may be of one polymer or a composite of two or more such as strong fibres embedded in a protective covering.

Choice of material depends on an assessment of likely corrosion, creep, durability in relation to the particular environment and application. Government specifications and guidelines have been issued on design criteria for highway earthworks. Specialist suppliers of reinforcement geogrids also offer design support including relevant computer software.

Granular fill material is favoured and the reinforcement provides a composite material which may even be constructed to a vertical face. The system is flexible, can accommodate settlement and can be constructed over soft ground. For a retaining wall the vertical face must be protected against erosion and the system of facing panels (often a proprietary method) is designed to settle with the soil wall and to provide this protection. The base of the front of the wall must also be deep enough to distribute horizontal forces. The panels are a facing skin taking only a nominal load. For slopes a geogrid is placed in layers of lm or less and soil is compacted over it; the sheet is then folded back over the front and a short way on to the top of each succeeding layer of soil to retain it. Seeding promotes grass cover to improve protection against sunlight which is adequate for the UK. If geotextiles are used in this way protection against ultra-violet light is essential and larger strain will occur for the material to mobilize its strength. There are many reinforced soil structures throughout the world but there is little information on railway embankments under high speed running lines. Present precautions are to limit embankment reinforcement up to a horizontal distance of 5m from the track. Placed across the base of a new embankment in a granular layer, the reinforcement permits steeper slopes with reduced land take. As the dimensions cannot be altered after construction, allowance must be made at design stage for cuttings for clearance to place OLE gantries, signals etc. Plastics reinforcement is preferable to metal in view of possible electrolytic corrosion caused by stray currents, particularly on lines where DC electric traction is used.

10.16.2 Soil Nailing

This is a form of soil reinforcement in which steel bars (typically 30mm diameter) are either driven into the ground or inserted into sloping boreholes and grouted. Adjacent soil is reinforced and there may also be a shear resistance (dowelling) where the steel crosses a slip plane (eg rail piles). Nails may be of the order of 5 to 10m long and the system is one of the few, apart from grouting, which can be used to repair a failed reinforced soil structure.

10.16.3 Ballast and Earthworks Stabilisation

There are a number of new ideas on the market to stabilise ballast and earthworks. An example is shown below courtesy of Balfour Beatty Rail Technologies Ltd.

Figure 10.9 Xi Track Installation (Courtesy of Balfour Beatty Rail Technologies)

CHAPTER 11

OFF TRACK MAINTENANCE

11.1 INTRODUCTION

In the UK there are more than 32,000 km of lineside infrastructure that needs management intervention. Track staff are in the forefront of the task of managing the land within the boundary and safeguarding railway property from encroachment.

This task, generally defined as Off Track Maintenance, embraces many activities in addition to track works and forms a significant part of the track staff duties. A wide range of knowledge is required, needing good awareness of the environment and a working understanding of current legislation affecting the countryside and pollution.

Care and thought in carrying out the activities described in this chapter can contribute much to the preservation of the landscape as well as meeting the primary object of achieving safe, well maintained track at a reasonable cost.

11.2 FENCES AND HEDGES

11.2.1 Fences

The principal types of barrier used for keeping animals off the railway and deterring trespassers are fences, which are most often these days concrete post and wire, but may also be timber post and wire, timber post and rail, unclimbable iron or wire mesh. Brick or stone walls and even occasionally hedges may be used.

A minimum height of 1350mm is required for all types of barrier. Sturdy construction is essential to keep out large livestock which can be a hazard to rail traffic.

In the UK, significant investment has now been made in palisade steel fencing of 2m or higher, especially in urban areas. On certain electrified lines, in particular those with third or fourth rail current supply, it is essential to provide fencing which young children cannot pass.

Figure 11.1 Railway Fence and Drain (courtesy D.Ratledge)

11.2.2 Hedges

Hedges are usually set at a distance of between 1200mm and 1800mm to their centre inside the railway boundary enabling the trackman to remain on railway property when cutting them. They should be trimmed in the autumn when sap is falling with the object of keeping the plant bushy. Main shoots should be cut at a height of about 1300mm and at about 300mm each side of the centre line. Gaps or thin places should be closed by layering longer shoots. The work can be done with a hedging knife or by portable powered hedge trimmers, electric or petrol driven.

The bottoms of hedges should be kept clear of long grass or weeds to encourage growth from the roots. Deep digging which might damage roots should be avoided.

11.3 DITCHES AND DRAINS

Many of the more troublesome track maintenance conditions can develop from neglected drainage systems. This is particularly the case with concrete sleepered track lengths.

If drains become blocked or damaged they will cease to drain the formation and standing water in the drains may even seep back through open joints or perforations and lead to softening of track formations.

All ditches, channels and piped drains should be inspected and cleaned out periodically, removing silt and other debris. Particular attention should be given to maintaining clear outlets. Ditches as well as requiring clearing may need spraying with chemical treatments to kill or slow the growth of reeds. This spraying must only be done by trained staff and after liaison with farmers to protect livestock on the downstream side of the ditch outlet. Machinery can be used economically to clear and extend ditch systems, often in co-operation with local Drainage Authorities.

11.3.1 Drain Cleaning

Piped drains can be cleared by a process known as "rodding". This consists of pushing through the pipe a block of wood or a soft "badger' which is trimmed to a loose fit in the bore. The block is fixed to the end of a rod, made of flexible material, and having screw threaded ends, so that further sections may be added to complete the pushing of the block through the pipe. When withdrawing the rods care must be taken not to unscrew rods while they are still inside the pipe.

Alternatives to rodding are winch-operated cleaning tools which can be drawn through the pipe, self-propelled nozzles driven by high pressure water, and power-operated mole cutting heads to clear complete blockages. The last two methods are available from specialist firms. Water jetting methods can be useful for clearing badly silted tunnel drains if an on track water tank supply is arranged.

Access to pipe drains for all these methods must be via catchpits or manholes which should be sited at intervals of no greater than 30 m. Some old drains may have catchpits too widely spaced or even lack them entirely. In such cases it is necessary to dig down to the drain and break into the pipe at suitable intervals for access. Catchpits for future access can then be provided on completion. Catchpit sumps must be cleared out regularly to avoid the pipe outlets becoming silted. Catchpit covers must be replaced when damaged both for safety of staff walking near them and to prevent debris falling into the catchpit. At least one grid section cover should be maintained to allow easy inspection of the drain flow.

11.3.2 Causes of Ineffective Drainage

A drain, however well laid, cannot perform its function if the water cannot reach it. Hence if a drain appears to be in good condition but is apparently not taking water from the ballast, the ballast should be investigated and any clay or other impermeable material impeding movement of water from the ballast to the drain must be removed and replaced with more suitable material. Where such a drain is deep it may be better to regard it purely as a carrier, and to supplement it by shallow collector drains in the form of channels or perforated pipes, outfalling into the carrier drain at catchpits.

Figure 11.2 Typical catch Pit Cleaning (courtesy D. Ratledge)

In clay type soils the porous filling around the drain may become clogged by fine soil particles and also the open collar joints or perforations in the pipes. This destroys the effectiveness of the drain. The only remedy is to dig down to the pipe, clear out the collars or holes and refill the trench with new filter medium or porous fill.

Occasionally a pipe may be broken and a collapse and blockage my result. If it is found impossible to rod through a section of drain a broken pipe is a likely cause. By measuring the length of drain rod that has been inserted it is possible to locate the site of the blockage. It is then possible to dig down to the drain and repair the broken length

11.3.3 Discharge from Drains

Track drains often discharge into streams or ponds outside railway property. Any diversion of a stream or filling in of a pond which may block the outlet of a drain or ditch should be reported promptly to the engineer. Also, any work observed outside the railway boundary on the high side which would result in an increased flow of water onto the railway should be reported. An example of such work is the construction of a roadway or paved area laid so as to slope towards the railway.

11.3.4 Slips affecting drains

If a slip occurs in ground through which a drain runs, the level of the drain could be disturbed to such an extent that water may be prevented from flowing through the drain. In such an event the drain should be opened up and re-laid to suitable levels. Drains laid in areas subject to mining subsidence are very likely to be affected in this way and may require repeated re-levelling. It is important that drains in subsidence areas should be maintained as effectively as possible, as the result of subsidence is sometimes to form hollows in the ground surface which may fill with water and flood the track if the drain becomes blocked.

11.3.5 Inspection of Drains

During regular inspections the flow of water through culverts under the track should be checked to ensure that they are clear and avoid the formation of deep ponds on the upstream side of embankments. On a high embankment such ponds could lead to slip conditions developing. Although the advice is unpopular, and indeed the PWSM may have other things to do at the time, the best time to examine drainage systems is during or immediately after heavy rainfall. This is what the drains are meant to cope with, and their behaviour under full load should occasionally be checked.

11.4   VEGETATION CONTROL

11.4.1 Vegetation Control Policy

It is recognised that good management of the lineside can make an important contribution to the landscape and provide a protected habitat for wild life. Grass cutting and burning should be confined to the minimum necessary for the safe operation of trains and the safety of staff. There is usually a Code of Practice on the Management of Lineside Vegetation to encourage awareness of environmental factors.

This minimum need can be met normally by controlling strictly the growth of vegetation on a strip 3m wide adjacent to the cess. Trees and shrubs on these strips should be cut down and regrowth controlled by selective herbicides aimed at preserving the strip as a rough, grassy zone. A similar strip can be maintained at the fence line where needed to keep access for fence maintenance and reduce encroachment of growth onto adjoining land. Grass cutting or burning on these strips can continue annually.

Within England and Wales the Wildlife and Countryside Act, 1981, enables the Nature Conservancy Council (NCC) to list Sites of Special Scientific Interest (SSSI). A number of these have been listed on railway property and methods of management of vegetation have been agreed with the NCC. Details of these sites are advised by the Track engineer and the aim is to cooperate fully with the interested outside parties, who may provide assistance in the work on site.

Figure 11.3 Severe Vegetation Growth on Main Line Railway (courtesy D.Ratledge)

11.4.2 Grass Cutting

Grass cutting should be done in June or July, according to locality, before it begins to seed. This will help to prevent the spread of grass etc to ballast and cesses. If this is done by hand, either scythes or reaping hooks can be used. On a slope the scythe is usually swung below the operator and men should work well apart. Near signal equipment the hook is best used and cables etc. held clear with a stick, not with the hand.

When the cut grass is dry it may either be burned where it lies or gathered up and sold. In some cases arrangements are made with local farmers to cut and remove the grass. There are a number of mechanical cutters available, powered by a petrol mix engine and easily used by one man. One of the more effective types described as a clearing saw can be fitted with interchangeable metal cutters or nylon strimmer head depending on the type of vegetation. With trained staff they can be operated safely and economically on slopes and close to equipment.

11.4.3 Scrub and Tree Clearance - Objectives

As mentioned in 11.4.1 above, the clearance of vegetation is limited to the minimum necessary to meet the needs of train operation, safety of staff, and the welfare of the neighbours. The overriding principles are:

- ensurance of good signal sighting
- prevention of encroachment on the structure gauge
- prevention of excessive autumn leaf fall.

Subject to these principles, conservation of the environment is the main objective influencing the way in which work should be done. This defines in clear, practical terms three lineside zones and describes the approach to control of vegetation within each zone.

11.4.4 The Cess Strip

Immediately alongside the cess is the cess strip, normally 3m wide, and the aim is to keep this strip free of substantial scrub and tree growth. Clearance of scrub can be by hand cutting or powered clearing saws, and chain saws can be used for the larger stumps. Future regrowth can be reduced by the chemical treatment of stumps immediately after cutting and spraying with selective herbicides. Spraying can be from hand operated back packs or from the weedspray train during its annual spray of the track formation.

Heavy scrub cutting within this strip can be carried out economically by mechanical flail cutters mounted on and powered by tractors. A number of firms have converted tractors with guide wheels to allow them to be driven along the track from a convenient access point, e.g. a level crossing. The machines can operate from the track and readily cut to 10m from the rail, both in cuttings and on embankments. A pre-planned possession of the line is needed for such work and normal rules must be applied for the protection of traffic operating over an adjacent line.

11.4.5 The Verge

Between the cess strip and the railway boundary is the verge. This is the zone which can benefit most from sensible management at minimum cost. Parts of the verge well away from track or structures can be left untouched for long periods to develop into woodland. Slow growing scrub such as dogwood, privet and ivy should be conserved as they help to prevent trees seeding into the same areas. Fast growing scrub such as hawthorn, ash, sycamore must be controlled before it begins to encroach into the cess strip. Control of this type of invasive scrub when needed should be by cutting and burning, followed by treatment of the cut stumps and regular chemical spray.

In addition to SSSI's, the Institute of Terrestrial Ecology publishes lists of Special Sites covering a number of lengths of the lineside where it is hoped the vegetation can be managed in defined ways. These lists are available to the engineer who can obtain advice on the management required from the relevant County Trust and decide the extent to which it can be applied by the track staff.

11.4.6 Tree Felling and Lopping

The need for felling or lopping of trees is dictated by considerations of the safety, either of the railway or of adjoining property. In addition to general vigilance throughout the year, an annual inspection of trees near the track is necessary. Any tree which is leaning towards the track or which appears to present a danger to the track in the event of a storm, must be reported and listed for felling or lopping. It should be borne in mind that elm trees are particularly likely to drop branches which appear healthy, and to be susceptible to hidden decay within the trunk.

Yew or other poisonous trees growing on railway property but spreading towards outside property should be lopped. Trees must be kept clear of telephone and other overhead wires, and trees and shrubs lopped or cut back to allow clear sighting of signals and level crossing approaches. Large trees on banks may be swayed by high winds and the roots loosened to an extent that would weaken the bank. Such trees should be felled.

The felling or lopping of large trees should be carried out by specialist forestry consultants and contractors, as they have wide experience and can advise the extent of work required, judge the condition of trees and are well equipped with power saws and lifting equipment for the work. When felling trees close to the line arrangements must be made to protect traffic in accordance with the rules. Certain trees are covered by Tree Preservation Orders and are so designated on maps. These trees and any others within Conservation Areas should not be felled or lopped without the consent of the local planning authority, except in an emergency. The track engineer's office will obtain consents for such work as required.

11.5 KEEPING WORKING AREAS WEED FREE

Weeds on and about the track are unsightly and, more seriously, cause the ballast to become choked with dirt and to lose its free-draining properties. Water held in dirty ballast leads to decay of track components, ballast attrition and the formation of washy areas around concrete sleepers. Growth around cableways is a severe fire hazard and must be prevented. Chemical herbicides, otherwise known as weedkillers, are used throughout the UK to control or eliminate weed growth from the track, the adjacent cesses and the wideways between tracks.

11.5.1 Weedspraying Trains

Throughout running lines the elimination of weed growth from the track and cesses can be achieved by the application of chemical weedkillers in liquid form using sprays from a train.

A number of specialist firms offer such a service competitively and undertake, if employed, to equip a train, supply the herbicides, and achieve a specified standard for weedkill as high as 98%, as defined by the engineer under the contract.

The train includes tank cars for water supply, a storage coach for holding liquids and solid herbicide, a spraying control coach and living accommodation. Pumps and mixing equipment are located in the spray control coach, and water is piped from the tanks. The chemicals are metered into the water flow to provide the required weed control relative to the speed of travel. Different chemicals can be applied simultaneously through a system of nozzles. These are arranged to cover finely, the four-foot, six-foot and cess, using any combination of cover as required. This system has been supplemented more recently by extra nozzles to include the 3 metre wide cess strip alongside the cess. Brushwood killers and growth inhibitors can also be sprayed from the train to control scrub growth.

The train location, speed and chemical flowrates are logged by computer at nine-second intervals.

Normally, one treatment annually from the train is adequate. The optimum time of year for spraying is during the Spring, when minimum regrowth has occurred, and certainly before seeding occurs. Trains operate at speeds up to 80 km/h for track spraying and 30 km/h for brushwood spray on banks, which needs more control.

Chemicals used are total weedkillers. These act either through the foliage, or remain in the ground as residual killers and act through the root system. Suppliers market weedkillers under many brand names, with varying combinations and strengths of the basic chemicals available. Each product has been developed for a particular range of applications, depending on method, site conditions and type of plant to be treated. The range available is large and continually being updated. Selection of product needs careful study of suppliers' information leaflets, which are very detailed, and much judgment and experience is needed to choose economically. All products used must be tested and approved for safe use and Network Rail restricts use to products on an Approved List, whether applied by railway staff or by contractor.

It is normal with train spray contracts, to allow the contractor to select the chemicals he wishes to use from the Approved List, according to conditions each day, and to decide the concentrations. Contractors can thus be expected to meet the specified degree of weedkill, and to respray any areas where their treatment has been unsuccessful. Deep rooted, persistent weeds such as horsetail often need special retreatment by hand. Spraying of slopes from the train to control scrub growth uses selective brushwood killers in varying combinations. These are developed to kill scrub, but do not affect grass and other small-leaved plants, which then re-establish themselves on the banks.

11.5.2   Weed Control Off-track

In sidings, wideways and localised areas of banks which cannot be dealt with readily from the train, it is normal to spray a similar range of chemicals by hand spray. Knapsack units carried on a man's back, and hand-held controlled droplet applicators which do not require water are available. Progress can be speeded up greatly if a platelayer's trolley or road vehicle can be used for larger containers.

The more recent innovation of CDA (Controlled Droplet Application) can speed up the rate of progress, as mixing of chemicals is eliminated, by using un-diluted herbicide which is precisely targeted on the areas to be treated. All spray methods of treatment must use well-trained staff, avoid risk of spray drift onto adjoining property in strong wind conditions, and avoid spraying at all in hot, dry weather. Employers must ensure good hand washing facilities after work is finished. It is normal to advise all lineside staff in advance of the spray train programme so that they can stand clear of the spray as it passes.

As an alternative to spray methods, dry weedkillers can be used, usually in the form of granules. These are relatively expensive and are only suitable for small areas, eg to suppress growth around mileposts, signal cabinets etc.

From time to time special treatments are needed using unlisted chemicals. An example is the use of sprays to slow the regrowth of reeds in large drainage watercourses. In such cases it is essential to obtain scientific advice in advance and to establish clearly the safe conditions of use.

11.6 RELATIONS WITH OUTSIDE PARTIES

Staff have long had a prime role both in keeping good relations with adjoining property owners and in safeguarding railway property from encroachment and conflicting developments by outside parties. This dual role is more important today than it has ever been, with the increased public awareness of their rights and the extension of legislation involving the relationship.

11.6.1 The Railway Boundary

Knowledge of the railway boundary is of first importance. Boundaries are usually clearly defined by standard fences. There are occasions when the land outside the fence is railway owned. Where hedges exist these are usually positioned a few feet inside the boundary. In such cases the true boundary is usually marked with rail posts at intervals and runs straight between them.

Types of encroachment which should be noted and reported to the engineer include the building of structures, new excavations, tipping of materials, road construction, erection of hoardings or fences, the laying of drains, pipes or cables to cross the railway without authority. Cases of trespass of public or vehicles should be prevented. Unauthorised crane working close to the track must be reported to the authorities. If known, full details of changes in land ownership on either side of private bridges or level crossings should be reported as the use of the crossing may be affected.

11.6.2 Avoiding Nuisance

When planning work, full consideration must be given to the likely effect on neighbours. Precautions for burning scrub and materials have been referred to in 11.4.2 and it is a legal requirement to avoid smoke nuisance or danger to neighbours or highways.

Noisy operations close to houses, particularly at night, require precautions to keep the noise produced below certain specified limits. Neighbours should be advised or press notices published outlining the duration of work and every effort made to comply with the Control of Pollution (Noise and Vibration) Act. Work during possessions must not be stopped, however, merely because of complaint and the engineer must be consulted if this is suggested.

11.6.3 Access

Access for road vehicles is often needed over private land. An important task is to keep co-operative relationships with farmers as good vehicle access can greatly reduce the cost of work. All staff must take care not to undermine these local agreements by damaging the farmer's land and fences or by leaving litter at the parking area. Common sense and consideration for neighbours when arranging work can prevent most potential problems arising.

11.7 EXAMINATION OF STRUCTURES

All structures for which the railway is responsible are inspected regularly and reported upon by trained examiners. The programme for maintenance, repair and renewal of structures arises essentially from these reports. The periodicity of such examinations varies and is determined by the Engineer or other senior authority responsible, but changes in the condition of a structure sufficient to affect its stability, and hence the safety of the railway and/or the public, can develop very rapidly. For this reason, track patrolmen are required to look for signs of rapid changes affecting bridges and structures when they carry out their own regular inspections of the track. Any recent changes likely to interfere with the safe running of rail or road traffic must be reported promptly to the senior works supervisor for the section concerned. In addition the PWSM should inspect the underside of each underbridge at frequencies specified by the engineer and under traffic whenever possible.

During or following floods or heavy rain a special examination by track staff should be arranged for all bridges over watercourses and culverts. This check should note and report specially any washing away of banks or riverbeds close to the structure which might undermine it and any obstruction of the waterway. After storms or high seas an inspection should be made of sea defences for any damage that could endanger the railway.

Any unauthorised use of space under bridges or arches must be reported and also any evidence that vehicles using bridges over the line may exceed the weight restriction shown on the warning notices.

In tunnels any movement of the lining, excessive scaling or movement in the rock face must be reported and if needed, immediate action taken to protect traffic until a structural examination is made.

The emphasis in all the inspections recommended for track staff to make is to look for signs of recent change since last seen. Note must be made of the development of new cracks, opening of old cracks, fallen or displaced bricks or masonry, movement of girders on bearings, signs of impact by vehicles, and storm damage. Distortion of arches or parapets, the cracking of arch rings or crumbling of mortar in the underside joints can indicate movement.

Loss of ballast, depressions in roadways, subsidence in cesses behind retaining walls and ground heave in front of walls can all be signs of failure. Walls out of plumb can indicate lack of stability or foundation failure. Lowering of ground level in front of walls can cause serious weakening, and must be reported promptly.

After periods of severe frost spandrel walls to arches and viaducts should be examined for signs of bulging and pushing due to the expansion caused when moisture freezes. The faces of soft bricks are often pushed off during these conditions and particular attention should be paid to brickwork over the track for loose or hanging sections.

Whilst modern bridges are waterproofed at deck level this is not the case with older ones. Hence water dripping through is not always a sign of failure needing quick attention. If the leakage continues into a long period of dry weather and is of some volume this should be reported as it may imply a damaged water main or sewer. Leakage in bridge decks above overhead wires in electrified areas can lead to damage to insulators and also to locomotive pantographs if icicles form. Cases of severe leakage should be reported. Movement affecting bridges can be caused by mining settlement, earth pressure, ground water changes and undermining by river erosion. Extra loads on walls and embankments can arise from developments outside the railway and lead to failure. '

Road vehicles frequently cause damage to parapets and pilasters of overbridges, undersides of arches and bottom flanges of metal girders of low bridges. Track staff should be on the lookout for such occurrences and report them. Damage to parapets of all types of bridge caused by vandalism should be reported. Immediate steps must be taken to protect the railway.

11.8 LEVEL CROSSINGS

At level crossings used by vehicles the roadway make up may vary from timber or ballast for field to field farm crossings to a tarmac or concrete surface for public roads. In all crossings used by vehicles the construction must include a clear flangeway 50 mm wide for the passage of wheels of trains.

Flangeways can be provided by fitting check rails, longitudinal check timbers or by design of prefabricated concrete units and stressed timber units. Prefabricated units are being used increasingly on heavily used lines and roads as they can be removed and replaced readily for track maintenance.

Essential work at level crossings must include the clearance of flangeways and the maintenance of a good standard of road surface repair. Potholes should be made good promptly. When installed, the vertical profile of the roadway across the tracks is designed to reduce the risk of grounding long, low bodied vehicles. The track must not therefore be lifted or recanted through crossings without the approval of the engineer, who will redesign the road profile to meet the required standards. At certain unmanned crossings datum posts are provided to indicate the required rail levels and these must be worked to where available. Whenever a crossing is opened up for repair or re-grading, the track components should be inspected and replaced as needed, fastenings tightened and sleepers tamped or Kango packed. If a rail joint exists within the crossing it should be re-sited clear of the roadway to improve future maintenance. This may be done by welding, fitting a longer rail or cutting in extra rail and adjusting.

At level crossings on third and fourth rail electrified lines provision must be made in the surfacing for the clearance of the current collecting shoes of electric stock. For this purpose no part of the surfacing should be above the plane of the rail head. This can be checked by using a suitably insulated straight-edge across the rails.

Staff inspecting the track should note the condition of road markings, gates, signs and fences and check the operation of any telephones provided. Defects must be reported and repaired promptly. Growth of trees and hedges on the road and rail approaches to the crossing must be controlled to keep good road/ rail intervisibility and to keep signs and equipment clearly visible.

At farm occupation crossings the farmer is under an obligation to keep the gates closed when not in use. Any failure to meet this obligation should be reported for action.

11.9 FOOTPATHS, ROADS AND CESSES

11.9.1   Roads and footpaths

The responsibilities of track staff for maintenance of roads at level crossings has been mentioned in section 11.8 above. At overline bridges the railway owner is normally responsible for maintenance of the road or footpath surface between the gates or stiles when provided or between the boundary fence lines otherwise. The railway is not normally responsible for the surface of roads or footpaths passing beneath underline bridges. These are maintained by outside parties or highway authorities.

Roads and footpaths maintained by the railway should be checked regularly in the course of inspections by track staff. Routine maintenance by track staff is limited to clearance of growth and temporary repairs to potholes before they become so large as to be dangerous to vehicles. Short term pothole repairs can be made with spent ballast or pre-packed, cold formed, bitumen based materials.

The need for extensive resurfacing of macadam, bitumenised or concrete roads should be reported to the engineer who will arrange this by his works staff or by a road contractor.

Roads and pathways should not be widened or have the surface level raised at overbridges or underbridges so as to alter parapet heights or headroom clearances, without the approval of the engineer. Any change of use, e.g. by heavier vehicles, should be reported promptly to the engineer so that the road category can be reassessed.

11.9.2 Cesses

Cesses are provided for the dual purpose of aiding track drainage and as a safe walkway and refuge for track staff during the passing of trains. The desirable standard at the time of writing is to maintain a minimum effective width of 500mm. This effective width should start at least 1.2m from the back edge of the nearest rail for line speeds up to 160km/h and 2.0m where speeds exceed 160km/h to ensure staff safety while trains pass. Where these safe widths cannot be provided continuously on at least one side of the line, consideration must be given for high speed routes to the provision of refuges at intervals of 40m.

The cess level must not be higher than the formation level for good drainage. This requires the cess surface to be maintained at depths below underside of sleeper varying from 250mm to 460mm according to the direction of formation crossfall. The surface should be maintained level and firm without raised tripping hazards such as cable duct covers.

All vegetation should be suppressed and all loose materials should be removed or stacked clear of the cess as work is completed. At all times the aim should be to keep the cess as a safe, clear walkway during the times when no men are working at the site.

11.10 LINESIDE TIDINESS

In concluding this Off Track Maintenance chapter, reference must be made to the increased importance now given to lineside tidiness. Tidiness alongside the track has long been accepted as a mark of efficiency, as a contribution to the environment and as a means of encouraging public confidence in the railway.

Vandalism risks today add urgency to the task of clearing away all unwanted rubbish and material so that it is not easily available to vandals to place on the track.

Regular clearance programmes should be made using recovery trains, self- propelled on rail type vehicles and scrap disposal contracts to supplement railway staff. The aim must be to clear all unwanted items including rubbish thrown onto the railway. Site work plans for both renewal and maintenance must include allowance for clearance and tidying the site on completion.

CHAPTER 12

GAUGING

12.1 INTRODUCTION TO GAUGING

Gauging, as it has always been, is the process by which safe passage
of trains is assured by providing adequate space between an operating
rail vehicle and the infrastructure through which it passes. The process
by which this is achieved is now somewhat different to the methods
used when railways were conceived and operated into the latter half of
the 20th century.

The original concept of 'structure gauging' was that rolling stock should
be built up to a vehicle gauge, and structures should be built out from a
structure gauge, there being a comfortable margin (generally around
a foot [300mm] of space between them). Thus were railways built.
However, before long, the success of rail meant that railway vehicles
became larger; longer coaches replaced carriages and locomotives
became more powerful and larger. This meant that additional
engineering was introduced into the gauges; as an example, gauges
were adjusted according to track curvature in order to provide for vehicle
overthrow where this occurred. The process of maximising the capability
of the available space had started.

Today, gauging very much relects that trend. We, in the UK are greatly
constrained by our Victorian infrastructure but have aspirations to
carry wide bodied passenger traffic, large containerised freight and
high speed tilting trains. Our gauging processes use sophisticated
measuring and computerised analysis tools to ensure safe passage,
whilst maintaining a managed level of risk consistent with that of the
railway system we are running. The environment of the 21st century
is one of expanding the capability of the network by a combination
of infrastructure replacement, understanding what capacity can
be achieved from the existing infrastructure and managing the
infrastructure in a way that the cost of achieving that capacity is
optimised throughout the lifetime of its components.

There is a belief, perpetuated by other texts, that a 'structure gauge' is central to the process of structure gauging. In practice, such concepts cannot provide operational clearance solutions to anything but new railways where space to operate modern trains is less of a constraint.

The process of gauging, as it is now performed, is one of a systematic understanding of the component systems, followed by the application of safety and management constraints to ensure that the relationship between permanent way and physical infrastructure is maintained to provide the safe passage of trains.

12.2 THE SIGNIFICANCE OF TRAIN DIMENSIONS IN THE UK

On the UK mainline railway, we carry large freight (9'6" high x 2600mm wide containers) concurrently with long (23m), wide bodied (2800mm) passenger rolling stock. The basic shape of such trains raises different issues; in general, containerised freight compromises clearances in arched bridges requiring tracks to be brought closer to each other to accommodate them, whilst wide passenger trains require sixfoots to be widened to allow them to pass safely. Simultaneously to this, we must ensure that platform 'gaps' comply with the increasingly demanding standards associated with the provisions of reduced mobility access to provide for the safe boarding of passengers.

When considering the issue of train size, we must consider how the size of these trains change when they are in motion. This size, known as the 'swept envelope' relates to the geometric overthrow of the vehicle centre and ends as it passes around curves, and the dynamic movements associated with sway induced by curving at high speed (or indeed at slow speeds on canted curves) and by the oscillations induced by track roughness. In the case of tilting trains, it is also necessary to consider displacements of the train body as a result of the compensation required for the curving forces seen by the passengers.

By considering these effects, we work to the current philosophy of gauging where 'knowns' that can be calculated are explicitly provided for in the system calculation rather than being accommodated in what has been loosely described as clearance in the past. Whilst the gauging process demands a high degree of sophistication and accuracy in these calculations, it is helpful for the practicing engineer to have a 'feel' for the magnitude of such effects in order to visualise the space requirement around the physical train.

A useful 'rule of thumb' for estimating vehicle overthrow on curves is the formula:

$$Overthow = \frac{125B^2}{R}$$

Where B is the vehicle bogie centres and R is the curve radius (both in metres). The overthrow calculated (in mm) is the approximate value at the centre of a vehicle, being an offset towards the inside of a curve. Generally trains are designed to have equal (but opposite direction) centre and end overthrow, although freight wagons tend to have reduced end overthrow as a result of the placement of bogies in relation to the loads. Thus, on a 200m curve, we would expect a 23m (typical Inter City) passenger vehicle (with 16m bogie centres) to have an overthrow of 160mm. A 20m (typical Suburban) vehicle with 14m bogie centres would have 123mm of overthrow on the same curve.

These calculations apply to circular curves. On transition curves, the calculations may be quite complex and are often simplified. Increasingly, accurate calculations of overthrow on transitions are sought, particularly in platforms.

Dynamic movement of the train, under the effects of curving and track roughness are also a component of the swept envelope. Vehicle suspensions vary, but a useful 'rule of thumb' would be to expect between 100mm and 200mm sway movement at the cantrail of a passenger vehicle (the level where the curved roof meets the body side) as a result of travelling at 100mph (160km/h) around a curve with 150mm cant deficiency. Movements at footstep level are likely to be up to 50mm in similar circumstances. Freight vehicles tend to sway less, although the latest generation of 'track friendly' bogies permit greater movement. It should be remembered that the energy that is not transmitted to the track is converted into sway motion and subsequently dissipated in the damping of this. In noting that vehicles sway outwards (in relation to a curve) under the effect of cant deficiency, it should also be noted that similar effects occur with slow moving vehicles where there is an excess of installed cant for that speed. This will cause vehicles to sway towards the inside of the curve. The effect is not at its worst when stationary (even though the cant excess is greatest). The combination of oscillation from track roughness (which is generally related to speed) and the balance of curving forces means that the movement towards the inside of a curve is generally greatest at a speed that has been known as 'trundle speed', now more correctly known as an 'inward critical speed', (this can be quite a high speed on shallow curves, and is line speed on 'straight' track).

Tilting trains introduce another component into the swept envelope; that of the tilt compensation. Modern tilting trains may develop up to 6 degrees of body roll around the tilt centre, which would be in the opposite direction to that induced by curving. A further consideration applicable to tilting trains is the ability to accommodate them in the rare event of tilt system failure (which may be in the wrong direction). Such 'failed' swept envelopes may be permitted to run with less clearance than normally required due to the low possibility of this occurring. Failed suspension (such as delated airbags) may also be accommodated with less than normal clearance for similar reasons.

Due to the complex shapes of trains, the swept envelope of the vehicle is usually calculated in relation to each and every cross-sectional profile of the vehicle in order to utilise the maximum possible vehicle envelope along its length. Often, items such as yaw dampers extend beyond the general body shape as they are fitted to the vehicle in areas that experience less overthrow.

Some gauging remains relative to vehicle gauges, freight being the most common. W6a gauge is considered to be the UK 'workhorse' freight gauge, although it is not, as often thought, a 'go-anywhere' gauge as some parts of the UK network cannot accommodate it. W7 to W12 gauges become progressively larger in either width or height of load permitted; W12 being considered to be the aspirational container gauge for the country. New gauges for passenger vehicles are also appearing, such as a 20m 'PG1' gauge for passenger vehicles.

As was noted earlier, gauges do not make best use of the available network space, since generalisation of shapes and dynamic characteristics against which clearance is determined cannot be as efficient as 'exactly' defining a vehicle. However, modern variants of the gauges (known as 3rd Generation) are more efficient than those traditionally defined as they are associated with dynamic suspension characteristics that adjust the gauge line according to the cant deficiency or excess at a location, in addition to the overthrow. Whilst generally requiring a computer simulation to apply them, they do permit a degree of network optimisation whilst simplifying and reducing the cost of new rolling stock introduction or freight train pathing.

12.3 DEFINITION OF CLEARANCE

Clearance is the gap between the moving train and the infrastructure. Its purpose is to provide a margin of safety, and to accommodate 'unknowns' (and some variability 'knowns') in the gauging system. As the technology of gauging progresses and more is known about the components of the system (for example how vehicles behave and how accurate our measurements are) it is possible to reduce the space to accommodate 'unknowns'. As a result of this progress, it is now normal to consider 100mm of clearance around a vehicle as being desirable which may be compared to the 200mm required in standards before 2003 and even greater in earlier years.

A particular parameter accommodated in the clearance is vehicle movement beyond that considered to be 'maximum'. In practice, the motion of a vehicle is statistical – there is no such thing as an absolute value of sway (in the same way that there is no absolute value of clearance). A measure of mean movement + 2.12 standard deviations is generally used as a 'definitive' value of train movements, although it is known that peak movements can be as great as mean +4.5 standard deviations can occur. Clearance accommodates such peaks, although in practice the aggregation of all system tolerances means that this component of clearance is unlikely to be used (all system tolerances conspiring together is not a credible scenario, even though it is generally analysed and sometimes gives indicative results that suggest a vehicle may not be capable of running even when it is known to).

12.3.1 Calculation of clearances

It is usual to use a computer program to calculate clearances between trains and structures. This process is referred to as 'absolute gauging'. Measurements of the infrastructure are entered into the computer program, which include both the shape of structures, the track geometry and the line speed. From this, the forces acting on trains passing the point are calculated (both at speed, and if running slowly) which are then used to calculate the swept envelope of any rolling stock passing. The clearances calculated are then compared with those mandated by Railway Standards in order to determine the level of risk associated with running those trains past the structure.

12.3.2 Measurement of infrastructure

Accuracy is paramount in the measurement of structures, in terms of using accurate systems, understanding how accurate they are, how frequently they sample the structure surface and the competence of the operators. Network operators generally have 'approved' measuring methods, for which accuracies are understood. If approved systems are not used (or such systems are used in an unconventional manner) then it will be necessary to provide evidence of the measuring system accuracy for those measurements to be considered acceptable.

For the track engineer, there are a limited number of tools that are now considered acceptable. For measuring platforms, a traditional platform gauge is considered optimal, although this must be of a robust, braced construction and used correctly to provide heights and offsets against the nearest rail running edge along the platform length. It should also be remembered that the platform edge may not be clearly defined if, for example, bull-nosed coping stones are used. The 'correct' platform edge is the intersection point of the top (horizontal) surface and the vertical struck from the lateral coper measurement tightest to the track – a point which may be imaginary, but from which regulatory stepping distances are calculated. Thus, if the actual shape of the coping stone is measured, for example by a cloud system, this measurement may not be acceptable for gauging purposes. The use of a calibrated 'shoe' would be required for such equipment.

Laser measuring systems are now the only acceptable tools for measuring structures such as bridges and tunnels; either discrete point systems (such as a 'LaserSweep') or as a point cloud. Discrete point systems must be used with skilled operators to ensure that the protrusions within irregular structures are accurately captured. Accuracies vary, but improve with each new generation of equipment. Some types of system provide a relative accuracy (i.e. between close points) of better than 1mm, and an absolute accuracy of 3mm.

Along with the structure measurements, it is also necessary to provide installed cant and curvature data; the latter often being measured by taking versine measurements from a chord strung between adjacent survey points, similar to that done for a Hallade survey. Again, accuracy is important, but equally important is the care necessary to ensure that the measurements taken relect the curvature of the track, rather than local misalignments of the track. For example, where track joints describe the curve as a series of straight lines. For gauging calculations to be accurate, design cant and curvature should be used or, if unavailable, suitably filtered site data. If raw data is used, additional uncertainty is introduced due to the double counting of geometric irregularities (these are accommodated in the vehicle modelling) resulting in over or under estimates of vehicle movement and resultant clearance calculations.

12.4   TRACK FIXITY

Track is often known as "Permanent Way" but it's anything but permanent. However, there is a belief that gauging measurements are associated with movement of structures when, in truth, the purpose of gauging is to manage the position of track in relation to a fixed structure. The concept of fixity is used to accommodate track movement within gauging calculations and is applied in an allowance to describe the possible variation of track position within a maintenance cycle.

In the absence of specific movement history information or of managed track position, it is assumed that normal ballasted track may move laterally by up to ±25mm, that the crosslevel may vary by up to ±10mm and that vertically it may vary within a 25mm band, generally assumed to be +15mm, -10mm from a nominal measured position if the design position is not known. Medium fixity is normally associated with undisturbed glued ballasted track and has a tighter lateral allowance of ±15mm and crosslevel of ±7.5mm. No vertical improvement is allowed. High fixity, associated with slabbed track, considers no movement allowance at all. Clearly, these are simplified cases. Allowances relecting known track movement envelopes may be used if those movements can be validated or controlled, for example by a managed track position regime. It is also possible that allowances may be applied asymmetrically, for example to relect track movement trends on curved track, or held against a platform wall.

As part of a gauging survey, track fixity should be noted (or at least track construction and anything affecting fixity), together with any observations of rail sidewear which is also an input to the gauging calculations.

It should also be noted that, as whole life costing models are introduced, allowances for fixity may be increased to allow for long term maintenance regimes. An example may be the inclusion of (say) 100mm possible upward movement to accommodate a lifetime of track lifting to accommodate ballast deterioration. Such allowances (and their effect on clearance) must be balanced against the cost of providing the clearance at the lifecycle start.

12.4.1 Passing clearance

The above calculations are equally applicable to the calculation of passing clearances between trains and the controlling of the track interval, or sixfoot. On multiple track sections of route, alterations to structure clearances will have a consequence on train passing clearances, which must also be safely maintained. Noting that in the traditional aggregation of tolerances it would be unlikely for the movement of one track to be in the opposite direction to its neighbours.

(i.e. that they are moving towards each other) a relaxation of fixity tolerances of 25% may be applied to this calculation.

12.4.2 Datum plates

Datum plates are installed in locations of managed clearance and form an important control measure. Horizontal, vertical and cant measurements of the track in relation to the datum plate provide a measure of how far the track may have deviated from its position when clearances were determined and the control regime established. Whenever work is performed on the track, and datum plates are present, the measurements should be checked against allowable track position tolerances and a record kept of the as-left measurements. Datum plates may be of various colours, each of which will have meaning in relation to a particular route and may signify a design position, a controlled position of a position established at a nominal point within a maintenance cycle.

12.4.3 The management of clearance

The track engineer inputs into the clearance management process, by providing some of the base data that ensures the continued provision of required clearances. Whenever track work is done where clearance is a constraint, suitable measurements must be taken and submitted to the infrastructure manager for logging. Track engineers may also be asked to undertake gauging surveys from time to time as part of a central program to manage clearances.

On the national network, managed by Network Rail, the provision of adequate clearance to all authorised traffic on a continuous basis requires a national program which also feeds into the processes necessary for accepting new trains and freight lows onto the network. Central to this process is the use of high speed structure measuring trains, a database of infrastructure records and processes to calculate clearances on a regular basis, which then feeds a register of tight clearances used to manage track movements at those locations.

High speed structure gauging trains have been available since the late

1980's and have used a variety of principles ranging from white light triangulation on the original BR Research Structure Gauging Train (SGT) through laser scanners and to sophisticated laser triangulation systems as are presently used. These systems produce huge amounts of data representing the shape of the network infrastructure at high point densities. Measuring trains run regularly, generally in the hours of darkness. Data is then processed to remove obviously bad data (for example, as caused by stray illumination or the measurement of airborne particles) and placed in a database, known as the National Gauging Database (NGD). From this, clearances are calculated for all rolling stock running at each structure location and where appropriate, a track position management routine established and provided to asset managers. The National Gauging Database is available to anyone having a requirement to understand the physical size of the network, for example in the design of new trains. It is also of great value to track engineers in the planning of track works to ensure that any potential issues of clearance are understood before track work commences. At present, the infrastructure is measured at a frequency consistent with the allowances for track fixity, and is updated every two months. Access to the database is generally provided through clearance calculation software.

12.5 STEPPING DISTANCES

The passenger / train interface (PTI), characterised by what is known as the stepping distance calculation, is becoming more significant as efforts increase to assist those with reduced mobility to use the rail network and to reduce the number of passenger / gap incidents. Providing a minimal gap conlicts with providing adequate passing clearance, especially when several types of rolling stock may need to pass (but not necessarily stop) at a platform.

Regulatory stepping distances are calculated by considering the lateral, vertical and diagonal components of a line strung from the nominal platform edge point to the step tip of a static train which is not subject to any suspension displacement and on track in its design position (i.e. without any consideration of tolerances or fixity). Platform / train combinations where the lateral dimension is less than or equal to 275mm, vertical 250mm and diagonal 350mm are compliant to GB domestic standards. However, this dimension is under constant review as increasingly European regulations for Persons of Reduced Mobility (the PRM TSI) strive to provide better than 50mm vertical and 75mm lateral.

In practice, the distance is theoretical. Site measurements will reveal a variety of variations even on a compliant interface as a result of track positional tolerance (as it moves through its lifecycle) and variability of the suspension settlements as the train comes to a standstill. For this reason, site measurements are no substitute for the correct calculation of this regulated measure.

12.6 FURTHER READING

This text has provided a brief insight into aspects of gauging relevant to a track engineer. However, the topic of gauging is one of complexity which only becomes clear once explored. Those entering the field with views of its obvious simplicity seldom leave with the same opinion, and occasionally never leave at all. Accordingly, there is much reading available which provides greater detail into some of the concepts raised, including the following documents:

1. The V/S SIC Guide to Gauging. The Vehicle / Structures Systems Interface Committee exists to provide continuity across the vehicle / infrastructure interface associated with the size and loading of rail vehicles running on the national infrastructure. Systems Interface Committees were set up in the aftermath of the Hatfield (2000) incident in recognition that a non-vertically integrated railway required regulation at its technical interfaces as well as at its commercial ones. The 'Guide to Gauging' was produced in response to the need for a simple guide for those wanting to know more about gauging processes used in the UK, for example those who were responsible for the introduction of new rolling stock.

2. Railway Group Standard GI/RT7073. This document, published by RSSB provides the core standards by which clearance is managed

on the main line railway network. This document prescribes required clearances against which maintenance regimes should be managed, and various parameters to be used in gauging calculations.

3. Railway Group Standard GE/RT8073. This RSSB document describes a series of vehicle gauges for the GB mainline network. Whilst not mandating that a vehicle is built to a gauge, this document prescribes the characteristics of a gauge if a vehicle is described as being compliant with that gauge.

4. Network Rail (UK) standard NR/L2/TRK/3203. This Network Rail standard prescribes their requirements for the recording of data associated with structure gauging. This includes aspects of frequency of measurement, approved measuring systems and their accuracies, etc.

5. Railway Guidance Note GE/GN8573. This RSSB document provides guidance on the background behind, and application of, the Railway Group Standards applicable to Gauging.

6. Railway Industry Standard RIS-2773-RST. This document, published by RSSB describes the presentation of gauging data for vehicles such that gauging calculations may be undertaken in a consistent manner. Calculations using the data are described in GE/GN8573.

7. BS-EN15273. This Euronorm regulates the wider context of gauging across Europe. At present, it mainly requires National Notified Technical Rules (Railway Group Standards) to be used to provide safe passage of trains. However, as an evolving document it will, eventually, become the standard that will be used in Great Britain, as well as Europe generally. Current UK methods are gradually being introduced to the document.

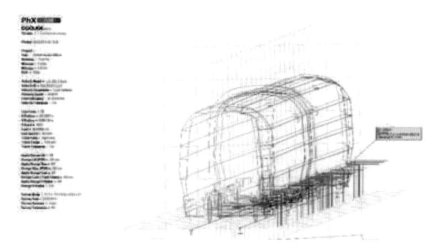

Figure 12.1 The output from a modern, computerised gauging analysis system.

In the diagrams, the relative size of rail vehicles operating in Great Britain can be seen:

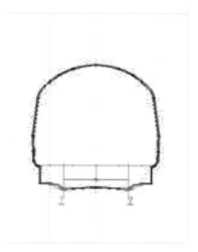

London Underground's Deep Tube network is illustrated by the static cross sectional profiles of 2009 Victoria Line stock, of 2897mm height.

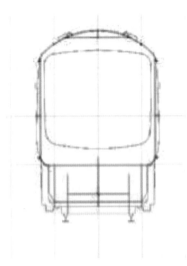

Typical mainline passenger rolling stock (including London Underground's Sub Surface lines) have a height of around 3775mm, as illustrated by the class 319 suburban Electric Multiple Unit (EMU). Such a 20m vehicle is likely to have a body width of around 2800mm.

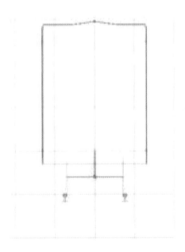

W12 is the largest GB freight gauge, having a height of 3965mm and a width of 2600mm.

Figure 12.2 Network Rail Structure Gauging Train
(Courtesy of Network Rail)

The train shown in figure 12.2, was first introduced in the 1980's using a white light triangulation system, with halogen projectors emitting a 'slice' of light from the slots in the centre of the vehicle.

The latest technology fitted, being the Balfour Beatty Rail Technologies 'LaserFlex' system, can be seen on the left hand end of the vehicle. This system works using laser light and digital camera triangulation in an integrated unit and is capable of operating at 75mph, giving a high degree of accuracy.

The LaserSweep, manufactured by Balfour Beatty Rail Technologies, uses a time-of-light laser system which can be pointed at parts of the structure which need to be measured. The device is ideally suited to smooth surfaces (such as underground tunnels), and measures both the structure profile and installed cant, which are stored on a Bluetooth connected portable computer.

**Figure 13.3 LaserSweep in use on London Underground
(Courtesy Balfour Beatty Rail Technologies)**

CHAPTER 13

TRACK MAINTENANCE

13.1   INTRODUCTION

Track maintenance is the key activity carried out on all railway track and guided way systems and society as a whole expects it to be carried out by competent professionally qualified staff. There are four reasons for this which relate to the operation of such as system and the delivery of absolute quality assurance, which is an expectation of the ultimate clients, the passenger and the freight customer:

- Guaranteed safety of the travelling public and the operational staff

- Performance of the railway in terms of zero delay and disruption to journeys

- The comfort of the ride at whatever planned speed, including the ability to write and drink

- A value for money efficient delivery where advanced innovative technology is promoted

This chapter describes the processes, both manual and mechanical, applied in maintaining the longitudinal profile, cross level gauge and alignment of track together with the tools and equipment employed. For information about maintenance of individual components, the reader is referred to the chapter dealing with that component.

The emphasis will be on describing the basic principles involved, followed by an account of the currently available techniques. It may be necessary to modify the methods described depending on the circumstances and environment, but the aim should always be to achieve good track standards and durability as this leads to a safe and efficient railway. It is also necessary to implement an organisation to deliver the maintenance activity. Track monitoring systems will clearly assist in providing information to achieve appropriate track geometry target levels. The systems for geometrical measurement are usually called "track geometry" and are described alongside the typical organisations needed for track maintenance.

13.2   MAINTENANCE OF LONGITUDINAL PROFILE (LEVEL or TOP)

Irregularities in longitudinal profile ("level" or "top") arise from differential settlement of the track, and the top maintenance operation consists in the correction of this differential settlement by lifting the over-subsided length of track so that it is level with the rest. The objective of top maintenance operations is to produce a quality of top which meets the standards laid down for the speed and weight of traffic using the line. However, the quality achieved must also be durable, and to achieve this, account must be taken of the state of the ballast and rails. The work to be done, and the equipment to be used, will depend on the circumstances, and the availability or otherwise of mechanical plant. Whatever the work, however, the fundamental principles involved are the location of those lengths of rail which have subsided relative to the remainder, the determination of the amount of lift required, the lifting of the rail with the sleepers attached, to the required level, and the insertion of ballast into the void thus created under the sleeper to support the track at the correct level. It is convenient to explain the principles involved in their application to manual work, and then to discuss machine maintenance operations.

13.2.1 Measurement methods

An isolated dip in the track (often termed a slack) may be located by eye, but it is not possible to identify by eye, errors in top where these are numerous, and/or have a long wave length relative to their amplitude, and/or extend over changes in gradient. Depending upon the length to be corrected and the method of correction, various methods of measurement are available.

Simple Line of Sight Methods

The simplest of these, which is also that used in measured shovel packing (MSP) and by many automatic track levelling machines, is based on the age-old method used as line of sight or colloquially known as "boning-in". In this method (see Figure 13.1) three Tee-shaped sighting boards called "Boning rods" all of the same height, are used. If two of them are placed upright on two points of known level, A and B, a distance apart, then by eyeing over the top of board A towards board B, and seeing when board C placed intermediately, just cuts off the visibility of board B, it is possible to determine by how much the intervening ground (or a rail) between them must be brought up or lowered so that its profile between A and B shall be perfectly straight. In permanent way work, A and B are almost always higher than C.

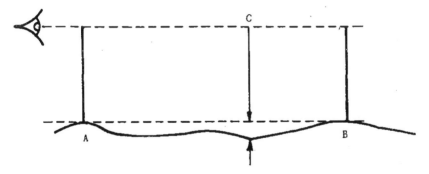

Figure 13.1 Line of Sight Method (Boning-In)

The boning rods may be very simple, even home-made articles, or they may be of more sophisticated design as a telescope with cross-wires to mark its axis of collimation. Such devices are available from suppliers of track maintenance equipment.

In automatic levelling machines the principle of boning is still applied. In early Plasser & Theurer machines the sighting end consisted of a light mounted on a trolley which moved in advance of, and in synchronisation with the main machine. The target end consisted of a photo-electric cell mounted on the main machine chassis. The intermediate board took the form of a shadow board supported on rollers which ran along the rail adjacent to the jacking assembly, so that it cut off the light to the photo-electric cell when the rail at the point of measurement was in the same plane as the light source and cell. This then stopped the lifting operation. In later machines the mensuration operation is carried out entirely within the wheelbase of the tamper, and typically the light is replaced by a stretched wire, but the same principle applies.

Optical levelling

The second method to be mentioned is the optical levelling survey, usually carried out to produce a redesigned longitudinal profile. The principle here is that a telescope with cross hairs is mounted so that it swivels about a vertical axis to define a datum plane which is horizontal and passes through the telescope at the point of observation. The level at any point within sight of the telescope can then be obtained relative to the datum plane by reading off a suitably graduated staff held at the desired point.

Lasers

Laser devices are available which will do the job either of the boning rods or the levelling survey, depending upon the design and application of the particular device. When using such devices, the makers instructions should be followed, particularly any which relate to avoidance of risk of eyesight damage which can occur if a person looks directly down a laser beam.

Electronic hand-pushed trolley

There are a number of lightweight trolleys available incorporating a stepping device and electronic equipment which measures the track geometry parameters over each successive step as it is pushed by hand along the track, or pulled along by a machine with which it is working. Real-time measurements displayed in 'easy-read' numerical formats on PDA include Gauge, Cross-level and Twist. The information is electronically stored and can subsequently be retrieved and processed to produce a longitudinal profile of the track.

Level correction

Having determined the amount of lift required by one of the means summarised above, it is necessary to raise the track and insert the necessary quantity of ballast to fill the observed slack or slacks. This may be done either manually or by machine. If done manually, the track must first be jacked up and then the ballast or other material is inserted by one or other of the methods outlined in the next clause. If done by machine, either an automatic tamper or stoneblower will be used as described below. It is worth pointing out here that in all manual methods of level correction it is necessary to take account of the possible existence of a void under the unladen sleeper, usually by the use of voidmeters. This correction is unnecessary when using on-track machines.

13.2.2 Manual Maintenance Techniques for Applying Top/Level Corrections

In the earliest days of railways, track was maintained by beater packing. The rails were lifted using long bars as levers and ballast was forced into the space under the sleeper using a kind of pickaxe with one end modified to a hammer head. This method, which is still in use in some countries and some of the heritage railways in UK, is the theoretical basis, if such it can be called, for both automatic machine tamping and for work with hand held tamping tools such as Kango hammers. Beater packing was superseded by shovel packing and then later by measured shovel packing.

Manual track lifting using jacks

It is possible to lift the track by using a long bar as a lever, but the most common method of manual lifting is by using track jacks. There are two main types, namely the mechanical ratchet jack and the hydraulic jack. It is the latter, being classed as obstruction-less, which must be used for work on track which is open to traffic without any restrictions. The ratchet type is primarily used in connection with track renewal work or where high lifts are required. Since a ratchet jack creates an obstruction it must be protected from trains. If a line possession is available a rail mounted portable powered jacking machine capable of lifting both rails at the same time may be used. The single jacking operation can be performed by one man although in general jacks are used in multiples either on a single rail or more normally on both rails opposite each other.

The following are some simple guidelines which should be applied when using jacks and lifting track

1.   Before lifting rails check that fastenings are tight on every sleeper to be lifted

2.   Use only an obstruction-less hydraulic jack of a type which is designed to close up if accidentally run over by a train.

3.   Ensure that all appropriate measures are taken to protect the safety of the staff on site before inserting any jack.

4.   Never leave a jack unattended and never leave the lever in place when not actually needed to work the jack.

5.   Place the jacks so that the lever sockets are on the "field" or "sixfoot" side of the rail and not facing into the fourfoot. Where an obstruction such as a conductor rail makes it impossible to follow this advice use any special equipment which may be provided or take such extra precautions as may be laid down.

6.   Never place a jack under a fishplated joint. This will strain the fishplates and in the case of an insulated joint the jack head could bridge the insulated endpost causing a temporary signal failure.

7.   If it is expected that a lift of more than 25mm will be required a temporary speed restriction should be imposed. Never lift more than 75mm on a track open to traffic.

8.   Lifting should normally be carried out towards the traffic, ie so that trains passing over the lifted track run "up" any ramp created by the jacking operation, onto the corrected track. On single lines or bi-directional tracks, the lifting should commence at the low point of any gradient and proceed up the slope.

9.   As we are dealing in this section with maintenance, it may be taken that the corrected track will line up "top-wise" with the existing track at the starting point of the work, and that a ramp connecting the uncorrected with the corrected track will develop as the work proceeds along the section of track being worked on. The length of this run down from corrected to uncorrected track where work is carried out without a speed restriction should not be less than 720mm per millimetre of lift. At any point where the lift reaches the limit of 25mm, the run down must be at least 18 metres long. This requirement clearly presents considerable practical problems in the management of the work and this is one reason why the limit of 25mm is placed on lifting track without a speed restriction.

10.   Both rails must be lifted together to ensure that no twist is imposed by the lifting process. Some twist may develop as a result of the difference in cross level between corrected and uncorrected track. Such twist must be limited to 1 in 240 (13mm per 3 metres).

11.   On curves it is usual for the low rail to be set at the correct grade or level and the high rail set at the correct cross level relative to it. Again, care needs to be taken to avoid introducing excessive cant gradients.

12.   When maintaining track under bridges the overhead clearances must be checked using the equipment provided to determine what lift if any can be safely applied.

Figure 13.2 Lifting and Patch re-sleepering on the Ffestiniog Railway at Porthmadog, Wales, UK. (Courtesy of Ffestiniog Railway Ltd.)

Manual Packing of Track

Shovel Packing

(1)   The rails are lifted to the required level and the crib ballast is pulled away from the side of the sleeper each side of the rail using the shovel. This exposes the bottom edge of the sleeper and the void underneath. The ballast in the crib is then 'punched' into the void from both sides using the bottom edge of the ballast shovel until no further stones can be accommodated.

(2)   This method will not adequately consolidate the packed ballast and will, therefore, not be durable under continuous heavy traffic loading. It is, however, a quick temporary measure for correcting poor rail top level and placing ballast under the sleeper where large voids exist. Its most common use is in forming large lifts following relaying or re-ballasting.

Measured Shovel Packing (MSP)

(1)   Introduction

Measured shovel packing is a means of applying simple levelling techniques to the correction of longitudinal level. Before the advent of mechanical tampers it was the standard method of track maintenance but it is now mainly used in the treatment of localised spots which deteriorate to an unacceptable degree between tamping cycles, or to tracks which are manifestly unsuitable for tamping, or for use on undertakings which for one reason or another do not have access to mechanical equipment. MSP requires apparatus consisting of a set of sighting boards, voidmeters, a cross level transfer gauge, one or more canisters to contain and measure the chippings, and special swan necked flat bladed spades to distribute the chippings under the sleepers.

The basis of the system is that the canister contains that quantity of chippings which when spread evenly over the whole width of the sleeper, for a distance of 15 inches (380 mm) either side of the running edge of the rail, will lift that sleeper by a sixteenth of an inch (1.6 mm). This is described as one "unit" of lift and the sighting boards, voidmeters, and cross level transfer gauges used in the system are all graduated in these units. This unit is implied in the descriptions which follow and it is to be understood that this is one area of work which is not metricated. The process involves two main stages, measurement and the placing of stone chippings.

(2)   Measurement

Before commencing the measurement of longitudinal profile, it is essential that the rails are seated firmly in the baseplates or rail seats on the sleepers and that all the fastenings are tight. The final quality of the work will depend on this preparation. In the computation of the required lift it is necessary to determine both the unloaded profile and also the depth to which the sleepers are depressed into voids under trains. The sum of the unloaded profile and the voids gives the dynamic top.

The unloaded profile is determined using a set of three sighting boards, comprising a sighting and a target board each with equal fixed heights, and one intermediate board with an adjustable stem and graduated vertical scale with moveable pointer.

The track maintenance team leader initially looks for high spots at which to fit the sighting and target boards onto the rail head. The maximum distance between the sighting and target board should not be more than 36 metres on straight track. It may have to be reduced on curved track in order to keep the view of the intermediate board within the length of the target board, but it should not be less than 10 metres. Before placing the sighting and target boards the leader should measure the cross level at the selected points with a track gauge and record any variation from design cross level. On straight track the rail seen as requiring the least correction is selected as the sighting rail. On curved track the low rail is usually used as the sighting rail. By this means, assuming that the cross level is approximately correct, it will usually be possible to achieve the required top by lifting both rails, rather than having to lower either. On curved track, and particularly in places where the cant is changing, such as transitions, it is advisable to check between the high spots to ensure that no adverse cant exists.

Having mounted the sighting and target boards on the rail head at the selected high points the intermediate board is mounted at the 4th sleeper from the sighting position, working towards the target board. It is important to ensure that all the boards are placed on the head of the rail with the board transversely level. The leader usually does the sighting through the slot of the sighting board and a second man at the intermediate board position adjusts the vertical stem until the leader can no longer see the lower portion of the target board through the slot in the sighting board. The reading on the graduated scale in units of lift is then marked on the sleeper position by the second person. The second person then moves the intermediate board four sleepers towards the target board and the process is repeated. This sequence of events continues until the intermediate board reaches its last position near to the target board. The readings marked on the sleepers represent the lift required to bring the rail up to the correct level at each mark.

When the readings have been completed on the "sighting" rail, a cross level reading is taken at each marked sleeper from which the number of units of lift required to produce correct cross level is calculated and marked on the opposite rail or sleeper end. The readings for the sleepers between the measured points are then obtained by proportion.

Example where the correct cross level is 32 units.

- At point A the sighting rail requires to be lifted 10 units, and the actual cross level is 34 units. Then the lift required on the opposite rail will be: *10 - 2 = 8 units.*

- At point B 4 sleepers farther on, the sighting rail requires to be lifted 14 units, and the actual cross level is 30 units. The lift required on the opposite rail will be: *14 + 2 = 16 units*.

The lifts at each jacking point will be:

Sleeper	Sighting Rail	Opposite rail
Sleeper A	10 units	8 units
Sleeper A + l	11 units	10 units
Sleeper A + 2	12 units	12 units
Sleeper A + 3	13 units	14 units
Sleeper A + 4 = B	14 units	16 units

The operation to determine the degree of voiding requires the fitting of void meters at the sleeper positions which have either been observed as moving under traffic or, in the absence of traffic, sound hollow when struck with a hammer. The readings taken from the void meters following the passage of a train are marked on the corresponding sleeper position. The accuracy of void meter readings is important to the final quality of top and it is advisable to fit void meters on as many sleepers as possible, always on joint sleepers and to repeat the readings where any doubt arises.

The dynamic profile or top is obtained by adding the readings obtained from the sighting boards to those of the void meter. This gives the final correction required in units of measured packing.

(3)   Placing of stone chippings.

The second part of the process is the placing of the chippings under the sleepers. The figures marked on the sleepers will indicate the number of canisters of chippings which must be spread under the sleeper approximately 375mm each side of the rail.

The special steel canister will when filled to a slot in its side represent the quantity of stone chippings required to raise the level of the track by one unit of measurement on plain track with standard width sleepers. Where wider sleepers are used or on 300mm wide S&C timbers, the canister should be filled flush to the top.

In order to place the chippings, it is necessary to remove ballast to expose the bottom edges of the sleepers. To do this it is normal for alternate bays or crib areas to be removed for half their length. Having done this, jacks are inserted under both rails and the track lifted sufficiently to enable the canisters of chippings to be spread under the sleeper by the special swan necked flat bladed spade. It is important to ensure that each canister is spread evenly over the whole area of the sleeper for 375mm each side of the rail.

Having placed all the chippings within the length being treated, the track is lowered down off the jacks. It is advisable before forking the ballast back into the cribs to observe traffic over the track and to check the correct bedding down of the sleepers. On being satisfied with the result it is normal practice to check the alignment and correct this where necessary.

Hand Held Stone Blowing

Arising from the research into and development of an alternative to the automatic tamping machine as a means of maintaining track, hand held equipment which will inject 20mm sized stone under the sleeper by means of a pressurised flow of air has been developed. Hand held stoneblowing offers a degree of mechanisation compared with the older measured shovel packing. More importantly it does away with the need to remove the crib ballast which was labour intensive even on timbered sleepered track and has become even more so with the advent of concrete sleepers.

The advantages of the process are:-

- minimum disturbance of ballast,
- economy in stone usage,
- good durability of the result.

The hand held stoneblowing process (HHSB) is seen as an intermediate maintenance aid between visits of on-track maintenance machines. It can be used effectively at fishplated or welded joint positions and in particular, insulated rail joints in CWR. Where wet spots develop, the HHSB is an effective remedy pending more radical treatment by, for example, ballast cleaning.

The procedure adopted in S&C is similar to that of plain line except that:-

a) The required rail top levels must take into account the long timbers carrying more than one pair of rails.

b) The amounts of stone required are blown at each rail bearing position on a timber.

c) In order to keep the initial jacking to 30mm a 14mm size stone is used instead of the 20mm stone used in plain line work.

Both the 14 and 20mm size stone should be of a hard and durable quality (Wet Attrition Value 4%) and angular in shape with all its dimensions near equal. Elongated shapes will tend to cause blockage during blowing. No more than 0.5% fines should be present.

The principal elements of the HHSB process are as follows

1) Survey and measurement of the section of track to be treated. Various methods are available, including conventional sighting boards as applied in the MSP process or optical

instrument survey As with other processes, the required lifts expressed as quantities of stone required, are calculated and marked for each rail bearing position. Practice has established that for each millimetre of correction to the rail top level required (loss of static top plus the dynamic void) 0.5 kg of stone must be placed under the rail. A suitable sized container can be used which contains a measured 0.5 or 1.0 kg weight of stone.

2)   Track Jacking. The track is jacked up by 30mm plus the correction in rail level to create a "packing" void. Any overhead clearances must be checked when doing this.

3)   Tube Insertion. Special steel blowing tubes are driven into the ballast against the sides of the sleeper, 225mm on either side of each rail, at the rate of 2 tubes per packing point. It is important to ensure that the tube is driven to the correct depth with the bottom of the tube 40mm below the underside of the sleeper.

4)   Blowing Stone. The quantity of stone as determined from the survey is fed into the steel tube and blown into the void under the sleeper.

5)   Reinstatement. Having blown the required amount of stone at each sleeper position the track is jacked down and the tubes are removed for re-insertion at the next position as required.

Hand Held Tamping Tools

Hand held power tools such as Kango hammers or similar have largely taken the place of the manual beater type tool of old. The Kango hammer is a relatively light weight electrically powered hammer in which a reciprocating or eccentrically supported weight driven by an electric motor strikes the socket end of a long swan necked beater. A Kango hammer is used in much the same way as a beater with the track being lifted to a level deemed satisfactory by eye or by reference to a pre-determined level established by survey and bench marks.

The crib ballast is first eased away from the sleeper side by shovel or fork in order to expose the underside edge. Then the track is jacked up. With the track lifted, the hammers are positioned close to the jack position. The hammer head is then used to force ballast into the void created and with its vibratory characteristic, pack and consolidate the ballast support under the sleeper for approximately 375mm each side of the rail heads.

The tools are generally used in sets of 2 or 4 and are powered from a portable generator or electrical power points at the track side. Each pair of hammers is used one on either side of the sleeper or timber so as to achieve the most rapid and effective consolidation. The user of the hammer will sense the point when adequate consolidation has been reached, and the hammer should be withdrawn before excessive breaking or crushing of the stone ballast occurs.

Devices similar to the Kango hammer are marketed by several other firms making track maintenance equipment. Hand held tamping tools are useful for local track maintenance operations notably in S&C and at joints. They are unsuitable for use if the ballast is very shallow and underlain by clay, and particularly at wet spots, as they will merely make holes in their formation which will fill with water.

Shimming

If the undersides of the sleepers are decayed, packing with ballast or chippings tends to be unreliable because the particles of stone penetrate the softened wood in the decayed part of the sleeper. One way of achieving a more reliable result is to use shims under the sleeper in place of chippings. The shims are made from waterproof hardboard 750 mm by 250mm and of suitable thickness (eg 3mm equates to one eight of an inch which equates to two units of measurement in the MSP system).

To use the shims, the track is first measured for lifting as for the MSP system, and the track jacked up and opened out in the usual way. Shims aggregating to as near as may be the calculated lift are then simply inserted in the void between the sleeper and ballast (making sure that no lumps of material are left adhering to the sleeper soffit) and the track finally jacked down and the ballast made good.

Adjustable fastenings

It is useful to include in the subject of manual maintenance of top some brief mention of the possibilities of adjustable fastenings. It is possible, with the more sophisticated modern spring fastenings (e.g. PANDROL as used in the UK and NAB LA as used on SNCF), to arrange so that a certain amount of adjustment can be built in between the underside of the rail and the rail seat area of the sleeper by means of a combination of varying thicknesses of pad under the rail and varying thicknesses of insulator between the rail and the clip holding it down.

13.2.3   Mechanical maintenance of top/level

The mechanical maintenance of top is usually carried out by one or other member of two families of machine; Tampers and Stoneblowers. These machines also correct line which is described later.

The tampers and stoneblowers typically combine the functions of correcting top, cross level and line on the one machine, and normally all corrections are carried out during the one pass of the machine. For convenience of presentation here however the top and cross level functions will be described in this clause, and the line correction function later. Top is measured and the necessary correction applied as described below, and as the tamper lifts the sleeper to the correct level, vibrating tamping tines are inserted into the ballast cribs adjacent to each sleeper and squeezed towards each other up to the sleeper's edges. The sleeper is then released (if the tamper signals that the correct lift and cross-level has been attained, otherwise it is retamped) and the tamper moves on to the next sleeper(s) for treatment. In the case of stoneblowing, chippings are injected at this stage.

The current situation in the UK is that mechanised maintenance is carried out by a combination of tamping and stoneblowing with both plain line and S&C machines being used. The geometrical principles remain the same in terms of improving the physical line and level.

Figure 13.3 Multi Purpose Stoneblower (courtesy Network Rail)

Correction of longitudinal level

Longitudinal level is measured and corrected by a mechanised form of sighting boards, in which the line of sight is replaced by a tightly stretched wire. Referring to Figure 13.1, board A (the sighting board) becomes a datum point on the tamper, to which one end of the wire is fixed. This will be positioned at a known, fixed height, above the *corrected* rail. The other end of the wire, which corresponds to board B in Figure 13.1, is fixed to a second datum point, the height of which above the *uncorrected* rail is adjustable. This is often called the "tower", or "front tower". The intermediate board, board C in Figure 13.1, which is also placed over *uncorrected* rail, becomes the lift control, and is conceptually the same height above the rail as the datum point. In operation, the uncorrected rail at C is lifted by roller clamps referenced to the intermediate board, until the board contacts the wire, when the lift cuts out, and tamping or stoneblowing is initiated.

The principle of control over the longitudinal profile rests on the fact that the distances between the datum, the lift point, and the tower, are fixed, so that the distance between the datum and the lift point (the "intermediate" length), is a fixed fraction of the distance between the datum and the tower (the "measuring base").

Smoothing

Consider the situation where the tamper/stoneblower is operated over straight, uncanted track having an irregular top, but a more or less steady gradient. Suppose also that the tower is kept at the same height above rail as the datum. The mechanism will improve the profile, since at any point where the rail at the lifting board is below a straight line (the "datum line") connecting the rail at the datum to the rail at the tower, the machine will lift the rail by the amount of that apparent depression. If we consider what happens between two high points, however, it will be noted, that whereas in MSP, the rail is lifted until its profile is a straight line between the high points, this does not happen with the machine. This is because, the distance between the datum and tower being fixed, neither of them will in general be at a high point. The result is that the corrected profile will still not be a straight line, and it will be lower than the straight line obtained by MSP. Furthermore, when the machine is so positioned that its measuring base straddles a high point, the fact that the corrected profile is lower than the straight line connecting the high points means that as the lift point approaches the high point, with the tower in the next depression, it is quite possible that the uncorrected rail at the lift point will be above the *datum* line and the track will be left uncorrected.

This can be prevented by carrying out the tamping/stoneblowing with the tower raised. The amount by which the tower should be raised to ensure that a lift always takes places is dependent on the initial roughness of the profile, and track recording car data may help here. If the standard deviation (SD) of top of the rail to be corrected is known, then using the rules of statistics, it will be about 99% certain that if the tower is raised by six times the SD, the rail will never come above the datum line (e.g., if the SD is 2.5mm, and the tower is raised by 15mm, then the uncorrected rail at the lifting point will almost never come above the datum line). Another way of determining the amount to raise the tower is to examine the Track Recording Car trace for the section being tamped, and to look for the worst irregularity.

It is often stated that apart from the consideration mentioned above, the slight track lift resulting from a modest heightening of the front tower improves the general performances of the machine. There will of course be a slight ramp at the start of the job, and the tower must be lowered below the height of the datum and lift boards to run out the lift with a gentle ramp down at the end of the work. This is the principle of smoothing.

It can be shown that, provided enough lift is given to ensure that the rail at the correction point never comes above the datum line, the roughness of the track will be reduced by smoothing tamping by a factor equal to the ratio between the measuring base and the intermediate length. This ratio, typically about 6:1, is hence sometimes referred to as the "improvement factor". Thus if the track before tamping has an SD of 2.4mm, its SD immediately post tamping should theoretically be 0.4mm.

Care must be taken when lifting, to avoid infringing any tight overhead clearances. Any such locations must be identified before work commences, and the front tower must be programmed so that lifting is run out before the tight point is reached.

Ramps and vertical curves in Smoothing

In smoothing tamping/stoneblowing, the machine takes the track "as found" and improves its longitudinal profile. However, since the measuring point is some metres ahead of the point at which the correction is supplied, the effect of the smoothing process is not only to reduce the magnitude of any slack by the improvement ratio, but is always to shift it backwards by a distance equal to the separation of the front tower and the lifting point. If the machine encounters a change of gradient, there will be a corresponding change in the elevation of the rail at the front tower, relative to that at the lifting point. The mechanism will see this as a defect (a slack if the gradient turns upwards and vice versa), and a "correction" will be applied, the effect of which will be to move the start of the gradient backwards. If the change of gradient is actually the start of a transition curve the effect will be to put the start of the cant gradient out of phase with the start of the spiral, and the result will be a cant deficiency defect which will be noticeable in the trains.

To allow for this the front tower has to be adjusted as the machine approaches any significant change in longitudinal profile. Typically, on approaching an upwards change of gradient, the tower must be lowered progressively, as it runs onto the new gradient, the amount of lowering reaching a maximum at the point at which the lifting point reaches the start of the new gradient, and returning to its normal value once the whole measuring base is safely on it. The amount of adjustment has to be calculated for each jacking location. The calculation uses the algorithm described below, and would be quite tedious if it had to be done by hand. However, provided that the necessary local data is available, modern machines have on-board computers which carry out the necessary calculations automatically. All that the operator then has to do is make sure that the appropriate information about gradients, vertical curves, cant, and the location, length, and gradient of transition curves or other places where cross level has to change is accurately input.

Correction of long wavelength defects

The efficiency of the smoothing algorithm falls off as the wavelength of any "top" defect extends beyond the length of the measuring base of the tamping machine. Various methods are now available for effectively extending the measuring base, and they comprise packages which are fitted to the machines, and work automatically once the necessary data has been put into the on-board computer. A number of software packages are used.

Design Tamping

In design tamping, the starting information will typically be a profile prepared for a re-levelling scheme. This will contain information about the existing and proposed reduced level at every sleeper throughout the section of track to be treated. Starting from the basic geometry of the measuring apparatus as described above, it can be shown by the application of the rules of similar triangles, that at any given machine position, the amount by which the front tower must be lifted or lowered in order to produce the required lift at the lifting point, is given by the following algorithm:

Let:

- N_d be the proposed reduced level at the datum point

- N_L be the proposed reduced level at the lifting point

- O_T be the existing reduced level at the front tower

- R be the improvement factor for the machine

- D_T be the amount by which the tower must be raised or lowered to give the required lift

Then: $D_T = (R \times N_L) - [N_d \times (R - 1) - O_T]$

D_T may be either positive (implying a raising of the tower) or negative (implying that the tower must be lowered).

As with smoothing tamping, it should be unnecessary when using a modern machine, to work out the front tower lifts by hand.

**Fig 13.4 Plasser and Theurer Tamping and Lining Machine (courtesy
Network Rail)**

Precautions to be observed to get the best results from tamping

In order to achieve the best results from automatic tamping the
following items should be checked in advance, depending on the
exact nature of the work:

a) Is the ballast quality suitable for tamping? Ash and very fine
    aggregates are unsuitable, as is badly fouled ballast.

b) Is there enough ballast? Sleeper cribs should be full of ballast
    to the sleeper tops, and the shoulders should extend to the full
    375mm and be well heaped up. If ballast is insufficient, or if it
    is expected that the track will be lifted far enough to expose
    sleeper sides, more ballast should be ordered. It is probably
    better however to have the ballast delivered after, rather than
    before, the tamping operation, as excessive ballast at the time

of tamping can, by obscuring the sleepers, lead to damage to the sleepers, and/or the machine itself.

c) Is there sufficient ballast depth? There should be about 225mm ballast under the sleeper soffit, to ensure that the tines cannot penetrate the subgrade. If this happens, the tines will dig holes in possibly soft, cohesive material, in which water can gather and initiate formation difficulties.

d) Is there a membrane in the ballast, which could be damaged? This can be either geotextile, on embankment or in cut, or a waterproofing layer on an underbridge deck.

e) Are the sleepers suitable for tamping? Decayed softwood sleepers can fail completely once fresh air is allowed to reach their undersides as a result of track lifting.

f) Are the sleepers suitably spaced and squared? Many machines tamp two or more sleepers simultaneously and are therefore unable to cope with uneven sleeper spacing. Tampers which tamp in multiple also need to be modified to cope with changes in sleeper spacing (e.g. from 26 to 30 sleepers per rail length). Most tampers cannot tamp out of square sleepers properly without moving the sleeper itself - to the detriment of the rail fastenings.

g) Check the sleeper depth. The depth of the tines has to be adjusted for differing depths of wood and concrete sleepers. If there are frequent changes in sleeper type, machine output may be reduced since the machine crew should conscientiously adjust tine depth for every variation encountered. If they fail to do this the result will be a poor job, and the bottom corners of the sleepers may be abraded or chipped.

h) Are the rail fastenings in good order? There is little benefit in lifting the rail if inadequate fastenings result in the sleepers being left behind in the ballast

i) Is the condition of the rail satisfactory? Lipping and sidewear may lead to inaccurate readings by the measuring systems of the machine.

j) Is the area where the work is planned free of obstructions? (Level crossings, barrow crossings, certain types of rail lubricators and switch heaters, S&T treadle mechanisms, hot axle box detector units, bond wires, point rodding, scrap rail, etc. etc.)

k) Track Geometry: Are the cant, transition length and curve radius clearly marked? Are limited clearances (vertical and horizontal) clearly marked?

l) Environment: Are the mileposts, etc, in position so that the worksite is easily located? Is the weather suitable for the proposed work? (e.g. frozen ballast, hot weather precautions). Has the possibility of noise complaints, etc, from lineside residents been considered? Is the proposed track possession suitable for the work proposed, and are any special safety precautions required?

The above is by no means an exhaustive list, but does serve to illustrate the problems that may arise if inadequate thought is given to preparation for mechanical maintenance. Indifferent results can often be traced back to indifferent preparation. It is emphasised that since mechanised maintenance forms the major proportion of the permanent way maintenance workload, it should receive a commensurate amount of attention when planning and preparing for site work.

It is critical to the tamping process itself that the tamping tines are not unduly worn, that they are inserted to the correct depth below the sleeper, and that sufficient squeeze pressure is used to ensure proper consolidation without either causing the sleeper to be pushed vertically upwards out of level, or physically damaging/moving the sleeper by exerting too much squeeze pressure.

Pneumatic Ballast Injection Machines - 'Stoneblowers'

With the advent of inertial track recording systems it was realised that vertical track geometry deteriorated rapidly under the passage of traffic following tamping. The resultant vertical profile was virtually identical to that which preceded tamping (a phenomenon known as ballast memory). Tamping machines cannot easily change this inherent ballast profile because, with small track lifts, no extra stone is introduced underneath the sleepers.

Stoneblowing can obviate this problem by introducing additional ballast (in 20mm chipping form) beneath the track sleepers - in effect an automated version of the original manual measured shovel packing, resulting in far greater durability of track 'top' than tamping.

The benefits to be obtained from stoneblowing include:

i. Track 'top' durabilities up to 8 times better than tamping have been recorded, with an average figure of 3. Thus far fewer stoneblowers are required to maintain track than tampers, with savings in capital expenditure, machine maintenance, manpower and track possessions/ disruption to rail traffic.

ii. Track suffering from sub-standard ballast conditions (e.g. ash ballast, wet/ clay spots, etc) can be satisfactorily maintained, and manual intervention between maintenance cycles reduced or eliminated. Further, speed restrictions can be removed or deferred due to the reduction in the rate of track profile deterioration.

iii. Conventional tampers tend to destroy ballast (reducing it to fines) entailing the tipping and spreading of 'maintenance' ballast and, ultimately, hastening the requirement for ballast cleaning itself. Because stoneblowers introduce all the requisite ballast (and due to the method employed do not reduce ballast to fines) stoneblower fleet implementation will result in considerable savings in ballast, manpower, ballast trains, ballast regulators and, ultimately, ballast renewal.

iv. Due to the stoneblower's computer system, only areas of track actually requiring attention will be treated. The computer system also ensures that the stoneblower is self-monitoring, and able to produce detailed records of individual site treatments.

v.   The stoneblower designs its own longitudinal track profile, onsite technical input is not required for surveys/setting out remedial treatment for long wavelength track profile irregularities.

vi.   Rail end straightening and weld straightening are reduced due to better track stability, and the individual joints and welds themselves straightened to a certain extent by the stoneblowing process.

vii. Partly-obstructed sleepers (e.g. rail lubricators, earth/bond wires, point rodding, hot axle box detector units, etc) can be successfully maintained provided tube access is available from one side of the sleeper, thus avoiding the need for manual packing of such problem areas.

Fig 13.5 Stoneblower injection Tubes (courtesy Network Rail)

The stoneblower is typically an 80 tonne, three-bogied articulated vehicle with a transit speed of 60 mph. The machine's hopper system contains up to 13 tonnes of stoneblowing chippings - sufficient for up to 4 shifts - the ballast (20mm granite chippings) being loaded into the hopper by means of the HIAB crane grab mounted at one end of the machine. The operations of track measurement (including sleeper voids), survey, design and treatment are all computer controlled, the track being maintained at a rate of up to 600 sleepers/hour by treating two sleepers simultaneously. The maximum design lift is 40mm, which requires injection of 44kg chippings beneath the individual sleeper. The machine is also capable of design or smooth lining using a system similar to that employed on a tamper fleet. The stoneblowing heads of this machine are illustrated in Figure 13.5.

13.3 MAINTENANCE OF ALIGNMENT (LINE)

13.3.1 Manual Maintenance of Line (Slewing)

Manual slewing is appropriate for dealing with localised misalignments which sometimes require to be dealt with in the course of day-to-day track maintenance, particularly when for any reason it is not convenient to bring in rail mounted plant immediately.

It is good practice to provide fixed datum points alongside the track in the form of pegs or monuments. Where these are provided an accurate gauging rod is needed to compare the position of the rail with its required location, and it is then merely a matter of slewing the track so that it matches the alignment of the monuments. If on the other hand monuments are not provided, the leader must use his visual skill to determine in which direction and by what amount the track requires movement in order to correct the alignment.

Before commencing realignment it is necessary to ensure that all fastenings are tight and that ballast conditions are adequate to provide good lateral restraint on completion of the work. The track gauge should also be checked, as irregular gauge will lead to the alignment differing as between the two rails. Where this is in curved track and no remedial measures can be taken at the time the lining should be carried out to produce the best alignment on the outside or high rail of the curve. Crippled rails or misaligned welds will also affect the quality of the work and where these cannot be improved by crowing or other means the defective item should be replaced.

The leader should send the team (which should comprise at least six persons equipped with bars) ahead for a distance of at least 2 rail lengths, or say 40 metres. One person of this group (the marker) will be directed to walk slowly ahead holding a bar on the head of the rail until he reaches a spot which the leader can see requires slewing. This point is then indicated eg by placing a particle of ballast on the rail head. Depending on the amount of slew it may then be advisable to ease the ballast away from the sleeper ends each side of this point marked on the rail to lessen the effort required to move the track.

Having been told in which direction to slew, the team divide into two groups and position themselves around the point marked on the rail. They then insert their bars into the ballast at an angle behind the rail with the bar against the rail foot and their feet either side of the bar so as to pull the rail in the required direction. The bar must be thrust well down into the ballast to provide good anchorage during the slewing. The marker calls for the pull so that all bars are exerting pressure in unison. Pulling continues until the desired alignment is achieved. On completion of the slewing the leader should ensure that the ballast is made good particularly where ballast has been removed to ease the slewing.

When slewing on underbridges the depth of ballast must be checked to ensure that the pointed ends of the bars will not damage the waterproofing layer below. Care should also be taken not to damage signalling cabling. Slewing may also strain cable attachments to rail and other signalling equipment.

When slewing in situations such as S&C and concrete sleeper track the work can be eased by using lifting jacks to take the weight of the track and also to provide additional horizontal pressure on the foot of the rail. Care should however be taken when doing this to avoid overlifting the track and causing ballast to fall beneath the sleeper.

Manual slewing can be done by fewer personnel if portable track liners (either mechanical or hydraulic) are available.

In the course of time and under the influence of traffic, the track gets pushed out of line due to repeated train forces and irregular lateral restraint by the ballast shoulders. Alignment irregularities on straight track are sometimes visible by the naked eye. Although with experience the ability to detect irregularities in curved track can be developed, probably the most certain way of finding and correcting alignment defects on curves is to use versines. A versine is the perpendicular distance from a chord to the centre of its curve. It is emphasised here that it is not necessary to embark on a long and complicated survey process in order to use the method. If a local defect is suspected (e.g. by reporting of an alignment exceedence from the Track Recording Car) a traverse of only six or eight half-chord points may suffice to locate the kink and indicate by how much the track must be slewed to correct it. Alignment correction may be done by hand or by machine. As already pointed out, machine realignment is normally done using a machine which maintains both top and line at the same time.

13.3.2 Mechanical Machine of Line

The Lining Process

Plain line alignment is measured and corrected by one or other of a number of geometrical algorithms which involve comparing the properties of a long segment of (curved) track with the properties of its shorter components. For example, if the versine on a long chord is measured, and compared with the versine on one or both of the two sub-chords, then the four or five points involved are only on a circle provided the versine on the subchords is exactly a quarter of the versine on the main chord. Alternatively, the versine on a chord may be compared with the offsets from one or both of the quarter points of the chord, in which case the sub-offset must be ¾ of the versine if the curve is to be circular.

The machine's control system identifies any mis-match, determines the required direction of slew, and actuates appropriate slewing jacks, the force from which is reacted across the wheelbase of the lining machine. Actuation ceases when the criteria for circularity are met.

The system is basically reliable but is prevented by the length of lining chord employed (usually about 20m) from detecting long wavelength misalignments. These can be removed by design lining (i.e. undertaking a manual Hallade curve survey and setting out the requisite slues) or indeed by the sophisticated Automatic Track Alignment system (see below).

Most lining problems arise either from poorly-calibrated machines, or indeed from track geometry that the machine cannot properly measure, e.g. lining adjacent to track obstructions, or track geometry that is not correctly marked up on site (if at all). Occasionally, problems are encountered with heavily- canted CWR 'sliding downhill' (especially during cold weather) whilst tamping. Under these conditions it is advisable for the operator to apply an 'uphill' adjustment to the front of the lining chord to counteract the phenomenon - reducing/increasing the chord adjustment over the length of transition itself.

Whilst lining and tamping can be done independently, it has been found that longitudinal track top deteriorates more quickly than horizontal alignment, making separate lining machines uneconomic, and virtually all lining is done using combined machines of the types described above.

Automatic correction of long wave alignment defects

With the drive for greater passenger speed and comfort in the UK, developments in methods have been made in improving the lining performance of its tamper and stoneblower fleet. With this system, the machine first records the horizontal track alignment versines, which are then fed into an on-board computer. The computer then redesigns these versines (using the Hallade method) according to the design parameters selected, e.g. maximum slue, length of transition, line speed, etc. Once the design and associated track slues have been calculated, the machine then maintains the track in a conventional manner, applying the design slues and tamping simultaneously. A further recording run can then be made so that the initial and final alignment quality can be compared. Most modern computer programs which combine both top and alignment algorithms have been developed.

Alignment Correction on long straights

Lasers have been used for some years now to improve the lining quality achievable on straight track. The laser emitter is mounted on a portable trolley, which is set up at a distance of up to 300m from the machine. The laser is aimed at a receiver mounted on the machine, and the front end of the lining system "locks on" to the laser beam. The machine then works towards the stationary trolley (the chord front of the lining system moving automatically to keep track of the laser) and the trolley is subsequently repositioned and reset when the machine is within 40m or so of it. Care needs to be exercised in use of this system. It can be adversely affected by fog and rain, which cause laser beam dispersion. Also, very large track slues can result if the long "straight" diverges significantly from a true straight line.

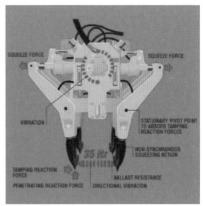

Fig 13.6 Tamping banks in detail (courtesy Network Rail)

13.4 MAINTENANCE OF GAUGE

13.4.1 Importance of Correct Track Gauge

The gauge is defined as the horizontal distance between the running edges of the track. The maintenance of correct and regular gauge is an essential ingredient in the attainment of satisfactory track geometrical standards, since it plays an important role in relation to the behaviour of the vehicle wheel, especially at high speed. Variations in gauge will cause vehicle oscillation, and this will in turn overload fastenings and induce uneven wear of the rail head. Tight gauge will also reduce the critical speed above which bogie hunting is liable to occur.

The standard gauge in most of the world on new plain line and "vertical" S&C layouts is 1435mm. Some designs of track were laid with a gauge of 1432mm in the latter part of the 20th century as a trial. The difference of 3mm is significant where lengths of track laid to the two gauges abut.

Gauge variation affects the alignment of track and it is wise to correct both gauge and alignment in the same (or at least sequential) operation. If this is not possible then when correcting gauge it is desirable to ascertain which rail has the better line. This rail is then used as a datum from which to set the corrected gauge.

13.4.2 Causes of Gauge Variation

Gauge variation is usually due to deterioration in some element of the rail support system. In general it is essential for good maintenance of gauge that all fastenings are kept tight and that correct seating of the rail in the baseplates and of the baseplates on the sleepers is maintained. In timber sleepered track, indications of movement in the baseplate under traffic loading can often be detected readily by eye and such sites should frequently be checked both for the tightness of the fastenings and with a track gauge until remedial measures are carried out. In concrete sleepered track insulators should be periodically inspected to ensure that no crushing or displacement has occurred. Gauge will be affected where sleepers are not square to the rail.

13.4.3 Correction of Gauge Variation.

When gauge is found to be incorrect there are various remedial measures which can be taken, including the following:

Fastenings in Timber Sleepered Track

Where fastenings have lost their holding power there are some proprietary devices which can be used to improve the gauge security as an interim measure. These include spike locks (requiring the withdrawal of the spike, insertion of the lock and re-driving the spike), and Vortok coils. On curved track in plain line and in S&C, gauge plates can be fitted to the outside edge of the baseplate.

Gauge Tie Bars

Where the track condition is such that gauge correction by rehabilitation of fastenings is impractical and replacement of components is inappropriate or not possible at the time (eg badly enlarged screw holes in decayed softwood sleepers), tie bars should be fitted. These should be checked periodically for tightness and gauge setting, and in track circuited lines they should be of the insulated type.

Concrete Sleepered Track

In the case of concrete sleeper fastenings wear or crushing of insulators is the main cause of gauge variation. The means of control is frequent inspection and replacement of worn and/or damaged insulators where necessary. Badly worn or damaged insulators can also cause track circuit failures. In a few cases the loosening of steel inserts in the sleeper or corrosion and breakage of these will lead to the early replacement of the sleeper.

13.5 MAINTENANCE OF ELECTRIFIED LINES

The mechanised maintenance of electrified lines poses many problems due to the restricted access and reduced clearance available for track maintenance operations.

Third and fourth rail electrified railways give limited access to the area around and beneath the running rails. This can restrict access to parts of the track for manual and mechanised maintenance operations. With tampers this problem is overcome by specially-modified machines that utilise twelve tines per sleeper (rather than the more usual sixteen): eight being located within the four-foot and only four (rather than the usual eight) on the outside of the two running rails.

All on-track machines require isolation of the third/fourth electrified rail for site operation. A further drawback is the multitude of earth bonding and feeder wires associated with the electrical supply - many of which are located across the 'four-foot' and impede access to the crib sides of the sleeper for tamping operations - although stoneblowing promises an elegant solution to the problem.

Overhead Line Electrification Equipment (OLE) creates a different set of problems for mechanical maintenance. Electrical isolations are required for certain mechanical maintenance operations, and staff access to the higher parts of machines is severely restricted to prevent any possibility of electrocution (note that electrical isolations are not required for tamping, ballast regulating, etc). The relationship between OLE and the locomotive's pantograph is critical, and therefore the amount of track slew or lift is strictly controlled between close limits: typically such data is marked on the OLE masts themselves.

The maximum amounts by which track may be allowed to vary from the correct position are given in Chapter 12. Some further notes on the effect of OLE on track maintenance generally are given in Chapter 1.

13.6 OTHER ON-TRACK MAINTENANCE MACHINES

The following section relate to a few of the more significant items of equipment in regular use.

13.6.1 Ballast Regulators

These machines are designed to regulate surface ballast to the requisite profile, and are available from several manufacturers. Typically, adjustable ploughs do most of the task, and rotating brushes provide the final finish. The ploughs are capable of transferring stone from the 'four-foot' to the sleeper end (and vice-versa) and also from sleeper end to the opposite sleeper end, with a horizontal reach up to 1.8m from rail, and up to 0.8m below rail level, dependent on manufacturer.

Speeds of 3-6 mph are typically achieved for each plough pass of the ballast regulator - the actual speed and number of passes required being dependent on the volume of ballast being regulated and actual model of ballast regulator; brushing speed being some 1 mph.

Ballast regulators may be fitted with an on-board hopper facility, typically of 12 tonnes capacity, so that excess track ballast may be collected, and stored for augmenting locations where ballast is deficient. These larger ballast regulators are mounted on two 4-wheel bogies, and are capable of transit speeds of 45 mph. Smaller, two axle regulators (usually without a hopper facility) tend to have a lower transit speed - typically 25 mph.

Ballast regulators may also be fitted with snow-ploughing equipment, and a variety of such machinery is marketed by several manufacturers for fitting to their machines.

13.6.2 Dynamic Track Stabilisers

Dynamic Track Stabilisers (DTS) manufactured by Plasser & Theurer, are utilised to simulate the effect of rail traffic by consolidating loose ballast. Normally, this will be in conjunction with ballast cleaning/track renewal.

However, normal tamping operations will also affect track stability, (particularly CWR) in very hot weather. Under these circumstances, tamping may have to be suspended, or special safety precautions, such as additional patrolmen or temporary speed restrictions may be required as described below. To some extent, these inconveniences may be avoided by use of the DTS which will consolidate the ballast and thereby increase the lateral stability of the track in the immediate post-tamping period.

The DTS operates by lateral vibration combined with vertical downwards pressure (up to 32 tonnes/sleeper) applied directly to the rails. Settlement of over 50mm can be expected in one machine pass, depending on ballast conditions, speed and frequency of machine. The DTS has an output rate of typically 500 m/hour.

A lining system is fitted to the machine, primarily to ensure that the horizontal track alignment is not altered by the operation of the machine. Each machine has a chart recorder fitted giving details of:

- operating speed

- vibration frequency (Hz)

- vertical downward pressure

- track 'twist', ie change in cross-level

- horizontal track versine

13.6.3 Ballast Removal Machines

There are a number of machines available in the UK for removal of ballast, mainly for heavy maintenance purposes rather than complete track renewal or ballast cleaning. Typical examples of the need for this type of machine could be under S&C or for isolated formation failure where ballast needs to be completely replaced over lengths of track up to 50 metres. The "Railvac" machine is a rail-mounted, self-propelled, high capacity vacuum ballast removal facility; in effect a full scale rail vehicle around the size of a large locomotive. The machine has a 250mm diameter tube with a tungsten tip connected to a hydraulic arm system with a remote control operating unit. The vacuum is a high power facility and can be used with a large variety of cohesive and large particle size non-cohesive materials. There is an integral hopper which collects the material and has side discharge chutes for the disposal of the spoil.

Figure 13.7 Railvac® machines working in the UK (courtesy Railcare SE)

13.7 TRACK MAINTENANCE OPERATIONS IN EXTREME WEATHER CONDITIONS

Buckling or thermal distortion of the track can occur when rail temperatures are unusually high and distortion due to thermal shortening of the rails on sharp curves can occur when temperatures are very low. Buckling is a very dangerous mishap which at the very least causes serious disruption to traffic, and is capable of causing catastrophic derailment of passenger trains with much loss of life.

For these reasons every case of track buckling or distortion is rigorously investigated, and these investigations have shown over time that buckles can be triggered by a very wide range of track features (e.g. wet spots, tight rail joints, changes of track material, shortage of ballast) which either tend to reduce the capability of the ballast to resist track distortion, or form localised points of weakness in the track structure.

Prevention of buckling thus depends in large measure on the faithful carrying out of appropriate maintenance work as described in this book, before the onset of hot weather, as well as, in the case of CWR, on conscientious execution of laid down procedures for rail tensioning. However, one significant cause of track distortion is the disturbance of ballast by the operations described in this chapter, and to prevent this the precautions described below are recommended in UK best practice.

13.7.1 Working temperature range

Maintenance work which reduces the stability of CWR must not be carried out when the rail temperature is above 32°C or below - 7°C. The person responsible for the completion of the work must frequently check the rail temperatures and stop the work and leave the track in a safe condition before either of these limits is reached.

13.7.2 High temperatures following maintenance work

If the rail temperature exceeds 38°C during the three days following any maintenance work which has disturbed the ballast, a trained operative must be appointed to observe over the site during the period of high rail temperature. Normally this precaution is all that is required, but the operative under the direction of a trained Maintenance Engineer has the duty to impose an emergency speed restriction or to block the line as a precautionary measure if they consider it necessary. Rail temperatures must be monitored so that the watch and/or the speed restriction can be withdrawn when the rail temperature falls below 32°C.

13.7.3 Precautions for maintaining lateral stability during maintenance work in hot weather

When it is necessary to pack sleepers manually, or to open out the ballast for any other reason (eg changing sleepers), not more than six alternate half-beds, or two full beds, may be opened out at the same time. Care must be taken to retain the top and line, and the ballast must be restored and consolidated around one group of sleepers before opening out another adjacent group. Defective sleepers may only be changed one at a time.

The replacement of individual rail pads (or any other task requiring the removal of rail fastenings) may be carried out provided that no more than eight consecutive pairs of clips on one rail are released at one time. This allows the rail to be lifted sufficiently to allow a pad to be removed and replaced without disturbing any sleepers.

13.7.4 Additional precautions to be taken during exceptionally hot weather

When exceptionally hot weather is officially forecast by the Metrological Office (under arrangements for notification), a set of instructions should be initiated, by 1600hrs on the previous day. When local circumstances may dictate a more restrictive precaution, then this shall be applied using the railway company's emergency procedures.

Level 1: Forecast air temperatures between 36°C and 40°C

Air temperatures between 36°C and 40°C can produce rail temperatures of between 53°C and 58°C. Whenever air temperatures are forecast to be between 36°C and 40°C, the following precautions should apply between 1200hrs (noon) and 2000hrs:

1) all passenger trains will be limited to a maximum speed of 90mph;

2) all container wagon trains will be limited to a maximum speed of 60mph;

3) all other freight trains will be limited to a maximum speed of 45mph;

4) track staff with rail thermometers to be positioned at all track locations which are considered susceptible to buckling, eg: North/South oriented cuttings sheltered from wind; wet spots or places where the ballast is inadequate or has recently been disturbed;

embankment slips or sites subject to subsidence; places where the track is restrained longitudinally, eg. underbridges or adjacent to S&C

Level 2: Forecast air temperatures above 41°C

Air temperatures in this range can produce rail temperatures of over 58°C, and in these circumstances the following additional precautions will apply between 1400hrs and 1800hrs:

1)   all passenger trains will be limited to a maximum speed of 60mph;

2)   all container wagon trains will be limited to a maximum speed of 45mph;

3)   all other freight trains will be limited to a maximum speed of 30mph.

13.8 SLEEPER CARE

It is almost always the case that provided proper care is taken of other elements of the track structure, ie, the rails, rail joints, top and line, fastenings and ballast, then the sleepers will take care of themselves, and all that is written in this book on those subjects is relevant to sleeper care.

Apart from these features, the following general advice can be given:

(a)   Sleepers must always be kept square to the rails and correctly spaced.

(b)   Sleepers on either side of a joint or weld must be of the same type.

(c)   The 375mm either side of the rail must be kept well packed.

(d)   Check and maintain the tightness of fastenings frequently and replace any which become defective.

(e)   Change promptly, any sleepers which develop defects before they have the chance to affect those on either side.

13.8.1 Concrete sleepers

Once installed, there is little preventative maintenance required for concrete sleepers, though the fastenings may need periodic attention.

Common faults and remedies are as follows:

- Abrasion of soffit due to excessive movement of sleepers or substandard ballast conditions. Exposure of reinforcement may eventually occur. This can only be prevented by attention to ballast condition and packing. Superficial damage due to traffic causes, eg, derailments, loose vehicle components. Such damage can be repaired by a suitable epoxy resin or similar synthetic bonding/grouting material.

- Major cracks and loose fastenings may occur. It may be possible to employ specialist repair techniques once the cause has been established but replacement is often the more economical solution.

The life of a concrete sleeper can be extended by relatively inexpensive but frequent attention to rail pads, particularly in the vicinity of joints and welds, to prevent abrasion or cracking of the sleeper surface in the rail-bearing area.

Fig 13.8 Damage to sleeper caused by ballast condition forming a Wet Spot/Wet Bed (courtesy Network Rail)

13.8.2 Timber sleepers

Timber sleepers may be subject to decay, splitting, indentation of chairs or baseplates, elongation of holes etc. If severe, replacement may be necessary. However, there are certain repairs or preventative techniques which can be applied to prolong the life of the sleeper. Splits in sleepers can be treated by the application of metal bands to close the open splits and prevent further propagation.

Enlarged or elongated sleeper screw holes can be treated by a maintenance plug or sleeving device such as the Vortok ® coil. This device consists of a helix made from aluminium alloy which is inserted in the old screw-hole, using a purpose-designed tool. Other tools are supplied by the manufacturers for removing the old ferrule from the baseplate, and for gauging the size of coil to be inserted.

Eventually it may be necessary to draw the sleeper through, far enough to enable a fresh set of holes to be drilled. If this is done, care must be taken to see that the new, bearing surfaces for the baseplates are plane and at the correct inclination, any adzed surfaces creosoted, and all redundant holes plugged.

A useful process to assess the life of timber sleepers in track, is the Sleeper Integrity Tester (SIT). SIT is a portable, battery powered apparatus which picks up and analyses the sound waves transmitted through a sleeper when it is struck by a rubber-ended hammer.

Sleepers graded as 3 or 4 by the SIT can be considered suitable for treatment with a proprietary timber repair system such as Pandrol "Timbershield" or similar products.

Under certain adverse conditions (eg on sharp curves), timber sleepers can suffer from gauge-spreading. Where this is considered to be a possibility, the use of gauge stops is advised. Where for any reason severe gauge-spreading does occur, arrangements should be made to change the affected sleepers, but to ensure safe running pending the replacement operation, temporary repairs can be effected by fitting gauge ties or tiebars. These must be insulated in track circuited areas.

13.8.3 Steel sleepers

Manual correction of line and level is a somewhat more onerous task with steel sleepers than with timber or concrete, owing to the hollow shape of the cross-section, and the spaded end. Satisfactory results can however be obtained with modern automatic tamping and lining machines.

The sleepers themselves are generally trouble free, until wear starts to occur around the fastenings, and it is around the fastenings that inspection must periodically be made for possible fatigue cracks.

It is often practicable to rehabilitate steel sleepers which have for any reason been removed from the track, by re-pressing them, repairing any loss of section by welding, and if necessary, by welding new plate onto the rail seat area to strengthen it.

13.9 MAINTENANCE OF RAIL JOINTS

13.9.1 Fishbolt tightening

All fishbolts must be tightened regularly, to the tightness noted previously in Chapter 5 and is usually prescribed for checking at regular intervals. It is also important if an ordinary fishplate is to work properly, that it is adequately lubricated, both because lubrication optimises the couple developed across the fishing surfaces from a given clamping force, and because the rail ends must be able to close up and open out to relieve thermal stress.

With the introduction of durable greases the use of "fishplate oil" has now ceased. Coupled with this the ultrasonic examination of rail ends has obviated the need for cleaning and visual inspection for rail defects. The use of spray grease has been introduced to this task and in general it is only necessary to slacken off the fishbolts sufficient for the grease to enter the top fishing surfaces.

It is normal practice for alternate fishplated joints to be greased each year in jointed track with the "uneven year" requirement being indicated by a fishbolt having its nut on the four foot side. Insulated fishplates and rail ends must not be greased.

13.9.2 Fishplate Shimming

Metal wear takes place at the interface of fishplate and rail on the fishing surfaces and is greatest at the extreme end of the rail. A joint worn in this way is unsatisfactory because it increases impact loads and causes further damage. It also puts the bolt holes in the rails out of alignment, preventing the joint opening and closing. The use of tapered steel inserts, or "shims", enables this loss of metal to be replaced and in so doing restores the joint to its proper level and normal functioning. The shims are made in six different sizes, varying in thickness and taper, and they are marked according to their maximum thickness.

A full length shim can be inserted for the complete length of the fishplate but half shims are favoured as these enable each rail end to be treated with its correct size. The wear at the extreme end of the rail end can differ between the running off and running on end of the joint.

To determine the correct size of shim to be used a 1 metre straight edge is placed on the rail top spanning the joint and a specially provided shim gauge is inserted approximately 18mm from each rail end between the straight edge and the rail top surface. Where the measurement differs by more than one half of a shim size, half shims should be used. The fishplates are then slackened sufficiently for the shim to be inserted between the top fishing surface of the fishplate and the upper fishing surface of the rail.

After re-tightening the fishbolts, the joint should be checked to ensure that no excessive bowing of the fishplates has occurred. Any voiding of the sleepers should be treated by packing at the time, ensuring however, that the joint is not humped. The joint should be reinspected after some traffic running and bolts checked for tightness. Insulated joints must not be shimmed as the electrical insulation integrity could be compromised.

13.9.3 Fastenings

Sleeper fastenings at joints may become loose or displaced and should be periodically checked and attended to. Particular attention should be given to any fastenings disturbed by any creep of the rail, or damaged by contact with fishplates.

13.9.4 Intermediate Packing (Dipped Joints)

Fishplated joints and weld positions have a tendency to require attention to rail top level at a higher frequency than other locations in the track. Where these locations are found to require attention between cyclic tamping by on-track machines following visual examination or as the result of track recording vehicle inspections, manual treatment is often the best way of restoring longitudinal level to acceptable standards. Where other features such as worn fishing surfaces have a direct bearing on the durability of this intermediate attention, they in turn must be dealt with otherwise the effort spent in packing will be wasted. The rail ends at fishplated joints and weld positions develop under traffic a vertical deformation which can be corrected by a hydraulically powered Rail End Straightening machine.

13.9.5 Changing components

Following periodic inspections the need to change one or more of the components in a fishplated joint may arise.

Fishplates

These may become cracked on either the upper or lower edges and should be changed before they break. They must be replaced with new plates, or at least with plates showing no differential wear in the fishing surfaces. Junction or lift fishplates must be fitted where change of rail section or differential depth exists. Insulated fishplates should be examined to determine any wear or defect in the insulating material. In the case of glued joints the standard joint can be reassembled using new components on the existing rail ends providing that these are not deformed. The solidified glue should be thoroughly removed from the rail ends before applying fresh glue.

Six-holed joints are designed so that four-holed fishplates can be used as an interim measure. Tight jointed fishplates are installed in welded S&C layouts and these can only be replaced by fishplates of the same design.

Fishbolts

The standard steel bolt should be used in all ordinary fishplated joints. These should be torqued up using an ordinary fishbolt spanner as already described in Chapter 5. In glued insulated joints, and S&C in welded layouts, all fishbolts are manufactured from high strength steel (Code V embossed on the bolt head), and must be replaced like for like. They must be tightened with a torque spanner set at 90 Kg.m.

Joint Sleepers

The sleepers on either side of the joint, particularly in older track, may require replacement before the rest of the track and it is normal in such cases to replace either the two or four sleepers around the joint at the same time.

When doing this, care should be taken to ensure that the old ballast bed is thoroughly scarified using a pick and any contaminated ballast removed. Following replacement the joint sleepers should be properly packed, and the rail top level reinstated to good standard. Care should be taken during the work not to disturb any signal cable connections in the vicinity of the joint. After some traffic running it may become necessary for further packing to be carried out. The standard ballast profile must be reinstated after any attention to the joint sleepers.

13.10 THE MEASUREMENT AND ANALYSIS OF TRACK GEOMETRY

The measurement and analysis of track geometry is carried out on a regular basis by most, if not all, railways as a means of obtaining data from which track maintenance can be scheduled. The quality indices obtained can be set against standards to determine priorities and thus ensure that work plans are cost effective. If resources are limited by financial restraints, track quality data can still help to identify those sections of track requiring the most urgent attention. The benefits obtained from limited resources can in this way be maximised.

A geometry recording train can also provide valuable data for use in fundamental research into the interaction between wheel and rail; in the past this has assisted the development of the improved bogies and suspensions which have made higher speeds and greater passenger comfort possible.

13.10.1 Principles of train borne systems

The inertial measuring principle makes use of the displacement arising from the inertia of a spring mounted mass in a sensor known as an accelerometer to sense and quantify acceleration. The data from the sensor can be integrated twice to give a very accurate electrical analogue of the displacement, or movement, of the surface to which it is mounted and when fixed to the body of a railway vehicle the arrangement provides a reference line from which vertical and horizontal track profile measurements can be made. To measure in the vertical plane, displacement sensors across the suspension transfer the measurement through the running wheels to the top of the rail (Figure 13.9). Because it is sensed through the running wheels, the measurement made is of the loaded profile at the point of contact. To measure in the horizontal plane, non-contacting optical sensors locate the position of each rail and transfer the measurement to the gauge face to obtain the horizontal profile.

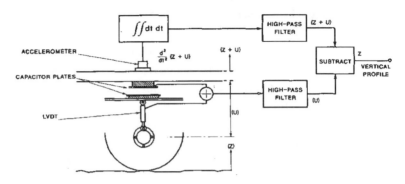

Figure 13.9 Vertical profile measuring system

To achieve an artificial datum of the desired form, the vertical and horizontal data are processed by electrical high-frequency-pass filters; these remove the long wavelength and steady state information, effectively creating a reference line based on a moving-data average. The wavelength at which the response falls to 70% is known as the cut-off wavelength, and this is normally set to 35 and 70 metres on mainlines in the UK. On London Underground (LU), a slightly different filter response is used, with a cut-off wavelength of 25m.

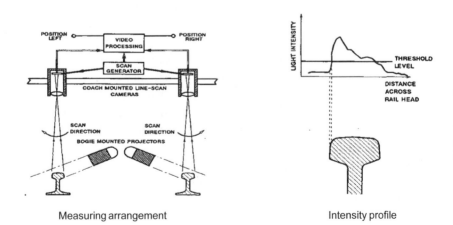

Measuring arrangement   Intensity profile

Figure 13.10 Optical sensing system

Chord offset systems have a disadvantage that measurements of short-wave alignment irregularities are not separated from long-wave curvature which is, of course, present in the track by design; this makes it difficult to process the data in a computer to generate quality indices. The UK system overcomes this problem by separating these two aspects of the data in the measuring equipment, and was the first system of its kind to make curvature-independent alignment measurements.

With the inertial system, measurements can only be made when the vehicle is moving at a speed greater than 15 km/h (l0km/h on the LU vehicle). This small disadvantage is outweighed by the absence of sliding probes or sensing wheels enabling high reliability to be achieved under intensive service conditions and the recording of data with good repeatability and minimum distortion. Below the limiting speed the recorded data tapers gradually to zero.

Additional instrumentation, consisting of a tiny gas bearing rate-of-tum sensor and accelerometers is used to derive crosslevel, curvature, gradient and corrugation amplitude. Twist, cant deficiency, equilibrium speed and passenger comfort indices are derived from these parameters. Curvature and cant deficiency are processed with low-frequency-pass filters to generate data from 35 metres to infinity. On-line computers are used to sample and analyse the data, prepare track quality reports and record on magnetic media for transfer to a central data base. Analysis consists mainly of deriving track quality indices using the statistical measure of variation "standard deviation". This form of analysis was adopted in the UK at the outset, and has since gained worldwide acceptance.

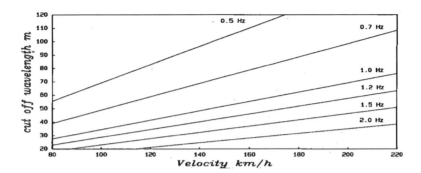

Figure 13.11 Relationship between wavelengths to be measured and line-speed and vehicle suspension frequency

13.10.2   Measuring wavelength range

Passengers respond to the acceleration environment of a vehicle in a complex way depending on the frequency and on whether they are seated or standing. Since the suspension acts as a low-pass filter and thus modifies the effect of irregularities in the track as a function of frequency and speed, deciding on the range of wavelengths to include in the measurement of track profiles is by no means straightforward.

The position is further complicated by the fact that if unnecessarily long wavelengths are included in the analysis, then since these will have larger amplitudes, the resulting maintenance work might be slower and more expensive, due to the requirement to provide greater lifts or slues.

Older tampers and liners can only correct irregularities of less than about 40m wavelength. Whilst modern developments enable a much improved long wavelength performance it is important to note that there is a direct conflict between, on the one hand, including all the wavelengths which might affect vehicle ride, and, on the other, avoiding the expense of unnecessary maintenance work.

It is important also, to distinguish between the shorter wavelength misalignments (less than about 35 metres) which increase in magnitude as a result of traffic, and which must therefore be corrected when an unsatisfactory situation is reached, and the longer wavelength misalignments (greater than about 40 metres) which exist perhaps as a result of past incorrect work or maybe unstable support conditions. When corrected, these misalignments may tend to recur much more slowly, if at all.

An approximate estimate of the measuring wavelength range can be obtained from a comparison of fleet vehicle characteristics and track recording filter characteristics. The relationship which relates cut-off wavelength in metres (l), speed in km/h (V) and suspension natural frequency in Hz (fr) shown graphically in Figure 13.11 is of the form: **l = 0.3 V/fr.**

Using this formula a wavelength range for vertical profile of 20 metres can be obtained for a suburban railway with a top speed of 120 km/h and fleet vehicles with a vertical suspension frequency of 1.8 Hz, and 60 metres for a high-speed (200 km/h) line with vehicles with a vertical suspension frequency of 1.0 Hz.

To avoid changing the wavelength range for different line speeds, a single filter of 25 metres cut-off wavelength is normally used on LU. On mainline UK two cutoff wavelengths are used: 35m for suburban and cross-country routes supplemented by 70m which provides additional information for high speed lines. The standards of acceptability are, however, varied with line classification; a higher speed line requiring tighter track standards and vice-versa.

When considering the wavelength range necessary for the measurement of horizontal profile or alignment irregularities, the position is more complicated because modern high speed passenger vehicles are designed with very soft lateral suspensions. Much longer wavelengths therefore need to be considered, perhaps as great as 150 metres, and there is great difficulty in distinguishing between "design" features, such as curves and transitions, and misalignments. Fortunately, measurements of curvature and cant deficiency together with analyses of their effect on passenger comfort, can be used to overcome this difficulty.

For the shorter wave measurements, reasonable correlation between track standard deviation and ride quality for seated passengers, can be obtained experimentally for both vertical and lateral directions.

13.10.3 On-board data analysis

The measured data is sampled at nominal intervals of 240mm and channelled to on-board computers which provide the following functions:

- Calculations of standard deviation, exceedence count, and mean, maximum and minimum value.
- Tracking of mile/km location, line speed and route identity.
- Provision of on board reports and summary data to magnetic media.

Standard deviation

Standard deviation is a universally used scientific measure of the variation of a random process. Track profiles have been found to have sufficiently similar statistical properties to random processes to enable a measure of the magnitude of track irregularities to be obtained from the standard deviation (SD) of the vertical and horizontal profile data. In the UK system this form of analysis provides the main track quality indices.

The standard deviation may be thought of as a measure of the energy imparted to a vehicle suspension due to the roughness of the track. A standard deviation of 0 corresponds to "perfect" track and the maximum reported value of 9.9mm corresponds to an extremely poor quality. A scale of intermediate values is thus available to specify the quality of all classes of line. To save space in the on-line reports, the left and right TOP SD's are combined, with the largest, or worst, value being printed. SD is calculated over Vs mile sections, Vs mile being a convenient subdivision of the V4 mile posted interval. Metric intervals of 200 and 100 metres are also available.

Standard deviation is calculated as the square root of the mean of the squares of the individual data samples, each of which represents a departure from an assumed mean whose value is zero.

Exceedence level

Two exceedence levels are used to identify isolated defects. The first (level 1) is used to derive an exceedence count which is reported with the standard deviation in the on-line report; the second higher level (level 2) is used to identify gross defects which need immediate attention and is linked to paint spray equipment which marks the track in order to assist in location. Irregular cant deficiency is also reported as a level 2 exceedence.

Mean, maximum and minimum value

The mean value of crosslevel, curvature, cant deficiency and gauge can be calculated and listed if required. The maximum value is used in the calculations for corrugation and passenger comfort indices.

13.10.4 Location and control

The operating philosophy adopted in the UK is that recording operations should be as automatic as possible, with no requirement for sighting of mileposts, so that the operation can take place at night or in conditions of poor visibility. The method of location uses pulses from an axle tachometer supplemented by position "fixes" from selected Automatic Warning System (AWS) magnets.

So that the recorded data can be accurately located with respect to the lineside posts, a route setting list is input to the on-board computer at the start of the run. The list, which is derived from a central data base, contains the following principal information:

- Region and area data to supply titles and headings to all recording devices.
- Mile or kilometre post sequences for the route.
- Line speed and permanent speed restriction details.
- Place names and comments to be printed on reports.
- Position of synchronization markers.
- Route and track identification, track type and category.
- Details of irregular quarter mile or 100 metre intervals.

Manual synchronization is needed at the first post to set up the sequence, thereafter operation is automatic.

Parameters measured

A list of available measured parameters is shown below:

PARAMETER	UNITS	FILTER
Top left	Mm	0.24-35/70m
Top right	Mm	0.24-35/70m
Alignment	Mm	0.24-35/70m
Gauge	Mm	—
Crosslevel	Mm	—
Cant deficiency	Mm	>35m
Twist 2 or 3	Mm	—
Twist 5 or 10	Mm	—
Curvature	Mm	>35m
Gradient	Degree	>70m
Equilibrium speed	km/hour	—'
Corrugation	Mm	—
Ramp angle	Mr	—
Cyclic irreg	mm	

Top, Left and Right

Top or vertical profile is measured at a single point using a running wheel as the sensor. On each side of the vehicle there are capacitive sensors and linear variable differential transformers (LVDT's) across the suspension together with a vertical sensing accelerometer in the body above rail position. Because the suspension movement is cancelled out in the computation, the measurement accuracy is independent of the suspension characteristics of the vehicle.

Alignment

Track centre alignment or horizontal profile is also measured at a single point using a lateral-sensing body mounted accelerometer and solid-state line-scan cameras mounted in the body above each rail to sense rail position. The line-scan cameras function by projecting an image of the rail head, which is illuminated by bogie-mounted projectors, onto a multi-element array of 256 photo-diodes which convert the light intensity into an electrical signal. This signal is processed to derive a measure of rail lateral position. The horizontal profiles of the two rails are then combined to give track centre line alignment.

The rail position located by this non-contacting optical sensor is approximately the gauge face point and for unworn rail is close to the point which would be located by a conventional contacting arrangement making a measurement 16mm below the rail crowns. For worn rail, measurement accuracy using this simple optical sensor is undefined although for modest wear it is not degraded significantly. The effect does not create a problem for the alignment measurement, because of the high-pass filtering which is used. The measuring technique which ensures a curvature-independent alignment measurement involves compensation for the effects of centripetal acceleration and body roll.

Gauge

Gauge is calculated by combining the data from each line-scan camera. The computation is the difference between the two position signals, and is affected by the lack of precision in the gauge face point on heavily worn rail referred to above. This causes no difficulty on a well maintained railway and the reliability and consistency of a non-contacting system outweighs this disadvantage.

Crosslevel

The data from a body-mounted lateral sensing accelerometer and a roll and yaw rate-of-tum sensor are combined in a novel way to produce a measurement of cant angle with a wideband response from 0-l00Hz. A measure of the crosslevel is then obtained by assuming a nominal rail centre spacing of 1510mm.

Cant deficiency

For a given track design speed (V) a so called 'cant deficiency' angle is calculated from:

$$\theta d = \theta - V^2/Rg$$

where R is the local radius of curvature, g is the acceleration due to gravity and ε the cant angle. The track design speed data is available from data held in the computer.

Twist

Twist measurements are obtained by differencing the crosslevel data in a computer over base lengths which have been selected to be nominally 2 and 10 metres for LU and 3 and 5 metres for UK mainline track.

Curvature

The curvature (reciprocal of the radius of curvature) is defined as the spatial turning rate of the track. A good estimate of this spatial rate over the range of wavelengths required, can be derived from the ratio of the measured rate-of-turn of the vehicle body and the measured forward rate of the vehicle. Filters are incorporated to remove wavelengths shorter than 35 metres and this also removes errors due to yawing of the body with respect to the track. The measurement is scaled to be the same magnitude as a 20 metre versine since this is the preferred form of curvature measurement used by the engineer.

Because the data is filtered, a small positional delay, of the order of 15 metres, is created. The curvature is referenced to the centre of the measuring vehicle which is about 5 metres from the measuring axle. So when the measuring axle is trailing the delay is about 10 metres and when it is leading the delay is about 20 metres.

Gradient

A longitudinal accelerometer, corrected by differentiated forward velocity is used for this measurement. The data is filtered to remove wavelengths shorter than 70 metres.

Equilibrium speed

Equilibrium speed V_e is calculated from:

$$V_e^2 = \theta g R$$

On straight track, where equilibrium speed is undefined, a zero value is substituted.

Corrugation

Rail head corrugations are classified as short-wave (wavelength range 35 to 80mm) and long-wave (wavelength greater than 80mm). In the UK mainly short wave corrugations are found.

The measuring system uses data from axle box mounted accelerometers and is valid only above speeds of 50 km/h. Output is in the form of the peak-to- peak value of the largest corrugation amplitude of the two rails. If the speed of the measuring vehicle differs appreciably from that of the rolling stock which causes the corrugation, then on curves, where there may be a different rolling line, the corrugation can be under-measured. It is important therefore for the vehicle to operate as close as possible to line speed for accurate corrugation measurements.

Ramp angle

An estimate of the impact force between wheel and rail can be obtained from the product of velocity and ramp angle, or change in slope of the vertical profile of the rail. The ramp angle can easily be obtained from the TOP data, and is therefore a convenient measure of the impact force. It is to be preferred over a direct measure of axlebox acceleration (used by some other railways) since it is independent of the speed of the measuring vehicle. The measured value can be tested against limiting values derived from line speed and tonnage as a means of preventing fracture damage.

In the UK, the system has been fitted to Laboratory 5 and used to identify situations at rail ends in bull-head track where there is an excessive ramp angle which could give rise to cracking of fishplates,

Cyclic irregularity

If dips occur in the track at the centre of the rails between the joints, the track then has a wave component with a wavelength of half rail spacing (a second harmonic of 9.1 metre wavelength). This can excite freight vehicle suspensions since the passing frequency can coincide with the vehicle vertical resonant frequency. If the wave in the track persists over several cycles, the suspension deflection may then build up due to resonance and derailment becomes possible. Laboratory 5 has a system which seeks these wave components in the "TOP" profile and then models the resonant build lip to identify sites where there is a potential risk. The system can also respond to the third harmonic of 6.1 metres.

Figure 13.12 Example of Cyclic Top Irregularities (courtesy Network Rail)

Calculation of passenger comfort indices

International standards defining comfort levels are applicable only to seated passengers and refer to "vibration", which can be interpreted as semi- continuous periodic-like accelerations. The isolated "jolts" which affect passengers who are standing or walking through gangways, and the steady state accelerations which cause discomfort through the need to continually readjust balance, must be analysed in a different way.

The accelerations arise because of the roll of the vehicle body due to crosslevel, centrifugal effects due to negotiating curves, and long-wave misalignments. Railway track is, of course, designed so that the first two of these accelerations tend to cancel and any lack of cancellation can be quantified in terms of cant deficiency, which is calculated and available on the recording car.

Since most of the long-wave irregularities which affect comfort result from misalignments, and since the measurement of cant deficiency has no wavelength filtering to attenuate long wavelengths, cant deficiency is very effective as an indicator of the presence of long-wave alignment defects. It is the "deficiency" which is measured, so the result will be less dependent on the design geometry and the difficulty of making the distinction between misalignment and design which would occur if analysis of a long-wave profile measurement was attempted, is conveniently avoided.

The processing regime adopted locates peak values of cant deficiency and then compares these to the mean or steady state value.

If the peak value exceeds the steady state value by more than a predetermined level then an exceedence is registered and reported. The level is selected from one of two values depending on whether the steady state value is above or below 110mm. The lower level is selected above 110mm which acknowledges that passenger sensitivity to these type of irregularities is greater when curving.

It should be understood that this analysis refers only to the discomfort caused by long-wave irregularities. It complements, rather than replaces, the standard deviation information from the alignment.

13.10.5 Application of track recording to freight lines

In the foregoing, the emphasis has been on the use of track geometry data to provide a satisfactory standard of comfort on passenger lines. Many railways, however, are principally freight carriers, often with heavy axle loads. The objectives here are somewhat different, the ride quality, as such, being much less important than ensuring that derailments are prevented, and that the forces generated in the track due to the roughness are not so great that the life of track components is seriously reduced.

Freight vehicles tend to have relatively stiff suspensions, and since speeds are normally less than 100 km/h, a measuring wavelength range of 25 metres, or even less, is adequate. Whilst standard deviation is still a useful measure of roughness, twist and top and alignment exceedences are much more important.

As well as the dipped joint and cyclic irregularity detection systems referred to earlier, a number of railways have worked on techniques to process data to highlight track sections where derailment could be more likely. Japanese National Railways have devised a technique called "alignment + 1.5 cant" which attempts to predict a combination of circumstances which can cause wheel unloading in freight vehicles.

These techniques are all of value in avoiding the often heavy financial consequences involved if a freight train mishap occurs, and can assist revenue by creating confidence which can allow line speeds to rise.

Sometimes, the derailment problem is worsened by the poor maintenance state of some wagons, and track faults are sought where none exist. It is essential that all derailments are properly investigated using sound scientific methods. The methods used are described in The BR Research Publication "A Guide to the Investigation of Derailments", by Duncan, McCann, McLoughlin and Poole (1982). The emphasis must be on establishing the cause so that future incidents can be avoided rather than apportioning blame. Where a pattern of track faults is established the recording car data can be used as a basis for locating other danger spots.

13.10.6 Future trends and developments

Railways have deservedly a reputation as a safe and reliable means of transport for people and commodities. The track geometry recording train is at the forefront in helping to maintain this reputation, and without doubt will continue so to do.

The rapid advances in high speed computing equipment open up considerable potential for much more powerful forms of on-board data analysis. The quantification of track quality in terms which are meaningful to passenger service operators can provide a business oriented framework for financing track maintenance from revenue, and begin to offer to the operator a guaranteed standard of quality in return for the financial support provided.

13.11 MAINTENANCE OF RAILS

13.11.1 Rail lubrication

Whatever grade of rail is used, the abrasive wear of the running face of the rail caused by guidance forces on curves can be reduced by installing grease dispensers on either the rolling stock or the rails. In the UK, the latter are usually used, and they are termed rail lubricators. Their principle of operation is that when activated by passing traffic, a charge of lubricant is deposited on that part of the rail which will come into contact with passing wheels, which then distribute the lubricant along the rail. The components of any lubricator system are a lubricant reservoir, a pump and one or more applicators.

The lubricant reservoir should be located clear of the track, and preferably clear of the ballast shoulder. It should be rechargeable without disconnecting any other part of the mechanism, and without exposing the attendant to any risk of being struck by passing trains. Depending upon the type, reservoir capacities can be up to 50kg.

Lubricant is fed to the applicator under pressure. This pressure is usually generated by the wheel via a plunger or treadle operated pump, but in more modern designs, compressed gas (eg nitrogen) can be used, the dispensing mechanism being triggered by a treadle or electronically. The latest technology in which the lubricator is driven by an electric motor/gear or pump, with electronic timer control, is proving very successful. Better control of grease application, longer periods between fitting and maintenance, and greatly enhanced safety of operatives being of major significance.

The usual method of grease deposition is by way of a long thin hollow blade secured to the gauge side of the rail. The pressure generated by the pump forces droplets of grease out of the open upper edge of the blade which must be adjusted to make contact with the wheel flanges. Other methods include a contactless system in which the droplets are projected at the rail from fine jets.

In the more usual method up to four applicator plates are installed, so as to give a uniform application of lubricant to the whole circumference of the wheel flange, and the unit is then supplied with regulating valves, to ensure an equal flow of grease to all the applicator plates.

There are several types of lubricator on the market, and it is not considered necessary or desirable to recommend any particular type here. The following are points which should be noticed in connection with the selection, installation, and maintenance of rail lubricators:

(1)   The equipment must be easily assembled.

(2)   Maintenance should be by straightforward component replacement. It should not be necessary to dismantle any component at the lineside for routine maintenance.

(3)   The equipment must be easily adjustable for lubricant flow and applicator plate height.

(4)   As far as possible, both recharging and maintenance should be able to be done without the attendant getting his hands covered with lubricant.

(5) The grease reservoir should have an external indicator to indicate the quantity of lubricant remaining without having to open the container.

(6) The method of attaching components to the rails should be "fit and forget" taking into account that they will be subject to intense vibration.

(7) The design of the equipment should be such that automatic tamping and lining can be carried out without dismantling any part of it.

Generally speaking the more dramatic savings in wear are obtained when lubricators are installed in sharp curves (say below 500m). It has been found by experience that measurable improvements to rail life can be obtained by installing lubricators on curves of up to 1000m radius, but it is not possible to lay down hard and fast rules for limiting radii above which they will no longer prove of value. Whilst the obvious place to site lubricators is at the start of the curve, optimising the siting of lubricators is largely a matter of observation and experience. On long curves two or more installations may be required.

Various vehicle mounted flange systems have been recently developed, and are in selective use. These devices spread lubricants on to the wheel flanges and may be activated manually, by curve sensors, or on a fixed distance pattern.

Rail lubricators may be used to apply grease to reduce the sidewear on the rubbing face of check rails. They should be sited as close to the entry end of the check rail as possible but clear of any entry flare. Maintenance and grease requirements for check rail lubricators are as for running rail lubricators.

Figure 13.13 Example of Sidewear of a rail on plain line track (courtesy Network Rail)

13.11.2 Turning and Transposing of Rail

It will be clear that eventually the outer rails on sharp curves will become sideworn to the extent that they may no longer remain in service. Since this damage is limited to the gauge side of the rail, it was traditional in the days of universal jointed track to arrange for badly sideworn rails to be turned end for end, thus presenting an unworn face to traffic. To deal with 18m rails in this way requires a crane with a lifting capacity of at least one tonne at around 13m radius, and with a jib whose proportions allow the rail to be turned end for end without fouling any part of the crane. This requirement was an important determinant of the traditional design of railway steam cranes of modest lifting capacity, but there are today many designs of hydraulically operated cranes with horizontally extending jibs which are suitable for this work.

In addition to the crane, a suitable lifting beam equipped with CAMLOK® clips or other approved rail lifting device must be available and the necessary number of operatives organised to remove, stack, clean, inspect, and replace fishplates and rail fastenings. When planning the work, the person organising it must ensure that a reasonable supply of replacement fishplates and fastenings are available, in case any of those taken out are worn out or damaged. This is particularly important if the operation includes any insulated joints.

If the operation takes place on a track circuited line, the work will involve the removal of track circuit bonds, whilst on an electrified line, not only will a current possession be required, but the work will involve removing traction current return bonds. The person in charge must carefully note how the rails are bonded and all bonds must be reinstated before handing the track back to traffic.

The operation is one of the few requiring the dismantling of rail joints, so opportunity should be taken to clean and make a visual examination of the ends of all the rails being turned. In case any incipient rail end failures are observed it is good practice to have a few spare rails available on the Engineering train. In addition to this, the person planning the work should look for any signs of active rail creep, and should correct this as part of the work, cutting and drilling rails as necessary.

The person in charge should ensure that when the rails are replaced, the correct expansion gaps are left between them, and the site must be left with all the fastenings in place, and all joints fishplated with the correct number of bolts inserted and fully tightened, not forgetting to replace any insulated joints with insulating fishplates in their correct locations.

If the high rail of a curve where the rails are continuously welded becomes worn, the equivalent of the operation described in the preceding section becomes that of transposing the two rails, so that the back face of the low rail becomes the running face of the high rail and vice versa. This work can be performed either manually or by means of specialist plant equipped with jibs which lift and reposition both rails at the same time, whilst being drawn slowly in the direction of operation.

VIEW SHOWING GDU & CLAMP IN POSITION ON 1131b RAIL

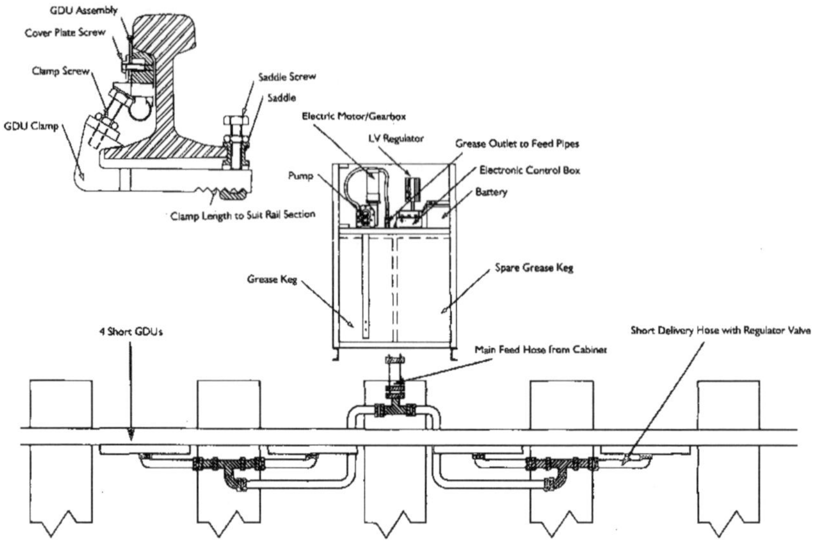

Figure 13.14 Typical Grease Dispenser Unit (GDU) and Control Box

The same preplanning is necessary for this operation as was described for rail turning, except that adjacent lines are not affected. Account must again be taken of the locations of any insulated joints. However there are one or two additional features to watch for:

(1)   The inner and outer rails will not fit back into their new places without some adjustment, due to the lead induced by the curve, and a suitable piece of rail must be available to cut in.

(2)   The rails will need to be re-welded into CWR, and the track affected must be re-stressed in accordance with standard instructions.

The process of rail transposition has a very important limitation to which attention must be drawn. It was observed above that on the low rail of curves metal flows from the gauge side towards the field side of the rail. As a result the shape of the top corner of the rail profile changes from the 12.7mm radius to which the rail is rolled, towards a sharp, rectangular corner, and even in bad cases, the rail head becomes lipped on the back face. This completely alters the relationship between the wheel and rail profiles. The effect is to produce unstable running at high speeds. Hence rails must never be transposed on tracks where train speeds exceed 120km/hour unless the work is associated with a re-profiling operation (see below).

13.11.3 Rail Reprofiling

The reprofiling of rails is done for one or more of three reasons, viz, to remove corrugations, to reproduce an "as new" shape on the head of a rail which has been turned, transposed or taken up and relaid, or to modify the shape of the head of a new rail to produce a profile which will give a longer life, particularly in curved track under very heavy traffic. The first and third of these operations are usually carried out using rail grinding trains or vehicles, whilst the second requires much more metal removal than can usually be achieved by grinding, and planing or milling machines mounted on suitable track vehicles are employed for this purpose.

Rail Grinding to Remove Corrugations

Elimination of corrugation was indeed the original initiative for the introduction of rail grinding trains. A typical rail grinding train consists of a number of vehicles permanently coupled together, capable of being driven from either end, incorporating its own prime mover which supplies power to drive both the train and the grindstones.

The propulsion system has to be capable of a wide range of speed from as near as possible the permitted line speeds of the client railway, to as low as 3km/h or less when grinding. Furthermore, this very slow grinding speed must be closely controlled, say to within ±10%. This is usually achieved nowadays electronically by a feedback system of some kind.

Since each grindstone can by the nature of the way it works produce only a plane surface, the production of an acceptable profile after grinding requires that the rail head surface is divided into a number of facets, each of which is attacked by one stone, and an important aspect of the design of the train is concerned with this facetting process, and in the positioning and driving of up to 30 to 40 stones at differing angles to the rail whilst keeping within the relevant loading gauge.

As may be imagined, a modern high performance grinding train when working, produces quite spectacular displays of sparks, and a fair amount of noise, whilst it leaves behind it a considerable deposit of steel filings. For the prevention of lineside fires and for environmental reasons the grinding stones are often shielded, whilst, particularly in tunnels it may be necessary to incorporate means to ingest the filings to prevent contamination of the track form and possible degeneration of the insulation of one rail from another.

The proper management of the operation requires firstly the identification of locations requiring attention. This should be done beforehand, the track being properly prioritised, and the running of the train preplanned. This aspect can be attempted by subjective methods (e.g. from noise observations) but it has been found that the inertial methods employed by BR's track recording trains will pick up corrugations and visual and/or statistical analysis of the record so produced will give an objective assessment of a route, identifying those places which would most benefit from the limited amount of attention that can be given.

The train should incorporate two sets of on-board mensuration equipment, situated at either end of the train. When operating, the leading set of equipment monitors the depth of corrugations, and this information is fed either directly to the actuators controlling the load on the stones and the speed of the train, or to a display panel by the use of which the train captain controls these variables. Either way the objective should be to ensure that the corrugations are completely ground out without wasting steel. The rail profile is monitored again by the trailing set of mensuration equipment to confirm success, or alternatively to indicate that a second pass is required. Since modern trains are capable of removing up to 0.4mm of metal per pass this control is of some importance.

The trains are usually designed to be self-contained, and are equipped with crew accommodation, workshop facilities, and stores for consumables etc. The design, construction, operation and maintenance of a grinding train is a specialisation which lends itself readily to contract operation.

The trains may either be owned by the contractor, or built to the specification of the client railway and leased, but in either event the train is usually crewed by the contractor. The staff have to be both highly skilled and dedicated, to work the unsocial hours demanded by the trains' operating schedules, and to endure the prolonged absences from home during the operational cruises.

13.11.4   Rail Adjustment and Straightening

As is mentioned, one of the difficulties in obtaining a long interval between track resurfacing operations is the so-called 'ballast memory'. In fact one of the elements assisting in the production of this 'ballast memory' effect turns out to be the shape of the rail, ie the natural longitudinal profile of the rail axis. If the longitudinal axis is badly distorted in the vertical plane at any point, then when in the unloaded state, and assuming that the sleeper soffits are all at the same level, the track will appear to be distorted, and there will be voids under some of the sleepers. This can only be corrected by introducing differences in level at some of the sleepers, which will then show as a track irregularity under load. Furthermore, the distortion of the rail leads to slight variations in rail seat reaction, and hence to differential settlement.

This can only be corrected by straightening the rails themselves. The most common location for rail distortion is at rail ends, either when the joints are fishplated or welded, hence the equipment used to correct this distortion is usually called a rail end straightener.

To correct the rail profile, it is necessary to bend the rail beyond its yield point, so that the steel becomes plastic, and does not return to its original profile on removal of the load. At the same time, the load applied must not be sufficient to break the rail. A manual three point bending machine (commonly called a jim-crow) fulfils this requirement, because as soon as the stress in the extreme fibres reaches the yield point the rail 'gives', and the load in the jim-crow is automatically relieved because it is unable to 'follow through'. What renders the manual operation virtually impractical is the difficulty in estimating, or guessing, the amount of bend to apply in order to finish up with a straight rail. If the operation is mechanised, eg by using a hydraulic jack, then the additional difficulty arises that the hydraulic actuator must be stopped before the jack either over-bends or even breaks the rail.

Rail end gaps in jointed track

In jointed track the build up of compressive force is prevented by building in expansion gaps at the rail joints when the track is laid. The size of gap varies according to the rail temperature at the time of laying. In the UK these should be:

Below 10°C	10mm
10-24°C	6mm
34-38°C	3mm
Over 38°C	nil

Special shims of the thicknesses prescribed in the table above (usually referred to as "Expansion Irons") are available and must always be used when laying jointed track. They are slipped over the ends of the rails as they are laid in, and the rails are pulled up tight against them before the fishplates are assembled and bolted up. Care must be taken to see that the irons are all collected up when the work is completed. A rail thermometer should be used to determine the rail temperature.

Rail creep and rail adjusting

In jointed track, particularly of older designs (eg BH, or FB track with Elastic Rail Spikes), traction and braking forces cause the rail to move longitudinally through the fastenings. This phenomenon is referred to as rail creep and its effect is to cause the rails to bunch up in some localities and open out in others. Where the rails bunch up there is clearly a risk of buckling in hot weather, and where the rails open out the fishplates and bolts are subjected to unnecessary and possibly deleterious tensile forces when the weather is cold. It is an important task therefore for track engineering staff when walking the track, to look for any signs of rail creep developing.

Indications that this is occurring include: fishplates bearing up tight against sleeper fastenings score marks where the rail moves against e.g. ERS sleepers becoming out of square to the track expansion joints closed up when the rail temperature is low.

Where rail creep is suspected to have occurred it may be corrected by rail adjusting. In planning such work the engineer must first note the position of all joints relative to the sleepers, and the state of the expansion gaps. This will determine whether the creep can be corrected by simply pulling back a few rails, or whether there is too much or too little rail in the length being considered. In the latter case, either rail must be cut out, or additional rail must be cut in, and the preliminary survey will enable the site at which the necessary cuts must be made to be determined, and what additional rail and equipment will be needed.

If the creep can be corrected by closing up some expansion joints and opening out others, then, particularly on FB track, it might be possible to adjust the rails in "between trains" possessions. In most cases however this will not be possible, and application of the full procedure for the obtaining and preplanning of an absolute possession of the line must be followed.

Typically the work under possession will involve:

- Unload rail and equipment at site

- Release track fastenings and rail anchors if fitted

- Create the gap into which the rail will be pulled and assemble the rail adjusting apparatus in the gap

- Loosen the fishplates one at a time over the length of rail to be pulled, starting from the end remote from the puller

- Commence pulling, and continue until the correct gap is obtained at the farthest joint

- Tighten this joint, and loosen the next one along repeat until all joints are correct

- Cut in the necessary closure, and retighten all fastenings clear the site and hand back to traffic

Opportunity should be taken during the possession to correct any sleepers which are out of square, renew defective fastenings, correct misalignments, and finally to fit new rail anchors in the hope of rendering further rail adjusting operations unnecessary.

Rail lifting and handling

The high carbon and alloy steels used in wear resistant rail renders rails especially sensitive to shock or impact loading, bruising, notching or scoring of the surface, and point or line loading. This is especially true of the longer rails being produced at the present time. All aspects of the storage and handling of rail from the moment of delivery into the recipient's responsibility, to the completion of the track laying process, must be thought through to ensure that there shall be as little chance as possible of inadvertent damage occurring. The notes below indicate some of the more important points to be observed.

Lifting

Straightness is very important in a rail, particularly if it is to be used in a high speed environment and when being lifted a rail is at its most vulnerable in this respect. Single point slinging incurs a high risk of bending and surface damage to rails, and in addition there are risks of injury to the staff in doing the work, so single point slinging should be strictly forbidden. It is essential that the rail be lifted at least two points, that spreader beams or similar devices are used to ensure that the lifting forces are applied vertically, and that the overhang of rail beyond the lifting points does not exceed half the distance between the lifting points. Table 13.1 gives an indication of the provision to be made for lifting BS113A rail.

TABLE 13.1 Lifting Points for BS113A Rail

Rail Length metres	Number of Lifting Points	Distance between Lifting Points metres	Maximum Rail End overhang metres
12-13	2	6	3.0-3.5
18	2	9	4.5
36	4	6 & 12	3

Rail Gripping Devices

In order to avoid damage to the rail, and risk of injury to operatives, the correct devices must be used to grip the rail when lifting. Lifting by magnet is the preferred technique (used in accordance with the principle outlined above), as it maintains the rails straight and horizontal, incurs little risk of surface damage, and permits several rails to be lifted at once. However, statutory regulations and technical limitations may preclude the use of magnets in many locations.

Rail head grips are suitable for most lifting operations provided that the rails are to be handled "head up". The necessary number of grips should be fitted to the appropriate length lifting or spreader beam. Care must be taken to ensure uniform engagement and disengagement of the load and to avoid sudden release and shock impact of the rails. Self-actuating rail grip devices such as CAMLOK® clips are much preferred to collar dogs or scissors grips. Indeed so many serious accidents have occurred over time to staff using collar dogs and scissors grips that their use has been prohibited for many years in the UK and this step is recommended to any organisation setting up rail handling operations.

For handling more than one rail at a time, the CAMLOK MULTI-RAIL CLAMP® has been developed. Up to five rails can be handled simultaneously in safety using this device, and it virtually eliminates any risk of surface damage to the rails during handling.

If none of the devices mentioned above is available, and slings must be used, then round link chains or hard wire cable slings must NOT be used as they cause surface damage. Fabric strapping (e.g. polypropylene) can be used, preferably with additional sleeves to protect the contact area between the sling and the edge of the rail foot. Alternatively flat profile sling chains with protective sleeving can be used. When using slings, rails must be set down clear of neighbouring rails in the stack, and landed onto suitable timber baulks ("dunnage") so that a clear space is left under the rail through which the slings can be drawn out. The final positioning of the rail must be done by manual levering.

It is permissible to handle rails beyond the reach of cranes with fork lift trucks of suitable capacity, provided that the rules for spacing of supports are respected (i.e. two or more trucks to be used for 18m rails and longer), and that the surface over which the fork lift trucks run is level and fairly smooth.

Stacking

A smooth, level, and firm base, free from protrusions, is essential. Concrete is ideal but a well compacted earth base is acceptable. Base supports or dunnage should be evenly spaced along the rails, with special attention being paid to the supports near the rail ends. The bottom layer of rails must be aligned with extra care, and rail flanges must not overlap. As assembly of the stock pile proceeds, the essential features are:

- Use rails of the same length. If this is not possible then the shorter rails must be laid nearer to the top.

- Keep the ends of rails of the same length vertically above one another.

- Do not cross layers without specific advice.

- Dunnage or spacers to be in identical positions along the rail length, so that when the stack is completed the dunnage is in good vertical alignment.

- Successive rail layers should be of the same or decreasing width.

13.11.5 Conductor Rail Maintenance

The following are the main points to be observed in maintaining conductor rail layouts.

(1) At the time of installation the height and lateral position of the conductor rail must be gauged and must be accurate within acceptable tolerances. The rail should be re-gauged at intervals as the running rail wears, and especially after any maintenance or renewal work.

(2) Any signs of excessive local wear or arc damage to the top contact surface or to the ramp ends should be reported.

(3) The clips holding the insulators in place must be kept tightly screwed down on to the sleepers

(4) Insulators must be clean to prevent current leakage. The insulators are designed to be self-cleaning, and should not

normally require attention, but heavily soiled insulators are also a potential cause of fire particularly on timber sleepers in dry weather, and should be cleaned or replaced as appropriate. Special attention should be given to areas of known pollution such as carriage washing plants, industrial sites, terminals, station platforms, etc.

(5) Displaced or broken insulators must be replaced at once.

(6) Fishplates - It is not usual to install steel conductor rail with fishplated joints. Where in situ, conductor rail fishplates must be maintained in the same way as running rail fishplates. When examining steel conductor rail fishplates the condition of the copper bond under the conductor rail joint must be checked to ensure that it is not loose, and that there are no broken or damaged laminations or strands. Failure of the bond will result in arcing across the joint and overheating. Any fishplated joints or bonds that have become defective should not be remade, but should be replaced by welding the joint as shown in the Electric Track Equipment Standards. The fishplated connections of conductor rails should be checked for signs of overheating or abnormal discolouration, and where this is observed, the joint should be broken, cleaned, re-greased and remade.

(7) The rails must be inspected regularly for signs of creep. Mid-point anchor insulator assemblies should be checked for signs of distress or displacement. Any sign of distress is a good indication that one or more insulator supports have seized to the conductor rail, and the section should be checked and released, pulling back rails as required. This is particularly important as if the rail creeps it will force the end ramp into a position where it may foul other equipment, e.g. a converging track, or level crossing cattle guards. In severe cases traction cables or conductor rail joint bonds may be damaged or torn off.

(8) Ballast etc. must be kept clear of the conductor rail cable connections and isolating switches, and ballast etc must not be allowed to encroach on the path swept by the disengaged

collector shoe. At switch and crossing work, level crossings and the like, it is wise to use a suitable gauge to check that hanging shoes are well clear of rails, and of any other fixed objects over which they must travel.

(9) Any maintenance or remedial work on conductor rail layouts during traffic with the rail energised must only be undertaken when the necessary safety precautions and the provisions of the Electrified Lines Working Instructions have been implemented. Care must be exercised to ensure that the alignment of the conductor rail is not altered during the course of the work in such a way as to cause damage to the collector shoes or any passing train.

Guard Boards

Where staff are required to work or walk regularly alongside a conductor rail, guard boards are provided on each side of the rail in order to prevent inadvertent contact. Traditionally, the boards were secured to a tall sleeper mounted bracket by small pocket type clips attached to the boards. This system worked well, but the tall brackets fouled tamping machines and therefore had to be removed and replaced each time the track was tamped. The latest design uses short brackets, combined with long clips, and only the board is lifted off leaving the short bracket clear of the machines.

When guard boards have been removed for machine maintenance or for other work to take place then they must be replaced as soon as the work is completed, for staff safety reasons.

The guard boards themselves are nominally 175mm deep and of varying lengths to suit the circumstances. Their width is approximately 35 to 38mm except where rolling stock provided with trip-cocks is used where it is reduced to a width of 25mm.

It is important that the boards are erected to the dimensions shown on the standard drawings, and special care must be taken to keep a good line on wooden sleepers or timbers. In all cases the correct height above conductor rail must be maintained.

For staff safety reasons it is essential that all guard boarding is kept in good repair and adjustment. Any boards that are twisted or any that become 'loose' should be renewed. If a clip falls off a board resulting in the board falling down at an angle then the board must be taken off and arrangements made for the clip to be re-secured as a matter of urgency.

In certain circumstances a single line of guard boarding is erected between the conductor rail and the running rail, to indicate to train drivers that single rail track circuits are in use at that location and that, in an emergency, the use of a short circuit bar will be ineffective. This use of guard boarding is now obsolete and is being replaced by a yellow plastic shroud clipped to the underside of the conductor rail and carries the words 'Do not use short circuit bar'. A further use for this shroud is at 'badger crossings' where it prevents the body of the badger from making contact with the conductor rail when passing underneath. In this case the shroud is painted green in order to advise Drivers that they may use a short circuit bar in the vicinity. It should be stressed that although this material is provided for the safety of badgers and other animals it must *not* be relied upon for staff safety purposes.

13.12 THE ORGANISATION AND PLANNING OF TRACK MAINTENANCE WORK

On larger railway networks and undertakings, the operation of track maintenance requires a complex and multi-functional input of human and material resources, and absorbs a very significant annual expenditure for the railway business. The management of track maintenance in an efficient and cost-effective manner has, consequently, considerable potential influence on the economic fortunes of the undertaking as a whole. The way in which the arising problems are tackled has varied with time, and with the objectives of the undertaking, as well as with the geographical extent of the network. It is common to identify the activity called "track maintenance" in a number of distinct categories each with dedicated staff who have had competency training:

- Track inspection
- Day to day manual maintenance or servicing
- Minor track repairs and component replacement
- Major track repairs and component replacement
- Replacement of Track
- Specialist Activities
  - Machine Operators, stoneblowers, tampers
  - Ultrasonic testing of rails
  - Welding to repair or join track
  - Earthwork and drainage repair
  - Off track repairs, eg. Fencing, access, steps

This section is devoted to a review of current practices on the two larger railway organisations in the UK; Network Rail and London Underground (LUL).

Figure 13.15 Completed Heavy Maintenance on a Double Junction (courtesy Network Rail)

13.12.1 Organisation

It is usual to divide a railway into areas with responsibility given to a professionally qualified track maintenance engineer (TME). The area can then be sub-divided into sections with a number of section managers who lead teams of inspection personnel and maintenance teams. The TME will also have a central support team of technical, planning and administration staff and specialist resources, such as welders.

Track Maintenance Engineer

The TME will be responsible to the railway authority for the safety of the line and employees, a maintenance budget, and standard of workmanship achieved. The TME is responsible for establishing safe systems of work for all track employees working on the area and observing the general state of the railway infrastructure and its surroundings and for taking appropriate action on these observations. The TME is required to carry out audits of the inspections made by reporting supervisory staff and team leaders.

It is usual practice for the TME to carry out personal inspections of the whole of his sub-area on foot and in train driving cabs at stated intervals, and to report on his inspections; to monitor patrolling and facing switch inspections; to take action on recording car output and track inspection sheets; to control his sub- area budget; and to contribute to preparation of renewal proposals. He will be responsible for safety procedures, and for general quality management on his area.

Track Section Manager

A typical organisation for a track maintenance engineer's area would comprise of a number of section managers, team leaders and operations with support staff providing technical and other resources. Activities would include track inspections, minor servicing and planned regular maintenance. It is also usual for each geographical area to have a rapid response service. The Track Section Manager (TSM) would, in conjunction with the TME, devise a programme of mechanical maintenance which usually comprises a combination of tamping and stone-blowing. A key requirement is to confirm specifications for technical compliance to fixed datum points or in some cases a local redesign.

Following the submission of on-track machine requirements for the Section, the TSM formulates a programme identifying the work that will be carried out on the Section and arranges all the necessary associated resources. Whilst any on-track machine is working on the Section, the TSM liaises with the an on-track machine manager responsible for monitoring and supervising track geometry quality input and arranges marking up the track. It is essential to ensure technical resources are provided, including the setting out for circular curves, transitions, fixed (or tied) points, cant and the existence of obstructions together with general track lifts and slues in advance of the machine work.

On-track machines, are usually fitted with computer controlled alignment and levelling systems and the TSM will ensure that all necessary track information is obtained prior to the machine working. Of particular importance are the location of points of restricted slues and/or lift, tied points and the value and direction of the slue or lift which can be permitted at each tied point. Also it is important to identify cables, traction, bonds and switch heaters that could be affected.

Typically, the TSM is required to inspect his section on foot. This task includes checking drainage, earthworks, fencing and lineside vegetation, switch heaters, conductor rails if any, level crossings, tracks on nonballasted underbridges, and structure gauge checking (this last includes checking permanent structures at six-monthly intervals and temporary structures on erection and then every two months). He must travel his section by cab riding, usually every two months, make reports on his inspections in the approved form, study the track recording car information supplied to him, and check and countersign "Track Inspection Report Sheets" and Patrolling Weekly Record Books. He is responsible for arranging inspection of facing switches jointly with the Signal Engineer's representative at three monthly intervals, and for arranging special inspections of track in tunnels at six-monthly intervals (note that these may involve technical staff and ultrasonic rail flaw detection (RFD) operators if the on-track rail depth measuring vehicle is not available). He must initiate appropriate corrective action on his observations as necessary. He initiates requests for S&C welding repairs, and has responsibilities in connection with private sidings. He initiates proposals for track renewal, submits notices for possessions and temporary speed restrictions on his section, and may be required to act as a track possession supervisor or crane supervisor, and in electrified areas to make isolation arrangements. He is responsible for safe custody and care of tools and equipment, stocks of maintenance materials (including detonators), and he has responsibilities for the road vehicles allocated to his Section. Finally, the TSM is responsible for maintaining registers of staff competencies, for the recruitment, induction, training, and discipline, of his staff, and for their safety, health and welfare.

Track Team Leader

A Track Team Leader is responsible mainly for supervising the work of a group of staff in carrying out work as instructed by the TSM. He is responsible for the general safety, health and welfare of the work force entrusted to him and, in particular, ensuring safe systems of work are in place when working on track. He is also required to carry out work as planned and report back to his Manager on the completion of the work and to submit details for input into the asset management system. A further responsibility is the safe custody and use of tools, material, equipment, transport, etc. allocated, the cleanliness and tidiness of accommodation and sites, and for track inspections as delegated.

Track Inspectors

The Track Inspector is responsible for the periodic examination of the entire extent of rail authority property on allocated lengths of route at frequencies laid down by the TME, either by himself or by an assistant. He records the content of his own inspections in a "Track Inspection Report Sheet", and countersigns other reports are made by his assistants. The objectives of these inspections are:

(1)   The identification of items requiring immediate or emergency attention, and responding to such situations (for this purpose he has special communication kit and carries equipment with which to halt rail traffic if necessary).

(2)   Identification of future work requirements affecting any part of the infrastructure for which the TSM is responsible, and recording details for input to the asset management system.

(3)   Monitoring the quality and quantity of work carried out to ensure that the required standards of work are maintained.

(4)   Contributing to the asset management process for diagnosing track deterioration profiles by liaising with the TSM and TME and the information from track recording vehicles.

Electric Track Maintenance Section Manager

This post is specific to routes electrified on the third rail system, and its holder is responsible for the specialist work associated with the conductor rail and the power supply connections to it. Details of the nature of the work involved may be gathered for Chapter 3.

13.12.2 The identification of Track Maintenance work

The first stage in the planning of any task is the identification of exactly what has to be done, and when. So far as track maintenance is concerned, this originally involved a dialogue between 'business managers', 'engineers' at various levels, 'inspectors' and other track staff. Since those early days, computer diagnostics are playing a major in asset management with information from a number of sources including track testing trains and manual inspections. Inspection and minor servicing work such as lubrication, tightening and replacement of single components, removal of wind borne material and debris is a key daily or weekly activity. Track maintenance itself usually involves a combination of on-track machine activity alongside activities requiring a small group of operatives using manual methods and had tools.

In a typical area, the track maintenance work would comprise of:

1. Track line and level repairs using Kango® packing or hand held stoneblowers.

2. Ballast replacement on a spot basis by manual techniques using hand tamping.

3. Ballast replacement assisted by specially adapted machines – there are a number of small machines that can excavate, an example is the Railvac® machine.

4. Ballast repair using chippings - this can be successful in bringing track geometry back into specification. (sometimes called "measured shovel packing")

5. Long timber and sleeper replacement – a key manual activity.

6. Timber repair through repositioning (sometimes known as "pulling through") – effective as an economic solution or where timbers were in good condition.

7. Replacement of S&C units such as crossings or sets of switches.

8. Replacement of steel keys with "panlocks" on bull head track track configurations.

9. Replacement of switch and crossing components including screws and bolts – an essential feature of minor refurbishment.

10. Weld repairs to all metallic track components including plain rail.

11. Welding up traditional sixty foot rail joints.

12. Stress reinstatement in areas of concern and weak points

13. Component replacement and thicker pad installation.

14. Realignment of track vertically and horizontally using lasers and advanced total station surveying.

CHAPTER 14

TRACKLAYING AND RENEWAL - PROCESSES AND EQUIPMENT

14.1   INTRODUCTION

The whole life costing of a railway asset will depend upon the required level of maintenance, half - life refurbishment and renewal. A comprehensive asset management process is needed. This will take the form of an asset register which considers not only the maintenance and performance history of the track and infrastructure asset but its failure analyses and a full suite of deterioration profiles that allow diagnostic and decision making to be modelled. This is against a backdrop of the actual train companies using the network with both a freight and passenger service pattern that will have a set of objectives over a variety of timescales including a high speed provision.

14.2   ASSESSMENT AND INSPECTION

Most railway authorities will have a system in place to identify, inspect and prioritise work on sections of track requiring complete and partial renewal. This is a continuous year on year process and produces a future workbank consisting typically of the following categories of work; complete relaying with associated ballasting, re-railing, re-sleepering, formation works, drainage, ballasting not associated with relaying, re-grading of embankments or the renewal of longitudinal timbers on certain types of underbridges.

Track engineering renewal activity proposals will have taken into account the following criteria:

- condition of the rails; present depth and rate of wear; presence of defects, gall and battered ends; sidewear or side cut; switch rails that are crippled or excessively hogged causing persistent detection faults, or crossings beyond economical weld repairing.

- condition of fastenings; loss of toe load; availability of replacement; resistance to rail creep; options and availability of special maintenance devices for the fastenings available.
- condition of sleepers; splits; rot; spalling of concrete sleepers and effects of ballast attrition on concrete sleepers; S&C timbers already pulled through and no new bearing areas available.
- condition of ballast inability to drain freely; tendency to attrition; of inadequate depth; incapable of being maintained by machine.
- condition of formation; poor drainage; tendency to slurry pumping through the ballast; cess heave.
- changes in traffic patterns; volume and type; effect of vehicles new to the route (or even new vehicles); effect of increased number of trains;
- materials policy changes associated with speed and tonnage changes; the present track construction may not meet current traffic flows for predicted business aspirations.
- delays in the work being done for many reasons; assess the deterioration from date of proposal to the probable year the work will be done and what will be necessary to keep the line open, albeit under a temporary speed restriction.

14.3   PLANNING BUDGETS AND PROGRAMMIMG

In its simplest form, the annual Programme of Work is drawn up from all the work which should be undertaken in the appropriate year, judged on its priority. There are, however, three important criteria which will affect this.

Firstly, it is desirable to have as constant a programme as possible from year to year on any given route to smooth out the requirements and allocation of resources. Peak years are difficult to accommodate within existing resources and availability of possessions, etc. especially when major projects are also being carried out and can last over a number of years.

Secondly, overall spending by the infrastructure owner is usually fixed for a financial period by regulation and scrutinised on a regular basis relating to improved performance and reduced cost whilst still carrying out volumes proposed. This is a difficult balancing act and detailed knowledge of the condition of infrastructure assets is essential to enable prioritisation to be effective whilst ensuring the optimum amount of maintenance and refurbishment is carried out.

Thirdly, other stakeholders need to be consulted in the local, regional and national transport arena as well as the aspirations of current passenger and freight companies. These will have significant influence on when work is done, particular for special possessions or speed restrictions. The track engineer plays a major role in offering methods of renewal solutions for infrastructure upgrades which meet an optimisation approach for performance and cost.

As mentioned above, the volume of track renewals should relate to whole life costing of the asset and the future aspirations of the railway and its part in the future of a country. A great deal of work was carried out in the UK in the early part of the 21st century to establish long term renewal plans. On the main line rail network in the UK, the current infrastructure owner, Network Rail has agreed with the UK government the following programme of track renewals work from 2014 to 2019:

S&C Full renewal with re-ballasting   1445 point ends
     Refurbishment, heavy and medium   3882 point ends

Formation Renewal and Drainage   171 km

Plain Line, using High Output Equipment
     Ballast Cleaning   856 km
     Rail and Sleeper Renewal   687 km
     Refurbishment, heavy   977 km

Plain Line, using Conventional Equipment
     Rail and Steel Sleeper Renewal   69 km

Complete Renewal (track and ballast)	824 km
Refurbishment, heavy	349 km
Refurbishment, medium, concrete & other sleeper types	1820 km
Rerailing	1478 km

The organisation of these volumes of work is a major task in terms of prioritisation and allocation of resources, as the completion of each individual item to its technical specification is essential if the projected life span of the track components under a normal maintenance regime is to be realised.

The task of eliminating a major maintenance problem by re-ballasting, or of relaying a major junction, in the space of a few hours requires planning and organisation of a high order, if completion to specification within the planned possession time is to be achieved. The price of failure so to achieve is often very high, not only in the direct cost of the work, but in customer confidence and satisfaction when their services are disrupted.

There is no one single preferred method to achieve the technical objectives on a renewals job; the following pages therefore do not seek to present a definitive method for a particular job, rather a range of proven options for the constituent parts from which can be chosen the combination which best suits local circumstances.

Manual methods for opening out the existing ballast, unloading the new materials, removing and loading the old materials, laying in the new materials, unloading/replacing ballast, lifting, packing, aligning and boxing in were extensively used until the mid-1950s.

There was a rapid decline in the use of manual methods from 1958 onwards with the advent of CWR, concrete sleepers and machine reballasting. Virtually all the handling of materials and components is now mechanised with manual work restricted to initial removal, final placing and adjustment of components.

Plain line renewal methods very much depend on the location and type of materials to be used. They can be broadly grouped into two techniques; the first employing High Output Equipment (HOE) and the other utilising many different variations of plant and equipment to suit particular circumstances.

The second technique is used for work in locations such as station platforms, tunnels, adjoining S&C, where side clearances are restricted, over some bridges and viaducts, and on routes electrified by 3rd of 4th rail systems. Explained further in 14.7, HOE is most economic when used to carry out an optimum length of renewal in the possession time available. Therefore, some work in lengths of less than 300m, although suitable for HOE, will be carried out using other techniques.

14.4   PLAIN LINE RELAYING METHODS

If complete relaying of rails, sleepers and fastenings is specified, new rail will be delivered to site as strings on long welded rail trains in advance of the programme dates for the work. Sleepers and fastenings are delivered directly to site on suitable wagons, usually on the date of relaying for immediate laying in. Occasionally they will be delivered in advance, in which case they will be unloaded to the side of the track.

Renewal of the old track is normally in panels of up to 18.3m in length by twin jib cranes, single line gantries (SLG) or suitably equipped and certificated road-rail vehicles (RRV) and loaded onto panel carrying wagons. Where the existing rail is to be re-used, it is first unclipped, and then moved to the side utilising an RRV with a rail handling device. The old sleepers are then removed by RRV's with suitable beams and either loaded into wagons or placed at the side of the track for subsequent disposal.

On the 3rd and 4th rail electrified routes, new conductor rail may be specified, in which case it will be delivered to site in advance together with its sleeper mountings. If the existing conductor rail is to be reused then it is initially moved to one side.

The next stage will normally be to carry out the reballasting work specified – see below for more detail.

14.4.1 Ballast Preparation

Before laying the new track, the ballast is prepared by levelling and consolidating to produce a top surface which is very close to design in respect of longitudinal level and cross-fall. This process is normally carried out using a tracked dozer with laser control and linked consolidating plates.

The new sleepers are then laid in using any of the items of plant previously used to remove the old track, depending on whether the sleepers are being taken directly from wagons or from the side of the track. Lifting beams are available which can accurately space and align the sleepers. Threading in of the rail follows, then the installation of the fastenings and the welding of the rail joints by the alumino-thermic process. The next step is to lay in the "boxing in" ballast which is unloaded from auto-ballast or hopper wagons, or by RRVs from open wagons, and profiled by an RRV with a ballast profile and sweeping device. On 3rd or 4th rail electrified routes, the conductor rails and supporting equipment is put in place.

The final job is to tamp and align the new track to its design alignment using OTMs which although may differ in type, design and performance rate, will produce the required track quality.

14.4.2 Reballasting Plain Line

This work is carried out with relaying, or as a stand-alone process where track components are not life-expired.

The work can be carried out using a Ballast Cleaning Machine (BC), in which case the track can remain in-situ or, if the track is to be renewed, by using a variety of RRVs. HOE is also extensively used, as with relaying, and is explained further in 14.7. Methodology commonly used in conjunction with the plain line relaying techniques is as follows:

a. Utilising a BC system to carry out the re-ballast as a separate process to the relaying, leaving the track below design level in the interim. This has the disadvantage of requiring a low TSR, and should not be used if the existing sleepers and fastenings will become insecure when disturbed.

b. Utilising a BC system to carry out the re-ballasting immediately prior to, or following, the relaying in the same possession. This carries risk in the event of BC system failure, but is an efficient use of plant and materials. This method can be used on single lines without double handling.

Ballast cleaning systems, utilising a medium output BC with automated spoil and new ballast wagons with laser control to the design alignment are extremely efficient and the mix of new and screened ballast is arguably the best way to produce quickly top quality track with minimum future maintenance intervention. However the systems cannot be used in a significant number of locations, such as station platforms, tunnels or where side clearances are restricted.

c. Utilising road rail excavators, carry out the excavation to the specified depth and cross-fall loading to open wagons. When using auto ballasters, or by RRVs from open wagons, unload new stone and level and consolidate using RRVs, tracked dozers and linked consolidating plates. All processes are laser controlled to design longitudinal level and cross-fall. This method produces very good quality, given experienced site personnel, but requires requires a multiplicity of machines and equipment and a considerable number of engineering trains. It can however by adapted for use on almost any double track site.

For re-ballasting not associated with relaying, the BC system will usually be the most efficient and economic method to use as the track can remain in situ, longer lengths can be carried out in the same possession time, and the line can be reopened with a high temporary speed restriction (TSR). These systems however, will be restricted to relatively open sites.

Where RR Excavations are considered to be the best option for reasons of site constraints and / or conditions, the track will of course have to be removed and then replaced. The whole process will involve virtually the same resources as for relaying and re-ballasting, and output will be less than that of a BC system in the same possession time, with a higher unit cost.

14.5   EARTHWORKS AND DRAINAGE OPERATIONS

These are remedial works required at sites where bearing strength failure has occurred, maintenance has become excessive, and speed restrictions have been imposed or are imminent.

The remedial work will be subject to detailed design to produce a long-term solution to the problem. The rebuilding of the formation with appropriate blanketing materials, reinforced with geotextiles and the provision of often complex drainage systems, all to exacting specifications, make these jobs amongst the most challenging to plan and execute in weekend possessions. In addition work on the cutting slopes may be necessary, together with work for a considerable distance beyond the boundaries of the job to provide drainage capacity to the outfall.

Techniques described in sections 14.3 and 14.4 are generally applicable. However, earthworks and drainage sites are usually relatively short and concurrent operations are not normally practicable. The drainage work usually proves to be the limiting factor, as attempts to do other work before the drainage is properly completed tend to produce a drainage system which does not conform to specification, particularly where the placing of filter material around the pipe is concerned. Sufficient time has to be allowed for the drainage work, and the logistics of excavating the trench and loading the debris. The unloading and placing of the filter material must be carefully considered. During the planning stage, it is important to produce a sequence of work that minimises the number of machine moves on the cut formation, as these cause rutting and pocketing of the formation and create the conditions for further failure in the future. Wherever possible, low ground pressure machines should be chosen.

Blanketing material is normally spread, levelled and consolidated in layers. Geotextiles are widely used to reinforce the blanket, and provide filter or impermeable layers depending on the failure or lack of strength problems to be solved. Membranes must be placed at the correct level and cross fall, and must not be rucked. Time and care *must* be spent in preparing the surface of the material on which the membrane is to be placed. The covering layer of material must be placed and levelled with great care to avoid displacing the membrane. There is no doubt that it is vital to spend as much time and expertise as necessary in this area to secure the long term solution to the original problem.

Use of the techniques described should also promote even and rapid consolidation and allow progressive raising of the temporary speed restriction to full line speed within a week or so of the job being carried out. Opening speeds in the order of 60-80km/h on jobs of this nature are now perfectly possible, a major step forward so far as the Businesses are concerned.

14.6 FOLLOW UP WORK

This encompasses the work required during the early life of the newly-laid and/or reballasted track prior to being considered in normal maintenance.

In all cases the volume of work to be done will be directly related to the quality produced at the conclusion of the main part of the job.

If at that stage the cut formation is parallel to the final rail level, the cross fall to the cut formation is to design and longitudinal levels, and crosslevels and alignment of the track are within the tolerances allowed for the route, then the follow up process will be simplified.

Immediate and subsequent consolidation will be uniform, predictable and manageable, and the best possible basis will have been laid for stable, quality track conditions with minimum maintenance requirements for the future. Should the requirements not be met, in particular the relationship of the cut formation to final rail level, then not only will the follow up process be more difficult but future maintenance will be more frequent.

Depending on the type, speed and axle loading of the trains using the route, follow up tamping will be carried out between 7 and 28 days after completion of the main job, then again at around 6 months. There should then be no necessity for intervention for at least 2 years, and more typically for up to 4 years.

14.7 HIGH OUTPUT TECHNIQUES

There is now a shift in emphasis from relaying and reballasting during possessions of two tracks lasting between eight and 54 hours over a weekend, to working in regular, short (6-9 hour) single line possessions on consecutive week nights.

Experience gained with ballast preparation techniques, laser control systems and "production-line" methods has been incorporated into the specifications for the high output, high efficiency equipment which can work in single line possession and without isolation of overhead equipment. Up to 75% of relaying and reballasting on UK routes is now be carried out by these systems.

High output techniques require the integration of equipment into units each capable of working at an average rate of 400m per hour, and requiring virtually no facilities other than possession of the line on which they are working. The availability and reliability of all the component units is crucial to the success of the system. To achieve this, dedicated teams of maintainer/operators are assigned to the units, to manage the systems. Purpose-built depots have been provided at specific locations at which the systems are maintained, emptied and reloaded on a daily basis.

14.7.1 Ballast Cleaning

The Ballast Cleaning System (BCS) has three matched components. First comes the Single Line Spoil Handling System (SLSHS) to receive the spoil from the ballast cleaning process. It is capable of self-discharge on site, at spoil transfer points, or at a tip. Next is the High Output Ballast Cleaner (HOBC). Lastly comes the Ballast Distribution System (BDS) which feeds new stone ballast to the HOBC to supplement that returned to the track from the ballast cleaning process. These three components combine to produce re-ballasted track to designed longitudinal levels, correct cross-level and boxed in with 10% excess, at an average production rate of 400m per hour, over the full range of normal re-ballasting depths below sleeper bottom, whether screening or carrying out total excavation. Following the BCS comes a suite of OTMs, including a DTS, to finish the track design.

At this rate the BCS will complete a job of 500m length, 280mm nominal ballast depth below sleeper bottom with 1 in 40 cross-fall, and average 50% return of cleaned ballast complete with the follow up work, the reconnection and testing of signalling and OLE equipment, and the line opened at high speed, within an 8 hour possession. Cleaned ballast returns of less than 50% will reduce the length of the job pro-rata, due to the capacity of the SLSHS and BDS systems.

No preparatory work to the track is necessary and the system is be able to work at its specified rate on any track curvature from straight to 400m radius and with up to 160m of cant, in any combination. It is capable of producing run-in and run-out ramps to the same degree of accuracy as the main cut.

Trains can pass the BCS on adjacent open tracks with a standard inter-track spacing under a Temporary Speed Restriction of 30 mph (this latter is required as a safety precaution for site staff), but need not be subject to any other restriction to normal running.

Their design incorporate comprehensive screening to protect staff from the live catenary, and the latter from damage by the machinery or the material being handled. Consequently on all but the most difficult and confined sites isolation of the OLE will not be needed. All elements of the BCS will conform to W6 loading gauge when travelling to and from site. When setting up or stowing away on site, and when working, all the elements of the BCS will conform to the Working Clearance Diagram of the Modified W6 loading gauge. Transit speed for the BCS is 60 mph.

Figure 14.1 NR P&T RM900 Ballast Cleaning System
(Courtesy of Rail Infrastructure)

14.7.2 The Relaying Train (RLT)

Relaying with associated re-ballasting is usually carried out in the following sequence:

(i)   Unload rail strings

(ii)   Relay with Relaying System

(iii)   Weld and stress new rail, restore line speed

(iv)   Load old rail subsequently re-ballast with BCS

(v)   Follow up and restore line speed.

This sequence enables high speed to be maintained between the various steps throughout the process. All the components of the relaying system conform to the basic requirements specified for the Ballast Cleaning System. Although any re-ballasting associated with the relaying process will in most cases be carried out in a separate, subsequent, possession, the processes are inextricably linked. The Relaying System consists of a Relaying Unit and a number of sleeper carrying vehicles. These last will incorporate handling equipment to deliver the new sleepers to the Relaying Unit and take away the old ones.

The Relaying Unit removes and loads the old fastenings, removes the old rail and places aside, removes the old sleepers, scarifies and profiles the existing ballast (including redistribution), lays in the replacement sleepers, positions the new rail on the new sleepers and inserts the clips. It then ploughs the excess ballast from the earlier operation into the beds and forms the shoulders..

A Mobile Flash Butt Welding Machine is incorporated in the system. These machines are capable of producing welds in a cycle time of five minutes or less including manipulation of the rail, and consistently meet the stringent weld specification. A typical night's relaying of up to 732m may require the making of 14 welds; in the time available only the flash butt welding process is capable of achieving the required productivity and consistent quality.

Following the RLT comes a suite of On Track Machines (OTM's), including a Dynamic Track Stabiliser (DTS), to finish the track to design. After reconnection and testing of signalling and OLE equipment, the line can be opened at high speed. This can be achieved in around an 8 hour track possession.

Finally, for both types of work, a follow-up with the suite of OTM's, including a DTS will be programmed 24-48 hours afterwards. This allows the line speed to be restored.

Figure 14.2 NR Matisa P95 Track Relaying System (Courtesy of Jörg Schnabel)

14.8 SWITCH AND CROSSING RELAYING

Whilst S&C layouts have in recent years become much less complicated, even in station approaches, the requirements for reballasting have become more onerous and the weight and nature of many of the modern components have created additional problems. Reballasting must be carried out to a very high standard. This is of crucial importance to the actual work on site, subsequent follow-up work, and the future maintenance of the layout. The techniques described in 14.3 and 14.4 will produce the required standard.

Virtually all new layouts are assembled at a manufacturer's yard to check that the components fit together properly, and that the overall dimensions are correct. The bearers are all numbered, a reference line is set out along the layout and marked on each bearer. Longitudinal bearer locations are marked on wooden laths, and rail joints are numbered. The layout is then stripped down either partially or completely for delivery to a point near the worksite. The degree of dismantling will depend on the relaying method chosen, but in any case consideration will have been given to the sequence of loading to ensure efficient handling at the delivery point. The following paragraphs describe some of the methods available, and the latest development of the Modular Switch System.

14.8.1 Crane and Trolley Methods

Where the new layout is to be assembled adjacent to or within a short distance of the site, it can be laid on the prepared ballast bed by high capacity rail cranes (Kirow Types 250, 810 or 1200) equipped with specially designed lifting beams in large sections. These cranes have an extensive range of duties, both in propped and pick and carry modes, and are capable of precise placing of the sections. They can work with similar duties with their main booms at any elevation between horizontal and maximum, which makes them suitable for deployment on most sites including OLE routes and other restricted areas, and can lay sections alongside and in line. In many cases, these cranes are also used to remove the old layout in large sections.

14.8.2 Modular Switch Methods

The Modular Switch system has been developed by Network Rail and implemented with its industry partners. The system enables the more simple layouts to be installed over a series of weeknight track possessions of c.8.5 hours, including re-ballasting. Layouts are designed to enable the installation the S&C units on one line in each possession; where "through" bearers to the adjacent line are required, these are provided by splitting the bearers in the track interval and then connecting them to their corresponding bearer in the adjacent line using bearer tie plates. Units are delivered to site on specially-designed tilting bed rail wagons, capable of transporting units up to 26.5 metres in length with up to 3.7 metre long bearers. The wagon bed is moved to the horizontal position for loading and unloading of the units, and tilted to become in gauge for transit in train formation.

Figure 14.3 Loading tilting wagon, bed horizontal, at Manufacturer's depot, with S&C panel prior to securing and tilting for transit (Courtesy of Rail Infrastructure)

Laying of the units on site is carried out by Kirow cranes. Each unit has lateral lifting beams attached to facilitate loading at the Manufacturer's depot, these remain on the unit for transit. The Kirow crane has a lifting beam, equipped with a load levelling and weight distribution device, which attaches to the lateral lifting beams. The switch unit has its point motor and associated equipment already fitted. A typical double line running or trailing crossover will have 5 units for each line.

This system and methodology is a real step forward to enable the installation of relatively simple layouts to be moved from long weekend to less disruptive short weeknight possessions.

Figure 14.4 Rake of tilting wagons conveying complete S&C layout on arrival on relaying site (Courtesy of Network Rail)

14.8.3 PEM and LEM

PEM/LEM machines (trolleys) are also extensively used for this work. These machines can handle a complete layout from the assembly ground and move it laterally into position, or transport it along the track to site. The system consists of individual self-powered units, and the dimensions, weight and movement required of a particular layout will determine the number of units that will need to be deployed. The units form an integrated system on site, controlled remotely, and allow precise handling and positioning of the layout. Use of these machines does entail considerable planning and physical preparation, including delivery to site and unloading, and removal following the work; it is also usual to remove the old layout using RRVs. However the method allows a large number of the rail joints to be welded prior to the work, offering considerable time savings and flexibility advantages during the installation.

Some smaller layouts can also be relayed using RRVs, but as lifting capacity is restricted compared with that of the Kirow cranes and the PEM/LEM machines the layouts need to be handled in much smaller sections or as individual units, thus increasing both installation time and the risk to the finished quality. This is however a viable option on some

Figure 14.5 KRC1200 with switch panel unloaded from tilting wagon, lifting beam incorporating load levelling device (Courtesy of Carillion)

Figure 14.6 KRC250s in tandem with central section of crossover lifted and ready to travel to installation position (Courtesy of Swietelsky Babcock)

sites, particularly where space is very constrained, as the RRVs are able to carry out a multiplicity of tasks, using a variety of attachments, concurrently rather than consecutively.

Figure 14.7 *Colas Rail Geismar PEM/LEM machines at New Cross Gate (Courtesy of Colas Rail)*

14.9 STRESSING OF CWR

The prevention of buckling of CWR in hot weather is ensured by a judicious combination of the following features:

- Heavy sleepers at close spacing

- Fully filled ballast beds

- Heaped up ballast shoulders

- Well consolidated ballast

- Fastenings having a good resistance to longitudinal rail creep, and to any twisting of the rail

- Limitation of longitudinal compression in the rail at high temperatures.

The requirements for the first five of these features are met by the standard provisions for sleepers, ballast, and fastenings described in the relevant chapters of this book.

In general terms, this objective is achieved by one of three methods. Whichever method is employed it is necessary to determine a rail temperature at which the rail must be free of longitudinal thermal load. This temperature is called the STRESS FREE TEMPERATURE (SFT) or NEUTRAL TEMPERATURE. The SFT must be chosen so that:

(a)   the compressive load at the highest summer rail temperature will not buckle the rail;

(b)   the tensile load at the lowest winter rail temperature will not initiate tensile fractures.

Having determined the SFT it is then necessary to ensure that when the rail is at that temperature it is actually free from longitudinal force.

The three ways in which this may be achieved are:

(a)   wait until the rail reaches the SFT, then release the fastenings, lift the rail onto rollers, vibrate it, cut off any longitudinal "growth", and finally re- secure.

(b)   release the fastenings, lift the rail onto rollers, and then deliberately warm it to the SFT before vibrating, cutting off "growth" and resecuring.

(c)   release the fastenings, lift the rail onto rollers, and then deliberately stretch it by a predetermined amount. Finally the "stretch" is cut off and the rail resecured.

Method (a) is commonly employed in parts of the world where the climate is suitable but in UK, method (c) is almost universally applied, and is the method implied in everything which follows.

14.9.1   Stress Free Temperature

The determination of SFT from first principles requires a knowledge of the climatic range affecting the terrain through which the railway passes. In the UK air temperatures rarely exceed 35°C or fall below about -12°C. Rail temperatures exceed ambient highs due to insolation, and go below ambient lows due to radiation giving a range of about +50°C to — 20°C and a year-on- year median of about 15°C. Since the prevention of buckling is the prime consideration the SFT is fixed substantially above the median. The allowed range for SFT on the UK network is 21 - 27°C.

It is emphasized that this value does not necessarily apply in other countries and it is known that where extreme "continental" type climates prevail stress ranges may be very large. This can influence the choice of rail section or even inhibit the adoption of CWR altogether.

14.9.2   Planning the Stressing of CWR

The stressing of CWR is the last operation in relaying, but it should be carried out as soon as possible after the long rails have been laid. This is particularly important on relaying done between mid-April and late September, due to the likelihood of high rail temperatures occurring during that period. Similarly any relaying/re-railing done during the winter must be stressed before mid-April.

The job plan must take into account:

(i)   the time of the year,

(ii)   the adjoining track type,

(iii)   location of the work,

(iv)   necessity for adjustment switches,

(v)   requirements for insulated joints,

(vi)   position of trap or catch joints,

(vii)   proximity of certain underbridges, viaducts, tunnels, other technical department's installations,

(viii) any significant alterations to longitudinal levels or lateral alignment,

(ix)   cant and curvature, gradients,

(x)   position of anchors, lengths to be pulled, and location of closing welds.

Prior to the date of stressing, the relaying should be completed to the extent that alignment and longitudinal levels are correct, ballasting to specification, and ideally tamping in conjunction with the DTS has been carried out.

Equipment needed for stressing will include rail thermometers, hydraulic rail tensors, support rollers, and side rollers. Specialist hand tools for removal/ replacement of fastenings are also required. This part of the process can be automated by using rail mounted clipping machines. However, these can only be used with PANDROL brand rail clips and whilst they can be extremely effective on well-laid track they are somewhat inflexible on site and do not allow reductions in site manpower in these circumstances. The number of welds to be made will dictate the number of teams of welders and alumino thermic welding equipment required, and indeed this part of the process will often be the deciding factor in the choice of stressing procedure.

The lifting of the rail to allow the placing of the under-rollers and adjustment of rail pads is carried out using ratchet or hydraulic jacks, or the specialist jacking trolleys which also have the advantage of being able to convey support rollers, side rollers, spare insulators, pads and clips.

14.9.3 Procedure for Stressing

The procedure to be adopted consists of the following steps:

   (i)   Preliminary adjustments

   (ii)   Establish 'anchor lengths'

   (iii)   Establish 'tell tale' points

   (iv)   Cut rail if restressing into existing continuous welded rail, release fastenings and mount on rollers

   (v)   Establish reference points

   (vi)   Note movement of rail at 'tell tale' points

   (vii) Establish average rail temperature, calculate and mark extension at pulling point and reference points

   (viii) Position tensors, and tension rail to calculated extension

   (ix)   Cut and weld final joints

   (x)   Remove rollers and fasten down the rail from the weld

   (xi)   Release and remove rail tensors

   (xii) Equalise stress

Preliminary adjustment

Before stressing or restressing it is essential to ensure that impediments to free movement of rail such as rail anchors are removed and expansion gaps are adjusted in adjoining jointed track.

Anchor lengths

For stressing and restressing purposes it is essential that an adequate anchor is provided at the fixed ends of the free rails.

The recommended minimum length to be fastened down to form an anchor is 100m. When extending continuous welded rail or restressing continuous welded rail this length should be increased to 150m if the fastenings, rail pads, insulators or ballast conditions are poor.

Certain parts of strengthened S&C may be included as part of an anchor length.

Tell tales

In order to monitor the movement of the rails to be stressed and the effectiveness of the anchor lengths, accurate, and reliable 'tell tale' points must be established for each anchor length.

Rail supports during tensioning

Fastenings must be released and commencing from the cut or free end of the rail lifted onto solid rollers positioned at intervals of 10 or 11 sleepers with pads removed to reduce friction. Side rollers or support arms should be used on curved track. To ensure that rails are free to move during tensioning, they should be vibrated or shaken by bars under the rail. Steel hammers must not be used to strike the rails.

Reference points

It is essential that the calculated extension is obtained uniformly throughout the length of the free rail. To enable this to be achieved, reference points must be established on each rail at equal intervals, e.g. 50m, or other similar dimension to suit easy division of the rail length. These reference marks are numbered 1, 2, 3, etc. consecutively from the first 'tell tale'.

Movement of rail

It is important that account should also be taken of any movement of the rail which occurs at the 'tell tales' when the fastenings are released. Movement at the inner 'tell tale' may be towards or away from the pulling point, depending upon the stress-free condition of the existing continuous welded rail including the anchor length, and also the average rail temperature. Movement at the outer 'tell tale' indicates that the anchor is not satisfactory and a new anchor length must be established.

Rail temperature readings, calculation and marking of extension

Thermometers should be placed on the rail foot or at the neutral axis (web of rail) on the shaded side of each rail and protected from the elements. At least three thermometers should be used on each rail. These should be located near to the pulling point, at the midpoint of the free rail, and near to the anchor length. Additional thermometers may be necessary, depending on site conditions, subject to a maximum spacing of 200m.

When the readings of all thermometers are reasonably uniform an average should be taken. If any reading is obviously inconsistent with the remainder then a check should be made by changing the thermometer/s concerned. If the rail temperature as measured proves to be above the required SFT, the work should continue to completion and arrangements made for restressing prior to the onset of winter.

The required extension of each rail to be stressed will be calculated according to a pre-determined formulae. The full extension should then be marked accurately at the free end (pulling point) and the proportional movement marked at each intermediate reference point in such a way that the rail movement may be monitored during tensioning.

Position tensors and tension rail

The tensors must be positioned on both rails to be stressed or restressed in accordance with the operating instructions.

The rail tensors must be used in accordance with the manufacturer's operating instructions. If the tensors do not have the capacity to obtain the calculated extension, the maximum pull must not be exceeded. In this case the length to be stressed should be reduced or the extension equivalent to the tensor capacity obtained and the rail restressed before the advent of hot weather.

During tensioning the rail should be extended to the reference mark at the pulling point and movement checked at the tell tale and at each established reference point. The force applied by the tensors should be monitored during tensioning to ensure that it is reasonably related to the required temperature difference. If this is not so, it is likely that the calculated extensions have not been obtained and the rail must be checked for possible obstructions to the free movement of the rail. The tensors, particularly the dial gauge, should also be checked for defects or slipping of the clamps.

In order to reduce the effect of frictional resistance and to ensure adequate stability during stressing, the maximum lengths of continuous welded rail to be tensioned in one direction are as follows:

For single items in straight track or curves flatter than 4000m a maximum length of 900m may be pulled in one operation. Provided that the overall length to be stressed does not exceed 1200m or double the lengths in the above table, and the tensor capability (load and extension) is adequate, it is desirable to obtain the required extension over the length by pulling both ways from a central pulling point.

When tensioning curved track, side rollers or support arms to suit the type of fastening installed must be used.

Cut and weld final joint

When the required total extension in each rail has been obtained (account being taken of any further movement at the tell tale which may occur during pulling) the overlap of rails between the tensor clamps must be carefully marked for the welding gap and cut off by the welder. When the two rails have been brought into line, the welder will check that the welding gap corresponds with the specification for the type of weld to be made, and trim it as necessary. About 10m of rail each side of the weld should be fastened in order to assist in lining and levelling at the weld. Immediately before removing the moulds and stripping off surplus weld material (three minutes from end of pour) an extra pull must be applied to the rails.

Remove rollers and fasten down

Fastening down will commence at the pulling point and work towards the anchor point. It should stop at about 36m from the anchor point to allow stress equalisation into the anchor. On sharp curves refastening should keep within 30m of the removal of each support arm to restrain the tendency of the stressed rail to leave the rail seating.

Release and removal of rail tensor

The clamps must remain in position under pressure for 20 minutes after the weld is poured to avoid the possibility of hot tears.

Stress equalisation

To achieve a smoothing out of any small stress difference which may occur during the tensioning sequences the fastenings for 36m on the anchor side of the inner tell tale should be released after tensioning. The rails for 30m either side of the inner tell tale must be lifted and then shaken or vibrated, and the fastenings replaced.

Technical control and record keeping

All stressing and restressing must be attended by an authorised person who will be specifically responsible for technical control of the site operation.

The authorised persons responsible for technical control of the operation will ensure that all relevant information concerning stressing or restressing is passed to the infrastructure owner for recording in the register of CWR.

14.9.4   Insulated Joints

Stressing may be carried through joints provided that the insulated fishplates are approved for use in continuous welded rail and are in good order. When planning stressing or restressing work it is desirable that anchor lengths are positioned close to insulated joints in order to minimise the longitudinal rail movement and thus keep joints centrally located between sleepers. If this is not possible, account must be taken of the amount of rail movement at the joint and, if necessary, respace and repack sleepers on each side of the insulated joint.

14.9.5   Switch and Crossing Work

Welded layouts which are incorporated within continuous welded rail and connecting lengths of welded plain line between switches and crossings should have their rails brought to a stress free temperature of 27°C following installation.

Built up obtuse crossings and switch diamonds are not designed to withstand the longitudinal forces produced by continuous welded rail and must in all cases be separated and protected from adjacent continuous welded rail by adjustment switches. Special consideration must be given to the installation of adjustment switches to protect S&C layouts incorporating rolled high manganese rail owing to the greater co-efficient of expansion of this material.

Precautions must be taken when stressing through turnouts to ensure that the machined portion of the stock rail is adequately stressed. In order to achieve this, provision must be made when stressing to allow this part of the stock rail with the timbers or bearers attached to it to move relative to the ballast. For this reason the length between the switch toe and the first heel block must never be used as part of the anchor length.

14.9.6 Adjustment Switches

Where anchor lengths abut adjustment switches which are to remain in position, the overlap of the switch rails must be set appropriate to the rail temperature at the time of stressing.

14.9.7   Special Precautions following Renewal of CWR

Following renewals work, consideration has to be given to the effects of high rail temperatures on the stability of the track, and a procedure for implementing appropriate precautions laid down.

For track which has not been stressed, the basis for the considerations is the Minimum Rail Temperature (MRT) recorded during the clipping down of the CWR. The Critical Rail Temperature (CRT) is then obtained by adding 10°, 15°, 20° or 25°C, depending on the track condition, to the MRT.

If the rail temperature approaches the CRT, then a TSR of 20 mph must be imposed until the rail temperature drops to 5° below the CRT.

This underlines the importance of carrying out stressing at the earliest possible time.

When reballasting with a ballast cleaner is carried out, and the CWR is not cut or otherwise disturbed, provided the track is returned at or near its original position (limits are clearly defined) and the stressing history of the length affected is properly documented and known to be in the range 21° - 27°C, then restressing will not be necessary. In this case, the CRT immediately following the work will be related to the track condition and in the worst condition will be 25°C when severe shortage of ballast exists. Where ballast is generally plentiful, but further work (particularly tamping) is required, the CRT will be 35°C. When full ballasting has been provided and the track is in good fettle, the CRT will be 41°C, and 43°C if the DTS has been used in conjunction with the tamping.

These conditions apply for five days following the provision of full ballasting and final tamping.

In the cases where restressing is deemed necessary following reballasting work, the CRT's will be 25°C when severe shortage of ballast exists, and 35°C when full ballasting is provided.

CHAPTER 15

NEW RAILWAY CONSTRUCTION

15.1   INTRODUCTION

The 21st century has seen a renaissance in new railway building
worldwide. This is not only in countries without existing urban or high
speed national networks but in areas with existing routes where
enhanced facilities are needed for economic growth. In the UK, there
are well developed plans for a high speed railway (HS2) linking London
and cities in the north. This chapter outlines the stages in planning,
designing and building a railway. The same principles apply whatever
the state of development of the territory through which the railway
passes, although the details may vary with the circumstances prevailing.

In each case, the need for a new railway has to be established. Next,
a business brief is prepared as a basis for the development of the
engineering specification. Feasibility studies will identify various route
options and a final choice has to be made, following which the railway
has to be designed in detail and eventually constructed. Of necessity
the process is an iterative one.

15.2   PRELIMINARY INQUIRIES

The idea that the construction of a new railway may be justified, usually
springs from the perception by a potential promoter of the existence of
a demand. If the promoter is a private undertaker, the demand will
be commercial (e.g., a specific commodity like iron ore, coal, oil, cattle,
or containers, etc between specific points of origin and destination). In
suitable circumstances, a new railway can be justified on one specific
traffic low (e.g., the Heavy Haul Railways of Western Australia exist
solely for the extraction of iron ore from particular mines). On the other
hand, where the promoter is a government or a government agency,
railways are more likely to be required for other reasons; for example,
they may form part of a long term investment plan for the improvement

of the national transport infrastructure (e.g., by providing high speed passenger train services between important cities or to provide commuter services to assist in the rejuvenation of city life) or for military purposes. Such a promoter is more likely to be interested in a mixed traffic railway, and is also more likely to be interested in linking up the new line with existing systems. Such an eventuality is the more likely, the more highly developed the area is in which the new facility is being considered, and in the ultimate, the proposal may turn out to involve the modification of an existing route rather than the construction of a completely new one. Whatever the circumstances however, the first approach comes from a promoter in the form of a question: here we have some potential traffic ... is it feasible to carry it by rail?

15.3   THE BUSINESS BRIEF

The form in which such a question is posed is the "Business Brief". Any transport system is expensive to construct, and before such a major investment decision is made, it is normal for the promoter to investigate all the alternative transport modes appropriate to the traffic on offer, such as roads, pipelines, conveyors, aircraft and ships, all of which can in the right conditions, be competitors with rail. The promoters will want to appraise the relative advantages of the possible alternative modes, in terms of their benefits to the community, as well as their relative capital and running costs.

The Business Brief will describe the nature and anticipated quantity of traffic likely to be on offer, the places to be served, and the general commercial requirements. It will indicate the timescale for construction, any predetermined conditions (e.g., track or structure gauge), and the fiscal and legislative framework within which the scheme is to be carried forward. This brief will be sent to a number of selected consultants, quite likely from different countries so that the ultimate comparison may be made, not only between one mode of transport and another, but between the solutions offered by railway technologies that differ from one another in detail. The brief will require the consultant to make a presentation on his proposals, to the promoter, at a stated time, and in such detail as the promoter may specify. This presentation will enable the promoter to compare alternative or competing schemes.

15.4   FEASIBILITY STUDIES

Each of the competing consultants will now have to arrange for an investigation to be carried out by a small team, usually including a civil engineer, a railway operator, and a traction engineer. Their task will be first to establish the practicability of constructing a railway at all, and then to determine how much it will cost to provide and operate it. In a less developed area, no survey work would be done at this stage. The study would be based on the best available maps, supported by aerial and satellite photography, and such geotechnical and hydrological data etc, as might be available. The route must be reconnoitred, preferably on the ground but if necessary by air. The general impression of the terrain and local features thus obtained, is vital to a proper understanding of the other data obtained.

Assuming that the construction of a railway is deemed practicable, the team will proceed to determine one or more routes in outline and will draw up proposals in broad terms as to how to meet the business brief. It is at this stage that the implications of alternative routes can be laid before the promoter, showing how first cost and running cost are affected by choice of ruling gradient, ruling radius, cant and cant deficiency, limitations on earthworks and tunnelling, etc. It may emerge that there are also viable alternatives in such matters as gauge or traction systems. On the other hand, building in a developed environment usually means building into an existing railway with known track and structure gauges, track design rules and components, types of traffic and rolling stock. If required speeds can be given, it is then a straightforward task to specify the minimum radius of curvature and maximum gradients.

Whilst drawing up the track specification may be simple, determining the route is not. If the environment is built-up, the choice of routes may be very restricted and each route in turn will probably be confined within very narrow limits. Even in country areas, restrictions will tend to be more political and social than physical, with the avoidance of private houses, industrial buildings, and environmentally sensitive areas, becoming a major problem. Gas, water, electricity, sewerage and other services must be accommodated or diverted. Suitable crossings of highways, other railways, canals etc, must be designed to suit their owners' requirements. Accommodation works will be required by people whose properties are cut in two by the railway.

Feasibility studies and route selection can largely be done using detail maps and geological and hydrological information which are usually readily available in developed countries. These will be supported by a relatively small amount of survey work and soil investigation carried out on the ground.

After the completion of the feasibility study, it would normally be several months before the next stage of the design procedure is put in hand. During this period, financial and economic evaluations are carried out and a decision reached, on the justification for further development of the project, and whether a railway is the preferred transport mode.

15.5   DESIGN

The next stage is the carrying out of more detailed studies with the objective of producing a railway design, a specification, and cost estimates of sufficient accuracy to serve as a basis for the eventual financial decision making.

At this stage, much more detailed and factual information will be required about the terrain which the proposed railway will cross. Detailed land surveys will be required, covering the topography of the corridor of land through which the railway will be routed. Geotechnical surveys will help to define the broad specifications for the railway earthworks, tunnels and bridge foundations. Sources of track ballast will be identified. Studies by an engineering hydrologist will help to determine the location and size of bridges and culverts and to contribute to the design of track drainage and flood protection.

The development of the final design is a continuation of the process which commenced during the feasibility study whereby many parties including planners, commercial specialists, operators, and engineers, work on the tentative alignments and gradient sections determined earlier (and the train performance statistics deduced from them). The effects of changes in each parameter are closely scrutinised so that eventually the most appropriate design is arrived at. Suitable computer programs greatly speed up this work and make possible the evaluation of a far larger number of variants than was ever the case when all calculation and drawing work had to be done by hand.

At the same time, the civil engineering specification for the scheme will be evolved. No two railway projects have identical specifications, and major differences both in principle and in detail may be expected depending upon whether the railway is being designed for developed or undeveloped territory. The major civil engineering decisions to be made are probably the following:

- Track gauge

- Number of tracks

- The alignment (ie, horizontal and vertical alignment, cant and cant deficiency)

- Structure gauge

- Track structure

15.5.1 Track Gauge

The track gauge will often be the gauge of the railway system of which the new railway will or could form a part. The most common gauges is 1435mm which exists in most of Europe, America and Australia. Other gauges in regular use are 100mm and 1520mm (The Spanish gauge). If there is no existing railway to which the new line would be likely to be linked, then the decision whether standard or one of the narrower gauges will be adopted, depends on a complex of factors including the type of terrain, the money and time available for construction, and the desired train speeds. A narrow gauge railway can be constructed within a narrower land-take than a standard gauge one, and this has advantages in mountainous territory where the track may run on a ledge on a steep mountainside. The quantity of material required to construct a narrow gauge railway is significantly less than that required for standard gauge. Hence the civil engineering works will be cheaper and quicker to construct. On the other hand, if a very high speed railway is required, it will be much easier to achieve and maintain an acceptable track geometry with a standard gauge than with a narrow gauge line.

In spite of these difficulties, it is noted that the operation of regular passenger train services at up to 160km/h are not unknown on narrow gauge lines. Narrow gauge railways form a significant part of the World's total mileage of railway, amounting to about half of the total outside Europe and North America, and to over 80% of railways on the African Continent.

Nevertheless it is probably true to say that in the absence of such overriding factors as those mentioned above, a new railway would most likely be designed in standard gauge.

15.5.2 Number of tracks

The choice of whether the railway will have one or more tracks will depend on the proposed density and pattern of train services and the signalling system to be used. If single track is to be adopted, the spacing, length and turnout speeds of the passing loops will also depend on the traffic and signalling factors.

15.5.3 The Alignment

The line that the railway is to follow, in plan and in profile, is sometimes referred to as the trace. Under this general heading are included comments on ruling gradient, curvature, cant and cant deficiency.

Ruling gradient determines the relationship between the locomotive power required and the weight and speed of trains. It also has an influence on operating practices, since if the track slopes steeply, precautions are required to ensure safe operation and prevent runaways. In this respect, practices have changed markedly over the last twenty years or so. It is now recognised that operation of powerful locomotives and automatic braking systems makes possible the combination of high speed passenger operation with quite steep gradients. For example, the SNCF route from Paris to Lyon, where the gradients are as steep as 1 in 28.6 (35 per 1000), has a design speed of 300 km/h.

Figure 15.1   Construction of a High Speed Line (HS1) Kent, UK (courtesy LCR Ltd.)

The significance of this can be appreciated by looking at one or two simple examples.

Suppose we have two trains whose locomotives are each putting out just enough power to overcome friction and windage, one travelling at

200 km/h, and the other at 300 km/h. The trains have to surmount a rise of 100 metres elevation. If the rise is surmounted by a gradient of 1 in 28.6, as in SNCF, then the length of track involved will be 2860 metres. The speed of the 200 km/h train will be reduced to 120 km/h by the time it has reached the summit, and the time taken over and above that required to cover the same distance at a steady 200 km/h will be about 14 seconds.

The speed of the 300 km/h train will however be reduced only to 254 km/h and the time loss will be only 3 seconds. If the same elevation is surmounted at a gradient of 1 in 100, the distance traversed will be 10000 metres, and the speeds of the respective trains will fall by the same amount as before. However, the time taken over and above that required to cover the same distance at a steady speed will be 44 seconds for the 200 km/h train and 10 seconds for the 300 km/h train.

Such a solution is only possible if the railway is dedicated to operation with vehicles of characteristics similar to those of passenger vehicles. Heavy axle load freight vehicles being normally operated at speeds around 100 km/h, do not mix well with 300 km/h traffic, and the power necessary to maintain 100 km/h on a gradient of 1 in 30, whilst it could be provided, may well be uneconomic.

Where on a predominantly freight railway, the direction of laden trains can be clearly identified, it is often possible to specify different ruling gradients for the loaded and the unloaded directions. A further refinement is to compensate for curvature by reducing the ruling gradient on curved track by an amount determined by the radius. An old rule of thumb states the relationship as follows:

The gradient, 1 in "n", that offers the same resistance as a curve of radius R, is given by the formula:

$$n = R(metres) \times 1.65$$

In the case of a new railway, it is important to recognise that the ruling curvature, cant and cant deficiency should be determined by a policy decision about the maximum speed of operation and not vice versa.

The rules governing the relationships between speed, curvature, cant and cant deficiency are discussed in the next chapter of this book, but in general terms it should be clear that the higher the speed is consistently required to be, the straighter must the route be. This can be expensive. A low cost railway will have an alignment which avoids heavy earthworks, bridges and tunnels, but in difficult terrain, a cheap alignment can entail a combination of sharp curves and steep gradients, which together preclude any pretensions to speed, as well as likely being considerably longer than the direct line between the places served. Thus, although cheap to build, the railway may prove expensive and inconvenient to operate. The final agreed alignment will be a compromise, one which meets as closely as possible the commercial and operating briefs, with optimum expenditure on construction.

15.5.4 Structure gauge

The structure gauge to be adopted, will be determined from the maximum desired cross-section of rolling stock plus allowances for dynamic movement of the stock relative to the track. The larger the structure gauge, the greater will be the cost of any structure over the track, and therefore it may be expected that there will be strong pressure to keep the structure gauge to as small a size as possible. This may prove false economy. The structure gauge should not be finalised without giving due consideration to the long term needs of the route, the possibility of traffic from other routes, and to the desirability of providing for the occasional exceptional load.

15.5.5 Track structure

The design of the track structure is a main subject of this book. All that can be done here is to indicate the nature of the choices to be made.

These concern principally:

- Rail section and metallurgy
- Whether to specify Continuous Welded Rail (CWR) or jointed track

- Sleeper spacing, type and design detailing

- Ballast-less track; concrete paved systems

- Fastenings

- Turnouts, Junctions , Switches and Crossings

- Ballast depth and specification

- Formation design

- Signalling systems, lineside or in-cab

- Maintenance

Guidance on these choices can be obtained from this book, from previous projects known to the promoters, from specialist publications including UIC leaflets, and from specific product manufacturers, but the circumstances of every new railway differ and engineering judgments have to be made in each case.

15.6   SURVEY

Surveys will be required, covering the topography and general characteristics of the corridor of land through which the railway will pass. The scale of the surveys produced will depend on the nature of the terrain. Built-up areas, the locations of large bridges, tunnels, and sites in difficult mountain terrain, need to be mapped at a comparatively large scale, perhaps 1:10,000 or greater, while lat, open deserts can be satisfactorily dealt with at 1:100,000. There are a number of techniques for producing maps and surveys.

A good starting point is aerial photography or photogrammetry. Survey stations have to be put into the ground, suitably located in the vicinity of the connections of the new railway with any existing network, and to any other pre-existing major features. They must form a closed traverse which can later be surveyed from the surface. Photogrammetry provides data in three dimensions which can be digitised and used in a computer to produce plans, cross-sections and other models, of the existing conditions. Increasingly, the source of data for such preliminary work is satellite photography, which has now reached a standard which is enabling the Ordnance Survey in UK to dispense with its triangulation stations.

Further data can if necessary be added from ground surveys using modem surveying instruments, and techniques are available to add ground collected data to that from photogrammetry stored in the computer. The objective is that once all the survey work is complete, a total and complete three-dimensional digital image of the area should exist for use in various software programs.

15.7   EARTHWORKS DESIGN

Modem design software is available which allows the designer to specify details such as standard cross-sections. When this is done, the designer can then input a proposed horizontal and vertical alignment and drawings will be produced illustrating the effects on the surrounding area, and will calculate the quantities of cut and fill very quickly. The designer is free to carry out this process any number of times until the optimum design is evolved. Once the alignments are finalised, all required plans, longitudinal sections, cross-sections, bills of quantities, setting out data, and other information as may be required will be available for preparing contract documents.

Traditionally in the design of a new railway, a major objective was to balance cut and fill. This avoided the need for borrow pits and/or spoil tips, at the expense of the cost of transporting the excavated material from one cut to another fill.

In today's context, a major factor in the design of a new railway is that once the line is open, maintenance must be minimised. This implies minimising settlement of the formation, and care must therefore be taken with the choice of filling material for embankments and in the methods of placing and compacting it. Railway embankments require much greater strength than highway embankments since the loading intensity is higher at formation level. Railways, particularly where speeds are high, will tolerate only slight defects in vertical alignment. A high quality fill material, properly graded and compacted is therefore essential, and material dug from cuttings is not necessarily suitable. The suitability or otherwise, of excavated material, will be determined by geotechnical investigation as the design proceeds. Despite the greater sophistication in earthwork specifications, the cost of earthworks relative to the total cost of the work, has declined in recent years with the development of earthmoving plant.

The designer must bear in mind the effect of local conditions on earthworks design. For instance, in hot arid countries, great difficulty may be experienced in obtaining sufficient water to achieve optimum consolidation of filling material. There may also be serious logistical problems in building a new railway across terrain where the existing transport infrastructure is inadequate or nonexistent. A railway requiring heavy earthworks in these circumstances can be disproportionately costly and slow to build.

15.8   THE UNDERLYING GROUND

In addition to carefully controlling embankment construction, the design process must ensure the competence of the underlying ground to carry the dead and live loads involved in the construction and operation of the railway. This is done by soil mechanics investigation during the survey and predesign work. Problem areas are identified and predictions of settlement made. If problems are anticipated with excessive settlement, various solutions are available, including surcharging, installing drainage, and in the last resort, replacement. Behavioural predictions are checked by instrumentation installed in the underlying material before work starts. Both settlement and stability must be monitored during construction, especially where progress is rapid. This is done by a combination of piezometers, inclinometers, and settlement gauges. Readings from these devices can be recorded by data collectors, permitting direct input to a computer which sorts the data, stores it and presents it in tabular form.

This allows the engineer to monitor and control his site construction with the aim of ensuring only minimum residual settlement when the line opens.

15.9   CONSTRUCTION

15.9.1   Setting out

Once all the formalities have been completed, the route must be established throughout as quickly as possible. Computerisation of survey and design brings further benefits at this stage. Information for setting out the route will often be provided in two complementary forms. Firstly, points on the centre line are defined by distance and bearing from the primary and secondary traverse stations. This procedure enables the establishment of principal setting-out stations at intervals of 400-500m, together with important locations such as intersection points for bridges. Points at 20m intervals are set out by distance and delection angle from the former. The advantages of using computer generated information are the rapid production of a large amount of accurate data, the facility to make rapid adjustments and corrections and the ability to introduce new setting-out points or survey stations into the system.

15.9.2   Infrastructure construction

Once the route is established, work will start at as many points as possible on the excavation of cuttings, the tipping and consolidation of embankments, and the construction of bridges. It is usual to control the planning and progress of the work as a whole, by means of Critical Path Analysis or a derivation there from, with the detailed plans for each self-contained section of the job being converted into Bar Chart form for the information and guidance of the engineers in charge of those sections. It is important that effective systems are established for the keeping of site diaries by all key personnel, and for control of the whole to be established and maintained through regular meetings. Where the client has a pre-existing organisation, many of the required systems will be found to be ready-made but any gaps in such a system must be identified and made good right from the start.

The actual construction process details will depend on the individual circumstances and cannot be detailed here, but it may be emphasised that whatever methods are used, construction traffic should be prevented from running on the finished formation or on anything above this level, to avoid rutting producing permanent damage.

15.9.3 Social impact

Invariably in developed countries there are restrictions imposed on construction in an effort to avoid or minimise damage, danger, dirt and noise which could adversely affect people working on the site and not least members of the local community. The routes by which bulk materials may be moved and the daily quantities of such materials over these routes, are frequently laid down. The railway authority becomes liable to pay for any damage caused to highways as a result of its activities. The contractor and the supervising engineer will have to arrange for any local "Health and Safety" legislation to be complied with.

15.9.4 Junction with existing works

Any new railway being built to fit into an existing network must connect into it at one or more points. The joining of the formations has to be done whilst traffic is running at line speed on the existing route and this requires special precautions. Some work may have to be done under possession. Where existing earthworks are interfered with, soil instrumentation as already described, will enable the behaviour of these earthworks to be monitored and constant watch must be kept on the line and level of the running tracks.

15.9.5 Track laying

The form of construction will depend upon the decisions made regarding ballasted or non-ballasted track. For ballasted track, there are normally mono-block or duo-block sleepers used. Some specialist track may be installed over bridges, structures, level crossings or tunnels

The laying of bottom ballast follows completion of the formation. Good quality, clean ballast, laid, and compacted to a level surface, is the basis of good track. Consideration is given elsewhere in this book to the details of the ballast specification and laying methods required to achieve these ends, and these should be adapted as necessary to suit the local conditions.

On non-ballasted track the structure will be set out exactly to surveying specifications prior to the installation of rails. Fully paved areas such as light rail tramways will have a specialist arrangement in place.

With proper control and monitoring of the earthworks, careful setting-out and laying of the ballast and track, and adequate attention to the final levelling, lining and tamping, the line can be opened for a fairly high speed and shortly afterwards, full line speed can be permitted with minimum future maintenance.

NOTES

NOTES

NOTES

NOTES

NOTES

NOTES

NOTES